Treatment of Genetic Diseases

TREATMENT OF GENETIC DISEASES

Edited by

Robert J. Desnick, Ph.D., M.D.
Arthur J. and Nellie Z. Cohen
Professor of Pediatrics and Genetics
Chief of the Division of Medical and Molecular Genetics
Mount Sinai School of Medicine
New York, New York

CHURCHILL LIVINGSTONE
New York, Edinburgh, London, Melbourne, Tokyo

Library of Congress Cataloging-in-Publication Data

Treatment of genetic diseases / edited by Robert J. Desnick.
　　p.　cm.
　　Includes bibliographical references and index.
　　ISBN 0-443-08773-3
　　1. Genetic disorders—Treatment. 2. Metabolism, Inborn errors of-
-Treatment. 3. Enzymes—Therapeutic use. 4. Bone marrow-
-Transplantation. I. Desnick Robert J.
　　[DNLM: 1. Bone Marrow Transplantation. 2. Enzymes—therapeutic
use. 3. Gene Therapy. 4. Hereditary Diseases—therapy. QZ 50
T784]
RB155.5.T74 1991
616'.042—dc20
DNLM/DLC
for Library of Congress　　　　　　　　　　　　　　　　　91-26797
　　　　　　　　　　　　　　　　　　　　　　　　　　　　　　CIP

© Churchill Livingstone Inc. 1991

All rights reserved. No part of this publication may be reproduced, stored in a retrieval system, or transmitted in any form or by any means, electronic, mechanical, photocopying, recording, or otherwise, without prior permission of the publisher (Churchill Livingstone Inc., 650 Avenue of the Americas, New York, NY 10011).

Distributed in the United Kingdom by Churchill Livingstone, Robert Stevenson House, 1–3 Baxter's Place, Leith Walk, Edinburgh EH1 3AF, and by associated companies, branches, and representatives throughout the world.

Accurate indications, adverse reactions, and dosage schedules for drugs are provided in this book, but it is possible that they may change. The reader is urged to review the package information data of the manufacturers of the medications mentioned.

The Publishers have made every effort to trace the copyright holders for borrowed material. If they have inadvertently overlooked any, they will be pleased to make the necessary arrangements at the first opportunity.

Acquisitions Editor: *Avé McCracken*
Copy Editor: *Bridgett Dickinson*
Designer and Typesetter: *Alice Cheung*
Production Supervisor: *Christina Hippeli*

Printed in the United States of America

First published in 1991　　　7 6 5 4 3 2 1

CONTRIBUTORS

Phyllis Acosta, Dr. P.H.
Director of Metabolic Diseases, Department of Medicine, Ross Laboratories, Columbus, Ohio

Patrick Auborg, M.D.
Clinic Chief, INSERM Unit 188 and Department of Pediatric Neurology, Hospital Saint Vincent de Paul, Paris, France

Henry J. Baker, D.V.M.
Professor, Department of Comparative Medicine, Bowman Gray School of Medicine of Wake Forest University, Winston-Salem, North Carolina

Norman W. Barton, M.D., Ph.D.
Chief, Clinical Care Unit, Developmental and Metabolic Neurology Branch, National Institute of Neurological Disorders and Stroke, National Institutes of Health, Bethesda, Maryland

Ann Bergin, M.R.C.P.
Fellow, Department of Pediatrics, The Johns Hopkins University School of Medicine; Postdoctoral Fellow, Kennedy Institute, Baltimore, Maryland

Edward Birkenmeier, M.D.
Staff Scientist, The Jackson Laboratories, Bar Harbor, Maine

Janet Borel, M.S., R.D.
Research Dietitian, Department of Neurology, The Johns Hopkins University School of Medicine; Research Dietitian, Kennedy Institute, Baltimore, Maryland

Roscoe O. Brady, M.D.
Branch Chief, Developmental and Metabolic Neurology Branch, National Institute of Neurological Disorders and Stroke, National Institutes of Health, Bethesda, Maryland

Xandra O. Breakefield, Ph.D.
Associate Professor, Department of Neuroscience, Harvard Medical School, Boston, Massachusetts; Associate Geneticist, Neurology Service, Massachusetts General Hospital, Charlestown, Massachusetts

Saul W. Brusilow, M.D.
Professor, Department of Pediatrics, The Johns Hopkins University School of Medicine; Staff Pediatrician, Department of Pediatrics, The Johns Hopkins Hospital, Baltimore, Maryland

Neil R.M. Buist, M.B., Ch.B., F.R.C.P.E., D.C.H.
Professor, Departments of Pediatrics and Medical Genetics, Oregon Health Sciences University School of Medicine; Director, Metabolic Program, Metabolic Birth Defects Center, Portland, Oregon

Sara Chaffee, M.D.
Assistant Professor, Division of Pediatric Hematology/Oncology, Department of Pediatrics, Duke University School of Medicine, Durham, North Carolina

David Cornblath, M.D.
Associate Professor, Department of Neurology, The Johns Hopkins University School of Medicine, Baltimore, Maryland

Neal A. DeLuca, Ph.D.
Assistant Professor, Department of Microbiology and Molecular Genetics, Harvard Medical School; Chief, Laboratory of Molecular Virology, Dana Farber Cancer Institute, Boston, Massachusetts

Robert J. Desnick, Ph.D., M.D.
Arthur J. and Nellie Z. Cohen Professor of Pediatrics and Genetics, Chief of the Division of Medical and Molecular Genetics, Mount Sinai School of Medicine, New York, New York

Konstantin Dobrenis, Ph.D.
Fellow, Department of Neuroscience, Albert Einstein College of Medicine, Bronx, New York

William A. Gahl, M.D., Ph.D.
Chief, Human Genetics Branch, National Institute of Child Health and Human Development, National Institutes of Health, Bethesda, Maryland

J. Victor Garcia, Ph.D.
Staff Scientist, Department of Molecular Medicine, Fred Hutchinson Cancer Research Center, Seattle, Washington

Mitchell S. Golbus, M.D.
Professor, Department of Obstetrics, Gynecology, and Reproductive Sciences, University of California, San Francisco, School of Medicine, San Francisco, California

James D. Goldberg, M.D.
Assistant Professor, Department of Obstetrics, Gynecology, and Reproductive Sciences, University of California, San Francisco, School of Medicine, San Francisco, California

Mark Haskins, V.M.D., Ph.D.
Head, Laboratory of Pathobiology, and Member, Section of Medical Genetics, School of Veterinary Medicine, University of Pennsylvania, Philadelphia, Pennsylvania

Michael S. Hershfield, M.D.
Professor, Division of Rheumatology and Immunology, Department of Medicine, Duke University School of Medicine, Durham, North Carolina

Peter M. Hoogerbrugge, M.D., Ph.D.
Department of Pediatrics, State University of Leiden, University Hospital, Leiden, The Netherlands

Kathleen L. Huntington, M.S., R.D.
Research Associate and Metabolic Clinic Dietitian, Metabolic and Nutrition Sections, Child Development and Rehabilitation Center, Oregon Health Sciences University School of Medicine, Portland, Oregon

Stephen G. Kahler, M.D.
Associate Professor, Division of Pediatric Genetics and Metabolism, Department of Pediatrics, Duke University School of Medicine, Durham, North Carolina

Michael Kaleko, M.D., Ph.D.
Group Leader, Genetic Therapy Inc., Gaithersburg, Maryland

William Kennedy, M.D.
Professor, Department of Neurology, University of Minnesota Medical School—Minneapolis, Minneapolis, Minnesota

Naoki Kodo, M.D., Ph.D.
Research Associate, Division of Pediatric Genetics and Metabolism, Duke University School of Medicine, Durham, North Carolina

Edwin H. Kolodny, M.D.
Professor, Department of Neurology, Harvard Medical School; Neurologist, Department of Neurology, Massachusetts General Hospital, Boston Massachusetts; Investigator, Eunice Kennedy Shriver Center, Waltham, Massachusetts

William Krivit, M.D.
Professor, Department of Pediatrics, University of Minnesota Medical School—Minneapolis, Minneapolis, Minnesota

Harvey L. Levy, M.D.
Associate Professor, Department of Neurology, Harvard Medical School, Boston, Massachusetts; Senior Associate in Medicine and Genetics, Children's Hospital, Boston, Massachusetts

Lawrence Lockman, M.D.
Associate Professor, Division of Pediatric Neurology, Department of Neurology, University of Minnesota Medical School—Minneapolis, Minneapolis, Minnesota

A. Dusty Miller, Ph.D.
Affiliate Associate Professor, Department of Pathology, University of Washington School of Medicine; Associate Member, Program in Molecular Medicine, Fred Hutchinson Cancer Research Center, Seattle, Washington

David S. Millington, Ph.D.
Director, Mass Spectrometry Facility, Division of Pediatric Genetics and Metabolism, Department of Pediatrics, Duke University School of Medicine, Durham, North Carolina

Ann B. Moser, B.A.
Assistant in Neurology, Department of Neurology, The Johns Hopkins University School of Medicine; Kennedy Institute, Baltimore, Maryland

Hugo W. Moser, M.D.
Professor, Departments of Neurology and Pediatrics, The Johns Hopkins University School of Medicine; Kennedy Institute, Baltimore, Maryland

Sakkubai Naidu, M.D.
Associate Professor, Departments of Neurology and Pediatrics, The Johns Hopkins University School of Medicine; Kennedy Institute, Baltimore, Maryland

Daniel L. Norwood, Ph.D.
Assistant Medical Research Professor, Division of Pediatric Genetics and Metabolism, Duke University School of Medicine, Durham, North Carolina

William L. Nyhan, M.D. Ph.D.
Professor, Division of Biochemical Genetics, Department of Pediatrics, University of California, San Diego, School of Medicine, La Jolla, California

William R.A. Osborne, Ph.D.
Research Professor, Department of Pediatrics, University of Washington School of Medicine, Seattle, Washington

Theo D. Palmer, Ph.D.
Research Associate, Department of Molecular Medicine, Fred Hutchinson Cancer Research Center, Seattle, Washington

Ben J.H.M. Poorthuis, Ph.D.
Department of Pediatrics, State University of Leiden, University Hospital, Leiden, The Netherlands

Berkley R. Powell, M.D., F.A.A.P.
Assistant Professor, Department of Pediatrics, Oregon Health Sciences University School of Medicine, Portland, Oregon

Annie P. Prince, Ph.D., R.D., L.D.
Research Associate and Nutrition Training Director, Metabolic and Nutrition Sections, Child Development and Rehabilitation Center, Oregon Health Sciences University School of Medicine, Portland, Oregon

Srinivasa Raghavan, Ph.D.
Associate Biochemist, Department of Biochemistry, Eunice Kennedy Shriver Center, Waltham, Massachusetts

Mario C. Rattazzi, M.D.
Professor, Department of Pediatrics, Cornell University Medical College, New York, New York; Physician-in-Charge, Human Biochemical Genetics Laboratory, Department of Pediatrics, North Shore University Hospital-Cornell University Medical College, Manhasset, New York

Marylynne Rice-Asaro, R.D.
Staff Research Associate, Division of Biochemistry and Genetics, Department of Pediatrics, University of California, San Diego, School of Medicine, La Jolla, California; Registered Dietitian, Department of Pediatrics, University of California, San Diego, Medical Center, San Diego, California

Charles R. Roe, M.D.
Professor, Division of Pediatric Genetics and Metabolism, Department of Pediatrics, Duke University School of Medicine, Durham, North Carolina

Takeshi Sakiyama, M.D., Ph.D.
Assistant Professor, Department of Pediatrics, Nihon University School of Medicine, Tokyo, Japan

Edward H. Schuchman, Ph.D.
Assistant Professor of Pediatrics and Genetics, Division of Medical and Molecular Genetics, Mount Sinai School of Medicine, New York, New York

Elsa G. Shapiro, Ph.D.
Assistant Professor, Division of Pediatric Neurology, Department of Neurology, University of Minnesota Medical School—Minneapolis, Minneapolis, Minnesota

Robert M. Shull, D.V.M.
Professor, Department of Pathobiology, College of Veterinary Medicine, University of Tennessee, Knoxville, Tennessee

C. Gail Summers, M.D.
Associate Professor, Department of Ophthalmology, University of Minnesota Medical School—Minneapolis, Minneapolis, Minnesota

Joo-Ho Sung, M.D.
Professor and Director, Neuropathology Laboratory, Departments of Neurology and Pathology, University of Minnesota Medical School—Minneapolis, Minneapolis, Minnesota

Rosanne M. Taylor, B.V.Sc., Ph.D.
Veterinary Policy Officer, Animal Welfare, New South Wales Department of Local Government, New South Wales, Australia

Arthur R. Thompson, M.D., Ph.D.
Professor, Department of Medicine, University of Washington School of Medicine; Director, Coagulation and Hemophilia Care, Puget Sound Blood Center, Seattle, Washington

Mary Anna Thrall, D.V.M., M.S.
Associate Professor, Department of Pathology, College of Veterinary Medicine and Biomedical Sciences, Colorado State University, Fort Collins, Colorado

Judith M. Tuerck, R.N., M.S.
Research Associate and Metabolic Clinic Coordinator, Metabolic and Nutrition Sections, Child Development and Rehabilitation Center, Oregon Health Sciences University School of Medicine, Portland, Oregon

Diane D. Waggoner, M.S.
Research Associate, Metabolic and Nutrition Sections, Child Development and Rehabilitation Center, Oregon Health Sciences University School of Medicine Portland, Oregon

Steven U. Walkley, D.V.M., Ph.D.
Associate Professor, Department of Neuroscience, Albert Einstein College of Medicine, Bronx, New York

David A. Wenger, Ph.D.
Professor, Departments of Medicine and Biochemistry and Molecular Biology, Jefferson Medical College of Thomas Jefferson University, Philadelphia, Pennsylvania

German Wiederschain, Ph.D.
Visiting Scientist, Eunice Kennedy Shriver Center, Waltham, Massachusetts; Head of Medical Sciences, Institute of Biological and Medical Chemistry, Moscow, Union of Soviet Socialist Republics

Catherine H. Wu, Ph.D.
Associate Professor, Department of Medicine, University of Connecticut School of Medicine, Farmington, Connecticut

George Y. Wu, M.D., Ph.D.
Associate Professor, Departments of Medicine and Physiology, University of Connecticut School of Medicine; Chief, Division of Gastroenterology-Hepatology, Department of Medicine, University of Connecticut Health Center, Farmington, Connecticut

Yan-Wan Wu, Ph.D.
Visiting Scientist, Endocrinology Branch, National Institute of Child Health and Human Development, National Institutes of Health, Bethesda, Maryland; Visiting Scientist, Cellco Advanced Bioreactors, Inc., Kensington, Maryland

Donald Zimmerman, M.D.
Associate Professor, Department of Pediatric Endocrinology, Mayo Medical School, Rochester, Minnesota

PREFACE

The 1990s—The Decade of Treatment for Inherited Diseases

The 20th century began with the rediscovery of mendelism by Correns, Tschermak, and deVries and with the recognition of the first *inborn errors of metabolism* by Garrod, the father of human biochemical genetics. These two seminal events established the foundation for the subsequent developments in biochemical, somatic cell, and molecular genetics that have provided the rationale for the treatment of genetic diseases. Garrod, an astute physician, described the first four inherited metabolic diseases, alkaptonuria, albinism, cystinosis, and pentosuria, in his famous Croonian Lectures of 1908. These disorders became the paradigms for the study of metabolic defects and biochemical individuality. During the next 50 years, the number of inherited metabolic diseases identified increased over 100-fold and the first edition of the definitive resource on inherited disorders, *The Metabolic Basis of Inherited Disease*, was published in 1960.

In the 1940s, two important discoveries further advanced our understanding of inherited metabolic disorders. First was the "one gene–one enzyme" concept of Beadle and Tatum that was so fundamental to the delineation of the basic biochemical aberrations in these diseases. Second was Pauling and Ingram's recognition that a single amino acid substitution in the hemoglobin molecule was the molecular basis of sickle cell disease. These two concepts set the stage for the second half of the 20th century, during which numerous metabolic disorders were delineated, their enzymatic and molecular lesions identified, and the first attempts to treat these disorders performed.

Bickel's report of the treatment of a child with phenylketonuria by dietary restriction in 1953 had a dramatic impact. This finding demonstrated that genetic disorders with severe neurologic manifestations could be effectively treated by metabolic manipulation. The success of dietary restriction for phenylketonuria sparked the initiation of a variety of strategies to treat other inherited disorders, including approaches to deplete the toxic accumulated substrate or replace the crucial deficient metabolic product. Strategies to decrease the concentration of a noxious substrate and/or its precursors and metabolic derivatives also involved the administration of chelators or other drugs that would bind the accumulated metabolites and mobilize their excretion. For the urea cycle defects, efforts focused on the use of alternative pathways to enhance the excretion of waste nitrogen in the form of metabolites other than urea, such as various urea cycle intermediates and amino acid acylation products. The recent advances in the treatment of various inborn errors by metabolic manipulation are described in Section I of this book.

The identification of the specific enzymatic defects underlying certain inborn errors provided the rationale for replacement therapy. Early efforts began in the mid-1960s and demonstrated the feasibility of enzyme therapy as well as the obstacles to be overcome before it could be therapeutically efficacious. These obstacles included the need to obtain large amounts of stable, nonimmunogenic human enzyme with high specific activity and the need to deliver sufficient quantities of the enzyme to the target cellular and subcellular sites of pathology. Studies also revealed that the blood-brain barrier precluded the delivery of an intravenously injected enzyme to neurons in the central nervous system for the treatment of inborn errors with primary neurologic involvement. Only recently has it been shown that these obstacles can be overcome. Various methods have been successfully employed to camouflage heterologous proteins from immunologic surveillance. Recombinant DNA technology has been used to produce large amounts of various human proteins with appropriate post-translational modifications. In addition, promising strategies are being devised to permit macromolecular therapeutic agents such as enzymes to cross the blood-brain barrier. The recent clinical successes of enzyme replacement in type I Gaucher disease and in severe combined immunodeficiency disease herald the coming of age for this therapeutic approach. These advances in enzyme replacement therapy are described in Section II.

An intriguing means for transferring normal genetic information into patients with selected structural and metabolic gene defects is allotransplantation. This approach exploits the grafting of normal, histocompatible cells, tissues, or organs for the production of active enzymes or other gene products in the recipient. Such cellular replacement therapy has already proven effective in patients with bone marrow-derived (e.g., severe combined immunodeficiency disease) or hepatic-specific (e.g.,ornithine transcarbamylase deficiency) defects. Current efforts are directed at extending this approach and determining whether other disorders can be improved by transplantation. Bone marrow transplantation is the focus of such endeavors currently underway in animal model systems or in humans. These endeavors are described in Section III.

Section IV is devoted to recent advances in gene therapy and in utero therapy. The strategies for gene replacement are reviewed and the early successes and limitations are presented. Various novel techniques are described for the insertion of genes into bone marrow stem cells, skin cells, hepatocytes, and neurons. In addition, the efforts of obstetric geneticists to correct or cure certain genetic disorders before birth are reviewed. The 1990s promise to bring specific and effective therapies to the patients and families suffering from a diverse array of currently untreatable genetic disorders. Certainly Garrod would be proud of the continuous progress made during this century and the prospects for the treatment and cure of these inherited disorders.

Robert J. Desnick, Ph.D., M.D.

CONTENTS

I. METABOLIC THERAPY

1. Nutritional Therapy in Inborn Errors of Metabolism   1
   Harvey L. Levy

2. Approaches to the Dietary Management of Hyperphenylalaninemia   23
   Neil R.M. Buist, Annie P. Prince, Kathleen L. Huntington, Judith M. Tuerck, Berkley R. Powell, and Diane D. Waggoner

3. Advances in the Treatment of Amino Acid and Organic Acid Disorders   45
   William L. Nyhan, Marylynne Rice-Asaro, and Phyllis Acosta

4. Therapeutic Applications of L-Carnitine in Metabolic Disorders   69
   Charles R. Roe, David S. Millington, Stephen G. Kahler, Naoki Kodo, and Daniel L. Norwood

5. Treatment of Urea Cycle Disorders   79
   Saul W. Brusilow

6. Therapy for Cystinosis   95
   William A. Gahl

7. Therapy for X-Linked Adrenoleukodystrophy   111
   Hugo W. Moser, Patrick Auborg, David Cornblath, Janet Borel, Yan-Wan Wu, Ann Bergin, Sakkubai Naidu, and Ann B. Moser

II. ENZYME REPLACEMENT THERAPY

8. Enzyme Replacement: Overview and Prospects   131
   Mario C. Rattazzi and Konstantin Dobrenis

9. Enzyme Replacement Therapy for Type I Gaucher Disease   153
   Roscoe O. Brady and Norman W. Barton

10. PEG-Enzyme Replacement Therapy for Adenosine Deaminase Deficiency   169
    Michael S. Hershfield and Sara Chaffee

III. BONE MARROW TRANSPLANTATION

11. Transplantation in Animal Model Systems — 183
    *Mark Haskins, Henry J. Baker, Edward Birkenmeier,
    Peter M. Hoogerbrugge, Ben J.H.M. Poorthuis, Takeshi Sakiyama,
    Robert M. Shull, Rosanne M. Taylor, Mary Anna Thrall, and
    Steven U. Walkley*

12. Bone Marrow Transplantation for Storage Diseases — 203
    William Krivit and Elsa G. Shapiro

13. Bone Marrow Transplantation as Treatment for
    Globoid Cell Leukodystrophy — 223
    *Elsa G. Shapiro, Lawrence Lockman, William Kennedy,
    Donald Zimmerman, Edwin H. Kolodny, Srinivasa Raghavan,
    German Wiederschain, David A. Wenger, Joo-Ho Sung,
    C. Gail Summers, and William Krivit*

IV. GENE THERAPY AND IN UTERO THERAPY

14. Human Gene Therapy: Strategies and Prospects for Inborn
    Errors of Metabolism — 239
    Robert J. Desnick and Edward H. Schuchman

15. Gene Transfer into Hematopoietic and Skin Cells — 261
    *A. Dusty Miller, Michael Kaleko, J. Victor Garcia,
    Arthur R. Thompson, William R.A. Osborne, and Theo D. Palmer*

16. Delivery and Expression of Genes in Hepatocytes — 273
    George Y. Wu and Catherine H. Wu

17. Herpes Simplex Virus as a Vector for Neurons — 287
    Xandra O. Breakefield and Neal A. DeLuca

18. In Utero Therapy for Genetic Diseases — 321
    James D. Goldberg and Mitchell S. Golbus

INDEX — 331

1

NUTRITIONAL THERAPY IN INBORN ERRORS OF METABOLISM

Harvey L. Levy

INTRODUCTION

It is not an exaggeration to state that nutritional therapy has been the major force behind the dramatically increased interest in the inborn errors of metabolism over the past two decades. The possibility of preventing mental retardation from phenylketonuria (PKU) by dietary treatment led to the development of newborn screening[1] and nutritional therapies for other inborn errors.[2] The result has been early diagnosis and treatment of tens of thousands of healthy and intelligent children throughout the world with inborn errors that would otherwise have left them mentally retarded. This veritable revolution in our ability to prevent mental retardation and to treat genetic diseases has stimulated physicians to devote their energy to the care of patients with inborn errors of metabolism, as well as other scientists, both physician and nonphysician, to investigate the clinical and molecular characteristics of these disorders.[3] In fact, the two major international scientific societies devoted to the inborn errors, the Society for the Study of Inborn Errors of Metabolism (SSIEM) and the Society for Inherited Metabolic Disorders (SIMD), sponsor of the Fifth International Congress of Inborn Errors of Metabolism, originated from the need to address questions about special diets for these disorders.[4]

BACKGROUND

The conceptual basis for nutritional therapy in the inborn errors of metabolism is the garrodian idea that an intermediary metabolite accumulates in the inborn error because of a block in its conversion to another compound (Fig. 1-1). Garrod developed this idea from (1) the observation by Wolkow and Baumann[5] that feeding tyrosine to a patient with alcaptonuria greatly

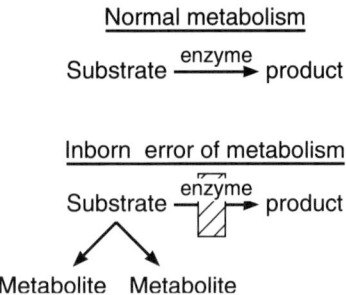

Fig. 1-1. Conceptualization of the inborn error of metabolism, based on the original idea of Garrod. (From Levy,[104] with permission.)

increased the urinary excretion of homogentisic acid and (2) his own observation that this did indeed occur in alcaptonuria but that normal persons never excreted homogentisic acid. He concluded that, in alcaptonuria, tyrosine could be converted to homogentisic acid, but no further.[6]

If administering the precursor of a metabolite increased the accumulation of the metabolite, decreasing the intake of the precursor might reduce metabolite accumulation. This was first attempted in the treatment of galactosemia.[7] Milk restriction improved or seemingly even cured infants with the disease, although the metabolic basis of this treatment was not at first understood.[8] The most notable and important demonstration of metabolite restriction for the treatment of the inborn errors, however, occurred in 1953, when Bickel and colleagues[9] showed that reducing dietary phenylalanine in PKU corrected the biochemical phenotype of this disease and led to clinical improvement. Within a few years, it became obvious that this dietary treatment could prevent the mental retardation of PKU if initiated during the neonatal period.[10] Thus, nutritional therapy for PKU converted the inborn errors of metabolism from curiosities of medicine to clinically significant diseases. From the experience with PKU have come nutritional therapies for many inborn errors of metabolism, and the list of disorders treatable in this manner continues to grow.

PRINCIPLES AND APPLICATIONS OF NUTRITIONAL THERAPY

Correction of Metabolite Imbalance(s)

A basic concept in the inborn errors is that the biochemical abnormality(ies) produces the clinical abnormality(ies). Garrod was convinced that the increased homogentisic acid in alcaptonuria was responsible for the ochronosis in that disease.[6] Fölling[11] believed that the phenylpyruvic acid he found in the urine of persons with PKU was probably the cause of their mental retardation. If so, correcting the biochemical imbalance might improve the clinical condition.

Substrate Accumulation

When there is an imbalance of metabolites the focus tends to be on the biochemical abnomality(ies).This usually means the accumulation of a substrate for either the primary defective enzyme or an enzyme that was secondarily involved (Fig. 1-1). In galactosemia, the substrate that accumulated was galactose observed in the urine of children who presented with severe neonatal liver disease or with cataracts, developmental delay, and liver disease later in infancy. The simple therapy involving the withdrawal of milk from their diet caused the disappearance of galactose from their urine and marked, often dramatic, clinical improvement, particularly in the liver disease.[8] In PKU, the observed biochemical abnormalities not only included phenylpyruvic acid in the urine but increased levels of phenylalanine in the blood and cerebrospinal fluid (CSF) as well.[12] By the early 1950s, it was clear that all these abnormalities could be explained by a block in the conversion of phenylalanine to tyrosine[13] and that restricting phenylalanine or even eliminating it from the diet might correct the imbalance (Fig. 1-2). But how could this be done? Since phenylalanine is a protein amino acid, it seemed that markedly reducing or eliminating its intake would require eliminating protein from the diet, a clearly impossible therapy.

Woolf and Vulliamy[14] attempted to circumvent this problem by giving glutamic acid to patients with PKU. Glutamic acid had been reported to reduce the blood levels of other amino acids, but it does so by enhancing intracellular transport[15]—a dubious benefit in PKU that is, in fact, potentially harmful. Nevertheless, these experiments failed. The levels of phenylalanine and its metabolites remained the same, and the clinical status of the patients was unchanged.

Two years after this failed attempt to treat PKU, another approach was taken. Bickel, then at the Children's Hospital in Birmingham, England, obtained and further prepared a casein hydrolysate that was free of phenylalanine. He administered this hydrolysate to a mentally retarded child with PKU and found that the blood phenylalanine concentration decreased to an almost normal level and that the phenylpyruvic acid almost disappeared from her urine.[9] Of even greater significance was the improvement in her behavior. She became more attentive and generally more responsive. Her development also

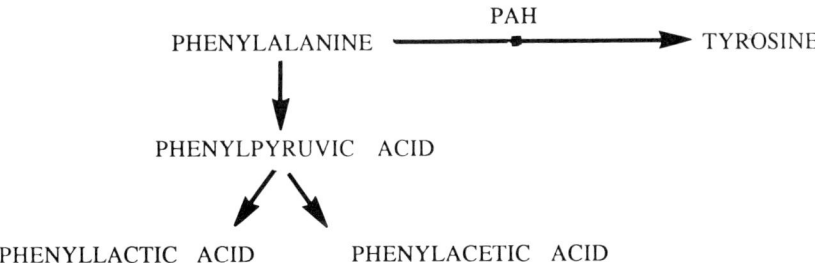

Fig. 1-2. Metabolic derangements in phenylketonuria (PKU). A defect in phenylalanine hydroxylase (PAH) results in a block in the conversion of phenylalanine to tyrosine, producing accumulations of phenylalanine and its metabolites.

improved.[16] Thus, therapy could be successfully designed to correct the metabolite abnormalities of a genetic disorder and at least begin to address a cause of mental retardation.

Despite the great significance of this demonstration, it was recognized that treating a mentally retarded child could at best probably only improve behavior. The mental retardation, an irreversible state, would remain. Investigators immediately asked whether the mental retardation and other neurologic manifestations of PKU, apparently postnatal in origin, could actually be prevented if the dietary treatment was to begin soon after birth and before clinical features appeared. Within 3 years of the original publications, this question began to be answered in the affirmative. In identifying PKU in newborn siblings of phenylketonuric children, Horner and Streamer[17] and Armstrong et al.[18] showed that initiating the diet in early infancy did indeed prevent mental retardation in PKU.

The potential of preventing mental retardation in all children born with PKU was at hand. However, this required the identification of all neonates with PKU, including those born to families in which PKU was unknown. This problem was solved by development of a bacterial inhibition assay that required only a simple specimen obtained by blotting a few drops of blood from the heel of a neonate into filter paper.[1] By the mid-1960s, routine PKU screening of neonates was performed in the United States and in a number of other countries. Many affected infants were identified and treated, and the results were definitive; the combination of newborn screening and early initiation of a dietary treatment that lowered the phenylalanine level and eliminated phenylalanine metabolites prevented mental retardation in PKU.[19]

The success of nutritional therapy for PKU led to the belief that the complications in any inborn error of metabolism could be prevented by a special diet designed to control the metabolite accumulation(s), provided treatment was instituted early in infancy. Accordingly, special diets were developed for maple syrup urine disease[20] (Fig. 1-3), classic homocystinuria (cystathionine β-synthase deficiency)[21] (Fig. 1-4), tyrosinemia,[22] and other disorders.[23] Table 1-1 lists the disorders for which nutritional therapies have proved useful, as well as the major metabolite abnormalities targeted for correction. In most instances, these therapies have been at least partially successful in preventing the clinical effects of such disorders. A notable exception is tyrosinemia I. This disorder, formerly known as hereditary tyrosinemia, produces severe and often fatal liver disease that only rarely responds to dietary therapy.[24] Unfortunately, the success of even "successful" substrate-limiting nutritional therapies must be somewhat qualified. Few children with inborn errors have attained their full potential, despite the earliest and best application of nutritional therapy. This problem is discussed in more detail later.

Product Supplementation

The conceptual model of the inborn error of metabolism posits not only substrate accumulation but also product reduction (Fig. 1-1). Total correction

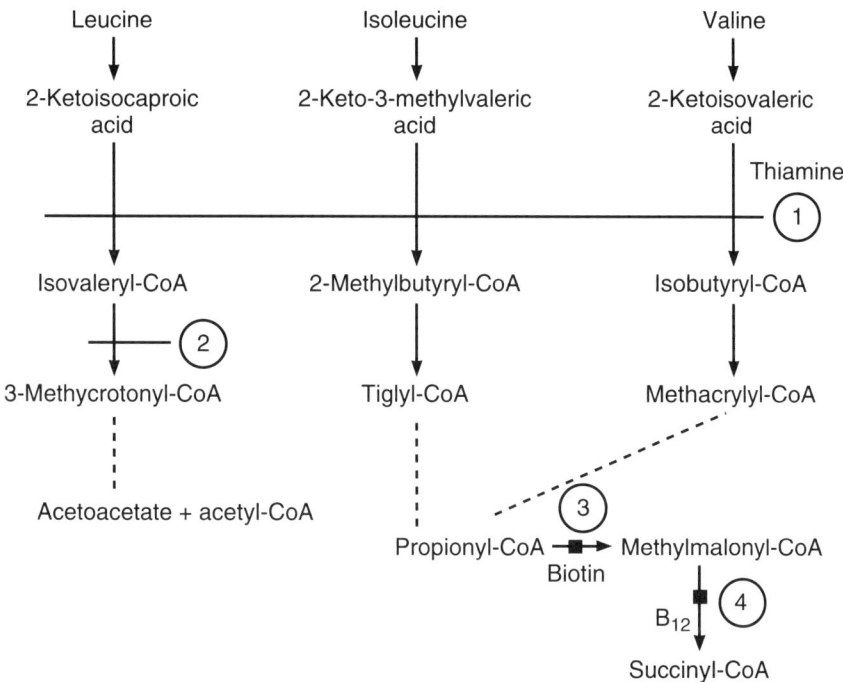

Fig. 1-3. Pathways of branched-chain amino acid degradation. Many of the organic acid disorders are caused by blocked steps in these pathways, including (1) maple syrup urine disease, (2) isovaleric acidemia, (3) propionic acidemia, and (4) methylmalonic acidemia. Vitamins thiamine (B_1), biotin, and cobalamin (B_{12}) relate to the disorders maple syrup urine disease, propionic acidemia, and methylmalonic acidemia, respectively.

of the metabolic imbalance would therefore require attention to the concentration of products distal to the blocked metabolic step.

This consideration was recognized in PKU almost from the beginning of dietary therapy. Unlike the increases in phenylalanine and its organic acid metabolites, which are obvious, the tyrosine levels in PKU are more difficult to interpret. They are not truly reduced but tend to be on the low side of the normal range.[25] This finding led to controversy over the importance of tyrosine in the pathophysiology of PKU. Some investigators maintained that the tyrosine level is of little if any consequence, while one physician went so far as to claim that tyrosine depletion is the sole cause of mental retardation in PKU.[26] The latter theory has been clearly disproved,[27] but there has always been some apprehension about allowing tyrosine levels in treated phenylketonuric patients to remain below the normal mean. Consequently, all the low phenylalanine and phenylalanine-free formulas used in the dietary treatment of PKU are supplemented with tyrosine so that the treated patient receives considerably more tyrosine per day than does someone on a normal diet.[28]

Recently, there has been considerable interest in the role of reduced levels of the tyrosine-derived catecholamine neurotransmitters in the brain damage

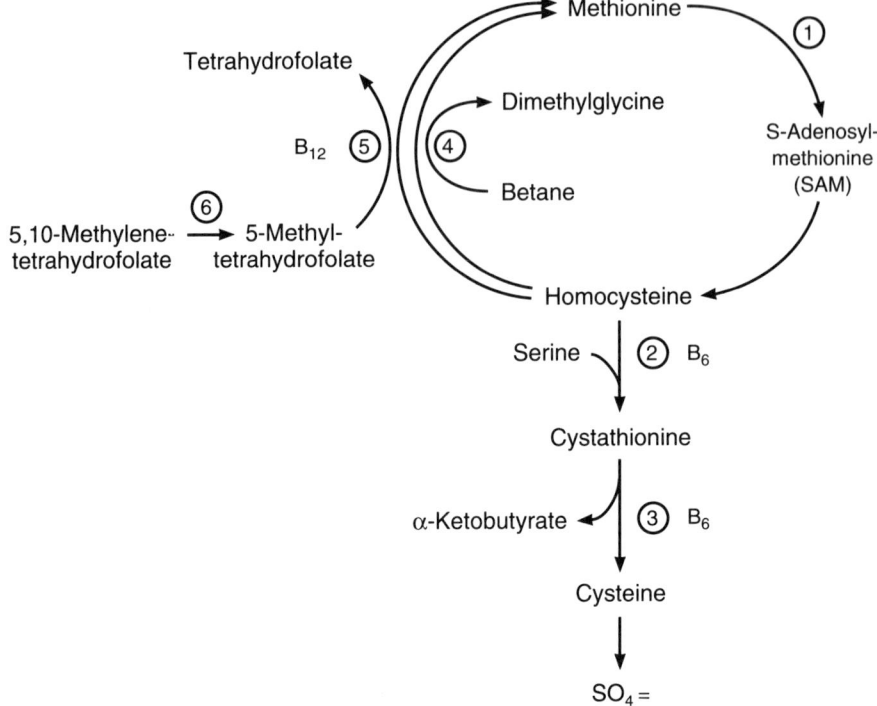

Fig. 1-4. Pathway of methionine metabolism. Enzymes that catalyze the key steps in this pathway include (1) methionine adenosyltransferase, (2) cystathionine β-synthase, (3) cystathionase, (4) betaine-homocysteine methyltransferase, (5) methionine synthase, and (6) 5,10-methylenetetrahydrofolate reductase. With the exception of betaine-homocysteine methyltransferase, each of these enzymes is associated with a known inborn error of metabolism. (From Avery and First,[105] with permission.)

associated with PKU. Monoamine metabolite levels in the urine and CSF may be reduced in untreated PKU,[29,30] possibly reflecting reduced neurotransmitter levels in the brain. Tyrosine supplementation can correct the CSF abnormalities and produce some clinical benefit, but not in all patients.[31]

Other inborn errors in which the product is supplemented are included in Table 1-1. In classic homocystinuria, for instance, cystine is almost undetectable in blood, requiring supplementation in an effort to attain metabolic homeostasis. Again, however, the evidence favors the accumulation of homocysteine, rather than cysteine depletion, as the toxic factor in the pathogenesis of at least most of the clinical abnormalities associated with this disease.[32] Nonetheless, cysteine depletion could be an important factor in the mental retardation caused by homocystinuria.

A recent ongoing controversy over dietary enhancement of metabolic product concerns galactosemia. It is now recognized that most children with galactosemia have long-term clinical complications regardless of how early milk avoidance therapy began and of how stringently this therapy has been applied.

Table 1-1. Inborn Errors of Metabolism for Which There Are Metabolite-Altering Nutritional Therapies

Disorder	Major Metabolite Abnormalities[a]	Substrate Restricted	Product Supplemented
Phenylketonuria	↑ Phenylalanine Phenylalanine metabolites ↓ (or low normal) tyrosine	Phenylalanine	Tyrosine
Maple syrup urine disease	↑ Leucine, isoleucine, valine Branched-chain ketoacids	Leucine, isoleucine, valine	—
Homocystinuria	↑ Methionine Homocyst(e)ine ↓ Cyst(e)ine	Methionine	Cystine
Tyrosinemia I (hereditary tyrosinemia)	↑ Tyrosine Tyrosine metabolites ↑ Methionine	Tyrosine Methionine	—
Lysinuric protein intolerance	↓ Citrulline	—	Citrulline
Urea cycle disorders	↑ Ammonia	Amino acids (protein)	Citrulline Arginine
Propionic acidemia	↑ Propionate Propionate metabolites	Isoleucine, valine, threonine, methionine	—
Methylmalonic acidemia	↑ Methylmalonate	Same diet as for propionic acidemia	—
Glutaric acidemia I	↑ Glutarate	Tryptophan Lysine	—
Galactosemia	↑ Galactose ↑ Galactose-1-phosphate ↓ UDP-Galactose	Galactose (lactose)	?Uridine
Pyruvic acidemia	↑ Pyruvate ↑ Lactate	Glucose	
Hereditary fructose intolerance	↑ Fructose ↑ Fructose-1-phosphate	Fructose	
Glycogen storage diseases	↓ Glucose	—	Glucose

[a] ↑, high; ↓ low.

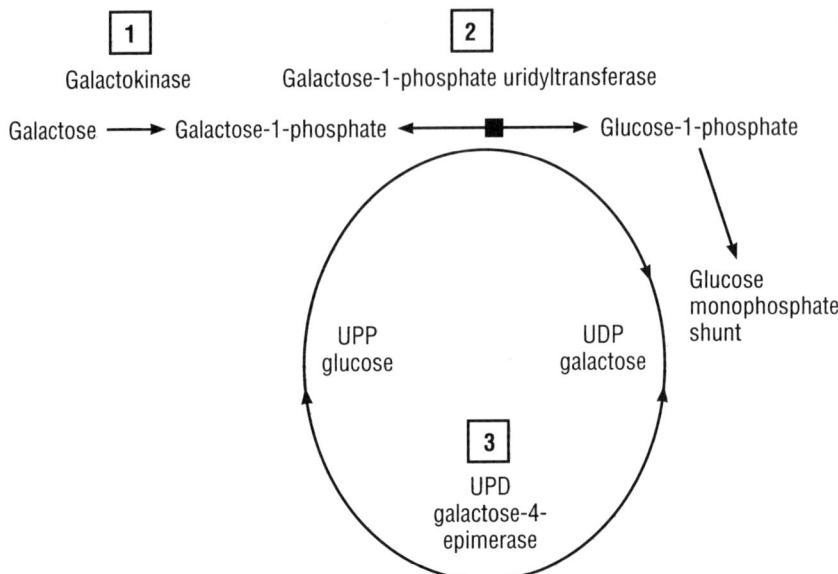

Fig. 1-5. Pathway of galacotse degradation. Galactose-1-phosphate uridyltransferase is the defective enzyme in galactosemia (block 2). This seems to result in a deficiency of uridine diphospate (UDP) galactose, as well as accumulations of galactose-1-phosphate and galactose. Other disorders in this pathway include galactokinase deficiency (block 1) and UDPgalactose-4-epimerase deficiency (block 3).

These complications include speech defects, learning disabilities, ovarian failure and, occasionally, neurologic deterioration.[33] The Los Angeles group has reported that uridine diphosphate galactose (UDPGal), a coproduct of the galactose-1-phosphate uridyltransferase (GALT) catalyzed reaction that is defective in galactosemia (Fig. 1-5), is substantially reduced in erythrocytes and cultured fibroblasts from these patients.[34] These workers have suggested that this reduction may be the cause of the observed chronic complications. Further study by this group demonstrated that UDPGal levels could be restored to normal in erythrocytes by exposure to large amounts of uridine and that this might be considered for therapy in patients with galactosemia.[34]

Several issues have arisen over these findings and their interpretation. The Philadelphia group[35] as well as Kirkman[36] have challenged the accuracy of the reported UDPGal values. Even if the values are accurate and it becomes clear that UDPGal is reduced in galactosemia, there is no direct evidence that this causes long-term complications. Finally, on the basis of their extensive studies of galactose metabolism in rats,[37] Segal and coworkers expressed reservations about the safety of giving uridine to galactosemic patients.[35] They are concerned that uridine might lead to an increase in the accumulation of galactose-1-phosphate by blocking residual GALT activity in these patients. Regardless, several studies of uridine supplementation in galactosemic infants and children are under way.

Relief of Metabolic Block

Enzymatic activity can be deficient for any of at least four reasons. First, the enzyme protein may be missing because of either lack of gene transcription or translation to protein or to instability of the protein. Second, the enzyme may be present but may not bind substrate or might bind it poorly (Fig. 1-6). This would result from a mutation that affects the substrate binding site or that affects conformation of the protein. Third, the apoenzyme might not bind the coenzyme or might bind it poorly (Fig. 1-6). A mutation that affects the coenzyme binding site would have this result. Finally, the apoenzyme may be normal, but there is an insufficient amount of coenzyme, resulting in reduced enzyme (holoenzyme) activity. This is caused by a defect in coenzyme production. In this instance, the block is a physiologic consequence of a primary defect elsewhere.

Defects in either the relationship of the apoenzyme to the coenzyme or to coenzyme production might be amenable to supplementation of the coenzyme or of a precursor of the coenzyme. This has indeed been effective in a number of inborn errors (Table 1-2). Since these coenzymes are usually vitamins, these disorders are often referred to as *vitamin responsive*. However, a few coenzymes involved in these disorders are not vitamins.

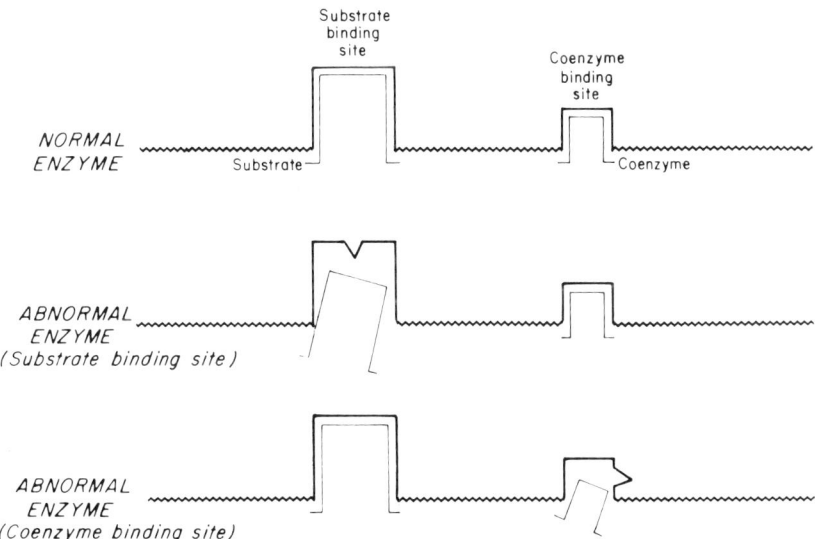

Fig. 1-6. Many enzymes contain sites that accommodate both substrate and coenzyme. A genetic mutation can produce a structural abnormality at either the substrate or the coenzyme binding site. Often the defect at the coenzyme binding site affects affinity for the coenzyme, permitting some degree of binding. In such instances, very large amounts of coenzyme at this site will support a sufficient degree or duration of binding to stimulate activity of the enzyme. (From Levy,[104] with permission.)

Table 1-2. Inborn Errors of Metabolism That Respond to Enhancement of Coenzyme

Coenzyme (or Precursor)	Disorder	Enzyme(s) Affected	Treatment
Defects in coenzyme binding			
Pyridoxine (vitamin B_6)	Cystathioninuria	Cystathionase	Pyridoxine HCl
	Classic homocystinuria	Cystathionine β-synthase	Pyridoxine HCl
	Hyperornithinemia with gyrate atrophy	Ornithine α-aminotransferase	Pyridoxine HCl
	Xanthurenic aciduria	Kynureninase	Pyridoxine HCl
	B_6-dependent seizures	Glutamic acid decarboxylase	Pyridoxine HCl
Thiamine	Maple syrup urine disease	Branched-chain α-ketoacid dehydrogenase	Thiamine
	Pyruvic acidemia	Pyruvate dehydrogenase	Thiamine
Defects in coenzyme synthesis			
Cobalamin (vitamin B_{12})	Methylmalonic acidemia (cbl A, B)	Methylmalonyl CoA mutase	Hydroxo (CN) cobalamin
	Homocystinuria and methylmalonic aciduria (cbl C, D, F)	Methionine synthase and methylmalonyl CoA mutase	Hydroxo (CN) cobalamin
	Homocystinuria (cbl E, G)	Methionine synthase	Methylcobalamin
Folate	Methylenetetrahydrofolate reductase deficiency	Methionine synthase	?Folates
Biopterin	Pterin defects (hyperphenylalaninemia)	Aromatic amino acid hydroxylases	Tetrahydrobiopterin
Biotin	Multiple carboxylase deficiency	Carboxylases	Biotin
Riboflavin	Glutaric acidemia II	Electron-transfer flavoprotein	Riboflavin

Coenzyme Binding

If the defect in coenzyme binding is such that affinity of the apoenzyme for the coenzyme is reduced, an excess of coenzyme at the binding site can support binding and can stimulate enzymatic activity. Frimpter et al.[38] first showed this for pyridoxine (vitamin B_6) in cystathionase deficiency (Fig. 1-4). Megadoses of pyridoxine (200 to 800 times the normal daily intake) administered to a patient markedly reduced the excretion of cystathionine. Subsequently, Frimpter[39] found that liver cystathionase activity was stimulated in vitro by supplementing the reaction medium with pyridoxal phosphate. We now know that the vast majority of patients with cystathionase deficiency are pyridoxine responsive.[32]

There are several other pyridoxine responsive disorders. Approximately 50 percent of patients with classic homocystinuria (cystathionine β-synthase deficiency) respond to pyridoxine supplementation with a marked reduction in the hypermethioninemia and elimination or reduction of urinary homocystine[40] (Fig. 1-4). Cultured fibroblasts from these patients display residual cystathionine β-synthase activity. This activity can be stimulated by pyridoxine when the cells are grown in a pyridoxine-free medium. By contrast, cells from pyridoxine-unresponsive patients have no activity, either residual or stimulatable.[41,42] Hyperornithinemia with gyrate atrophy of the choroid and retina (ornithine α-aminotransferase deficiency) also has a pyridoxine responsive form[43] as does xanthurenic aciduria.[44] All patients with vitamin B_6-dependent seizures have been responsive to pyridoxine[45] seemingly because of stimulation of glutamic acid decarboxylase.[45,46] Rare cases of maple syrup urine disease and pyruvate dehydrogenase deficiency have been responsive to thiamine.[47,48]

Coenzyme Production

Deficiencies of coenzymes, in the absence of nutritional or nutritionally related deficiencies, are caused by defects in coenzyme production. These defects may also be treated by providing the coenzyme as a nutritional supplement or by administering large amounts of a more readily available precursor, which is effective presumably because the defect is incomplete and permits some synthesis of coenzyme. The first demonstration of a defect in coenzyme production was in cobalamin (vitamin B_{12}) metabolism. Mudd and colleagues studied an infant with failure to thrive, microcephaly, and respiratory disease who had homocystinuria, cystathioninuria, hypomethioninemia, and methylmalonic aciduria.[49,50] This biochemical profile could be explained by deficiencies of the two coenzymatically active cobalamins: methylcobalamin and deoxyadenosylcobalamin. The infant died, and studies of his tissues disclosed reduced activities of both methionine synthase, which requires methylcobalamin (Fig. 1-4), and methylmalonyl-CoA mutase, which requires deoxyadenosylcobalamin (Fig. 1-3). In addition, deoxyadenosylcobalamin levels were reduced in the tissues. Thus, the explanation of the biochemical phenotype was supported by postmortem studies. Since that discovery, a number of defects in cobalamin coenzyme formation have been identified. Two, known as cbl A and cbl B, limit the synthesis of deoxyadenosylcobalamin alone. These defects cause methylmalonic acidemia and severe ketoacidosis.[51] Three others, known as cbl C, cbl D, and cbl F, limit the synthesis of both methylcobalamin and deoxyadenosylcobalamin.[52,53] Finally, two defects that limit only methylcobalamin synthesis have recently been reported. These are known as cbl E and cbl G.[54,55] All these cobalamin defects are treated with supplements of cobalamin precursor (vitamin B_{12}), usually cyanocobalamin or hydroxocobalamin, as well as with supportive metabolic therapy. Specific methylcobalamin supplementation may be more appropriate for the cbl E and cbl G disorders.[56]

Defects in the metabolism of other vitamins may also lead to deficient coenzyme production. A block in the reduction of methylenetetrahydrofolate (Fig. 1-4) results in methyltetrahydrofolate deficiency and reduced methionine synthase activity.[57] Folic acid therapy may provide some benefit for these patients, albeit minor. A defect in biotinidase, a key enzyme for intracellular recycling of biotin, has been discovered to be the cause of late-onset multiple carboxylase deficiency, with its clinical complications, including developmental delay, seizures, eczema, alopecia, and hearing loss.[58] Newborn screening identification and treatment with supplementary biotin prevent the clinical consequences.[59,60] Reduced production of coenzymatically active riboflavin may also account for patients with glutaric aciduria type II who have responded to riboflavin supplementation.[61]

Defects in the metabolic pathway of a nonvitamin coenzyme, tetrahydrobiopterin (BH_4), have been very prominent in the hyperphenylalaninemias. Although only a rare infant with hyperphenylalaninemia has BH_4 deficiency rather than PKU, the clinical features of BH_4 deficiency are so striking that much attention has been given to these disorders of pterin metabolism in recent years. Affected children have severe neurologic deficits with a progressive downhill course that usually leads to death during the early or middle childhood years.[62] Since BH_4 is the coenzyme for phenylalanine hydroxylase, the hyperphenylalaninemia is prominent. Dietary treatment with control of the blood phenylalanine levels, however, will not prevent the brain damage in these defects as it does in PKU. The reason seems to be that BH_4 is also required for the activation of tyrosine and tryptophan hydroxylases, which are necessary for the production of monoamine neurotransmitters. Administration of BH_4 as a dietary supplement may be therapeutically beneficial for some of those patients.

Stimulation of Alternative Pathways

In the homocystinurias, whether attributable to a defect in transsulfuration, such as cystathionine β-synthase deficiency,[32] or to a remethylation block,[52] the major aim of therapy is to reduce the accumulation of homocysteine, putatively the toxic metabolite.[32] Often, therapies such as methionine restriction in cystathionine β-synthase deficiency or cobalamin supplementation in the remethylation defects are only partially successful in meeting this need. Furthermore, adolescents and adults with homocystinuria caused by cystathionine β-synthase deficiency may be unable to tolerate the very difficult methionine-restricted diet.[32]

A successful approach to this problem has been to provide betaine as a nutritional supplement to stimulate the alternative pathway for homocysteine methylation catalyzed by betaine-homocysteine methyltransferase (Fig. 1-4). Betaine supplementation has been shown to reduce the levels of free homocystine and total homocysteine substantially in affected patients.[63-65] As might be expected, methionine levels are increased by this therapy, but this effect is probably not pathogenetically important in cystathionine β-synthase deficiency[32]; in fact, it may be an advantage in remethylation defects.[23] Notably, betaine has produced marked clinical improvement in many patients with cystathionine β-synthase deficiency.[63,64] Betaine therapy may be particularly important in preventing thromboembolic complications in these patients.

Detoxification of Toxic Metabolites

A rather novel approach to reducing the amount of a toxic metabolite is to bind the metabolite into a nontoxic and readily excreted conjugate. This approach may be necessary when the toxic metabolite is a nonessential compound that cannot be controlled by diet alone (i.e., can be synthesized so that the sources are nondietary as well as dietary) or when the diet required for effective control would be so stringent as to threaten normal growth and development.

One example of this approach is in the treatment of nonketotic hyperglycinemia (NKH). This inborn error of glycine metabolism has devastating neurologic consequences.[66] As shown in Table 1-3, the putative toxic compound is glycine, which accumulates in the brain as well as peripherally. Glycine is believed to produce the neurologic damage by excessive inhibition of postsynaptic neurotransmission[67] or by excessive activation of the N-methyl-D-aspartate type excitatory receptors,[68] or perhaps a combination of both actions. Reduction of dietary glycine only partially lowers the levels of glycine in blood and urine and may have little or no effect on the level of glycine in the central nervous system (CNS).[69] However, benzoic acid (as benzoyl-CoA) will form a conjugate with free glycine. This conjugate, hippuric acid, is a normal, nontoxic constituent of the body that is readily excreted in the urine. Treating patients who have NKH with large mounts of benzoate has resulted in lowered levels of glycine in blood and CSF and improved seizure control.[70] Unfortunately, this treatment has not produced general neurologic improvement or prevented the neurologic deterioration that accompanies this disease.

A more beneficial use of conjugating therapy has been in the administration of glycine to treat isovaleric acidemia (Table 1-3). This organic acid disorder, often referred to by the sobriquet *sweaty foot syndrome* because of the distinctive odor imparted by isovaleric acid to the body, produces episodic ketoacidosis and leads to mental retardation.[71] The toxic metabolite appears to be isovaleric acid, which accumulates as the result of a metabolic block in the oxidation of isovaleryl-CoA[72] (Fig. 1-3). Leucine is the amino acid precursor of isovaleryl-

CoA. A low-leucine diet has been effective in partially controlling the accumulations of isovaleric acid[73] (Table 1-1); however, this is a difficult diet. An alternative is to provide glycine as a supplement, either alone or in combination with reduced intake of leucine.[74] Glycine conjugates with isovaleryl-CoA to form the *N*-acyl derivative isovalerylglycine, a nontoxic and easily excreted metabolite. Thus, isovaleryl-CoA (and isovaleric acid) is detoxified and removed. This therapy with or without the low-leucine diet has been effective in controlling the isovaleric acid levels and in preventing ketoacidosis as well as neurologic complications in a number of patients.[75]

No therapy aimed at detoxifying the toxic metabolite, however, has had the wide application of carnitine (Table 1-3). Carnitine is a naturally occurring amine with the primary role of facilitating the mitochondrial transport of long-chain fatty acids by forming esters with these acids. In fact, carnitine can be esterified by fatty acids of most chain lengths as well as by certain nonfatty organic acids. This feature has been used in therapy for many organic acid disorders (see Ch. 4). For instance, carnitine is esterified by isovaleryl-CoA to form the nontoxic isovalerylcarnitine, presumably reducing isovaleryl-CoA and isovaleric acid in isovaleric acidemia, as does glycine.[76] In propionic acidemia

Table 1-3. Disorders for Which Detoxifying Therapy Is Used

Disorder	Toxic Metabolite(s) (Presumed)	Therapy	Nontoxic Conjugate
Nonketotic hyperglycinemia	Glycine	Benzoate	Hippuric acid
Isovaleric acidemia	Isovaleric acid	Glycine	Isovalerylglycine
Propionic acidemia	Propionic acid	Carnitine	Propionylcarnitine
Methylmalonic acidemia	Propionic acid Methylmalonic acid	Carnitine	Propionylcarnitine
Medium-chain acyl-CoA dehydrogenase deficiency	Hexanoic acid Octanoic acid	Carnitine	Hexanoyl- and octanoylcarnitines
Glutaric acidemia I	Glutaric acid	Carnitine	Glutarylcarnitine
Glutaric aciduria II	Glutaric acid Butyric acid Isovaleric acid Hexanoic acid Octanoic acid	Carnitine	Various acylcarnitines
Urea cycle disorders	Ammonia	Benzoate Phenylacetate Phenylbutyrate	Hippuric acid Phenylacetylglutamine

and methylmalonic acidemia, administration of L-carnitine results in the clearance of propionyl-CoA as the nontoxic and readily excretable propionylcarnitine[77,78] and perhaps has been clinically beneficial.[79] Some investigators believe that carnitine is especially useful in the treatment of medium-chain acyl-CoA dehydrogenase (MCAD) deficiency. This defect of medium-chain fatty acid oxidation produces episodic nonketotic hypoglycemia with metabolic acidosis and may be a cause of sudden infant death syndrome (SIDS).[80] As in other disorders, presumably toxic intermediates such as hexanoyl-CoA and octanoyl-CoA are removed as carnitine esters with, also presumably, the replenishment of intramitochondrial coenzyme A.[81]

Table 1-3 lists these and other metabolic disorders in which conjugation therapy may be useful. Those most clearly benefitted by this type of therapy are the urea cycle disorders. In fact, the administration of benzoate and phenylacetate (or phenylbutyrate) to conjugate glycine and glutamine, respectively, and prevent these free amino acids from contributing nitrogen to the ammonia pool represents a major advance in our ability to treat these diseases (see Ch. 5).

Blocking Substrate Production

An important recent discovery with the potential for wide application in metabolic treatment is oleic acid therapy for the peroxisomal disorders. Oleic acid blocks the synthesis of C26:0 very long chain fatty acids (VLCFA) in fibroblasts from patients with adrenoleukodystrophy[82] (see Ch. 7). While dietary restriction of VLCFA alone does not lower blood VLCFA levels in these patients, oleic acid therapy in combination with dietary VLCFA restriction does reduce blood levels of VLCFA.[83] Erucic acid may be even more potent as an inhibitor of VLCFA synthesis and is now being investigated in the treatment of these disorders.

Correction of Secondary Imbalances

The inborn errors or their therapies may create secondary biochemical imbalances. These imbalances could be important in the pathophysiology of the disorder and may require correction. In the former category are the reduced blood levels of a number of amino acids unrelated to phenylalanine metabolism in patients with PKU.[84] The origin of this amino acid imbalance is unclear, but it may result from inhibition of amino acid exodus from cells by high levels of intracellular phenylalanine.[85] To compensate for this, the diet in PKU contains larger amounts of amino acids than does the usual normal diet. This maintains blood non-phenylalanine amino acid levels within a normal range.[86]

In the latter category, patients receiving dietary treatment for PKU and maple syrup urine disease have had reduced serum levels of trace elements, such as zinc[87] and selenium.[88] The origin of these deficiencies seems to be a reduced bioavailability in children receiving elemental diets. The physiologic significance of these findings is unclear. Nevertheless, there is concern that subclinical manifestations of these deficiencies could be chronically harmful. Accordingly, elemental and other formulas for the dietary treatment of inborn errors are fortified with trace elements.[28]

OUTCOME OF NUTRITIONALLY TREATED INBORN ERRORS

The promise of nutritional therapy for inborn errors has been the prevention of disease, especially with early treatment made possible by newborn screening. There is no question but that early nutritional therapy has greatly improved the outcome in many inborn errors. Mental retardation from PKU has been almost eliminated.[89] Neonatal deaths from sepsis in galactosemia[90] and profound ketoacidosis in maple syrup urine disease[91] are rare. Dislocation of the ocular lens occurs later and the intelligence quotient (IQ) is higher among patients with homocystinuria who are treated early than among untreated patients.[40]

Despite these advances, nutritional therapy accomplishes only part of the task of prevention. While patients with PKU who are treated early are not mentally retarded, most have cognitive and psychological deficits. The cognitive difficulties include a lower IQ than normally expected,[92] as well as learning disabilities[93] and perceptual-motor difficulties.[94] The psychological deficits are often emotional in nature, such as deviant behavior[95] and, in adolescents, increased emotional lability,[96] attention-deficit disorder, oppositional disorder, and phobias.[97] Continuation of dietary therapy for PKU, rather than its discontinuation during childhood, seems to alleviate some of these problems but does not prevent all of them.[98] Furthermore, continuing diet therapy, perhaps for life, creates major difficulties in compliance for affected persons.[99] Improvements in taste and other factors of acceptability of the special medical formulas used in the nutritional treatment of the inborn errors[100] do not entirely satisfy problems of compliance.

Nutritional therapies for inborn errors other than PKU have the same limitations. For example, early-treated homocystinuric patients experience dislocation of the ocular lens (albeit later than untreated patients) and, although they may not be mentally retarded, they do have lower than normal IQ scores.[40] Children with maple syrup urine disease who are treated early also have lower than normal IQ scores.[101] The intellectual outcome of children with even the earliest-treated urea cycle disorders is well below normal. A number of these children are mentally retarded despite the best nutritional therapy available.[102] Nor does early dietary therapy prevent diminution of IQ, learning disabilities, or ovarian failure in galactosemia.[33] In fact, no person with an inborn error completely escapes some degree of adversity, despite nutritional therapy.[103]

CONCLUSION

Nutritional therapy has been a major force in advancing our understanding of the inborn errors of metabolism and in our ability to treat these disorders. Yet we may now be approaching the peak effectiveness of this mode of therapy and are beginning to encounter its limitations. Some of these limitations are of a technical nature and will likely be surmounted in the near future. For instance, improvements in taste and variety of special medical foods will certainly continue and will lead to increased acceptability and compliance. Other limitations are social in nature and are far more difficult to address. These include the limitations in normal social interaction and the attendant alterations in health perception imposed by a special diet.

More basic treatments for the inborn errors, such as enzyme replacement or gene therapy, are needed if advancement in preventing the complications of these genetic diseases is to continue. The chapters in this book that discuss these new frontiers of therapy are a taste of what lies ahead in this exciting field.

REFERENCES

1. Guthrie R, Susi A: A simple phenylalanine method for detecting phenylketonuria in large populations of newborn infants. Pediatrics 32:338, 1963
2. Guthrie R, Murphey WH: Microbiological screening procedures for detection of inborn errors of metabolism in the newborn infant. p. 132. In Bickel H, Hudson FP, Woolf LI (eds): Phenylketonuria and Some Other Inborn Errors of Metabolism: Biochemistry/Genetics/Diagnosis/Therapy. Georg Thieme Verlag, Stuttgart, 1971
3. Levy HL: Genetic screening. Adv Hum Genet 4:1, 1973
4. Committee on Nutrition, American Academy of Pediatrics: Special diets for infants with inborn errors of metabolism. Pediatrics 57:783, 1976
5. Wolkow M, Baumann E: Über das Wesen der Alkaptonurie. Z Physiol Chem 15:228, 1891
6. Harris H: Garrod's Inborn Errors of Metabolism. Oxford University, London, 1963
7. Göppert F: Galaktosurie nach Milchzuckergabe bei angeborenem, familiärem, chronischem Leberleiden. Berl Klin Wochenschr 54:473, 1917
8. Mason HH, Turner ME: Chronic galactemia. Report of case with studies on carbohydrates. Am J Dis Child 50:359, 1935
9. Bickel H, Gerrard J, Hickmans EM: Influence of phenylalanine intake on phenylketonuria. Lancet 2:812, 1953
10. Baumeister AA: The effects of dietary control on intelligence in phenylketonuria. Am J Ment Def 71:840, 1967
11. Fölling A: Über aussheidung von phenylbrenztraubensaure in den harn als stoffwechselanomalie in verbindung mit imbezillitat. Z Physiol Chem 227:169, 1934
12. Jervis GA, Block RJ, Bolling D, Kanze E: Chemical and metabolic studies on phenylalanine II. The phenylalanine content of the blood and spinal fluid in phenylpyruvic oligophrenia. J Biol Chem 134:105, 1940
13. Jervis GA: Phenylpyruvic oligophrenia deficiency of phenylalanine-oxidizing system. Proc Soc Exp Biol Med 82:514, 1953
14. Woolf LI, Vulliamy DG: Phenylketonuria with a study of the effect upon it of glutamic acid. Arch Dis Child 26:487, 1951
15. Christensen HN, Streicher JA, Elbinger EL: Effects of feeding individual amino acids upon the distribution of other amino acids between cells and extracellular fluid. J Biol Chem 172:515, 1948
16. Bickel H, Gerrard J, Hickmans EM: The influence of phenylalanine intake on the chemistry and behavior of a phenylketonuric child. Acta Paediatr Scand 43:64, 1954

17. Horner FA, Streamer CW: Effect of a phenylalanine-restricted diet on patients with phenylketonuria. JAMA 161:1628, 1956
18. Armstrong MD, Low NL, Bosma JF: Studies on phenylketonuria. IX. Further observations on the effect of phenylalanine-restricted diet on patients with phenylketonuria. Am J Clin Nutr 5:543, 1957
19. Kang ES, Sollee ND, Gerald PS: Results of treatment and termination of the diet in phenylketonuria (PKU). Pediatrics 46:881, 1970
20. Westall RG: Dietary treatment of a child with maple syrup urine disease (branched-chain ketoaciduria). Arch Dis Child 38:485, 1963
21. Perry TL, Dunn HG, Hansen S: Early diagnosis and treatment of homocystinuria. Pediatrics 37:502, 1966
22. Gentz J, Lindblad B, Lindstedt S, et al: Dietary treatment in tyrosinemia (tyrosinosis). Am J Dis Child 113:31, 1967
23. Rohr FJ, Levy HL, Shih VE: Inborn errors of metabolism. p. 391. In Walker WA, Watkins JB (eds): Nutrition in Pediatrics. Little Brown, Boston, 1985
24. Michals K, Matalon R, Wong P: Dietary treatment of tyrosinemia Type I. J Am Diet Assoc 73:507, 1978
25. Partington MW, Lewis EJM: Variations with age in plasma phenylalanine and tyrosine levels in phenylketonuria. J Pediatr 62:348, 1963
26. Bessman SP, Williamson ML, Koch R: Diet, genetics and mental retardation interaction between phenylketonuric heterozygous mother and fetus to produce nonspecific diminution of IQ: Evidence in support of the justification hypothesis. Proc Natl Acad Sci USA 75:1562, 1978
27. Batshaw ML, Valle D, Bessman SP: Unsuccessful treatment of phenylketonuria with tyrosine. J Pediatr 99: 159, 1981
28. Committee on Nutrition, American Academy of Pediatrics: Task Force on the Dietary Management of Metabolic Disorders. American Academy of Pediatrics, Evanston, IL, 1985
29. Krause W, Halminski M, McDonald L, et al: Biochemical and neuropsychological effects of elevated plasma phenylalanine in patients with treated phenylketonuria. J Clin Invest 75:40, 1985
30. Lykkelund C, Nielsen JB, Lou HC, et al: Increased neurotransmitter biosynthesis in phenylketonuria induced by phenylalanine restriction or by supplementation of unrestricted diet with large amounts of tyrosine. Eur J Pediatr 148:238, 1988
31. Lou HC, Lykkelund C, Gerdes A-M, et al: Increased vigilance and dopamine synthesis by large doses of tyrosine or phenylalanine restriction in phenylketonuria. Acta Paediatr Scand 76:560, 1987
32. Mudd SH, Levy HL, Skovby F: Disorders of transsulfuration. p. 693. In Scriver CR, Beaudet AL, Sly WS, Valle D (eds): The Metabolic Basis of Inherited Disease. 6th Ed. McGraw-Hill, New York, 1989
33. Buist N, Waggoner D, Donnell G, Levy H: The effect of newborn screening on prognosis in galactosemia: Results of the international survey. Am J Hum Genet 43: A3, 1988
34. Ng WG, Xu YK, Kaufman FR, Donnell GN: Deficit of uridine diphosphate galactose in galactosaemia. J Inher Metab Dis 12:257, 1989
35. Palmieri MJ, Rogers S, Berry GT, Segal S: Uridine diphosphate glucose and uridine diphosphate galactose in galactosemia. J Pediatr 117:839, 1990
36. Kirkman HN Jr: Uridine diphosphate glucose and uridine diphosphate galactose in galactosemia. J Pediatr 117:838, 1990
37. Rogers S, Bovee BW, Segal S: Effect of uridine on hepatic galactose-1-phosphate uridyltransferase. Enzyme 42:53, 1989

38. Frimpter GW, Haymovitz A, Horwith M: Cystathioninuria. N Engl J Med 268:333, 1963
39. Frimpter GW: Cystathioninuria: Nature of the defect. Science 149:1095, 1965
40. Mudd SH, Skovby F, Levy HL, et al: The natural history of homocystinuria due to cystathionine β-synthase deficiency. Am J Hum Genet 37:1, 1985
41. Uhlendorf BW, Conerly EB, Mudd SH: Homocystinuria: Studies in tissue culture. Pediatr Res 7:645, 1973
42. Fowler B, Kraus J, Packman S, Rosenberg LE: Homocystinuria: Evidence of three distinct classes of cystathionine β-synthase mutants in cultured fibroblasts. J Clin Invest 61:645, 1978
43. Berson EL, Schmidt SY, Shih VE: Ocular and biochemical abnormalities in gyrate atrophy of the choroid and retina. Ophthalmology 85:1018, 1978
44. Tada K, Yokoyama Y, Nakagawa H, et al: Vitamin B_6 dependent xanthurenic aciduria. Tohoku J Exp Med 93:115, 1967
45. Scriver CR, Whelan DT: Glutamic acid decarboxylase (GAD) in mammalian tissue outside the central nervous system, and its possible relevance to hereditary vitamin B_6 dependency with seizures. Ann NY Acad Sci 166:83, 1969
46. Lott IT, Coulombe T, DiPaolo RV, et al: Vitamin B_6-dependent seizures: Pathology and clinical findings in brain. Neurology 28:47, 1978
47. Zhang B, Wappner RS, Brandt IK, et al: Sequence of the E1α subunit of branched-chain α-ketoacid dehydrogenase in two patients with thiamine-responsive maple syrup urine disease. Am J Hum Genet 46:843, 1990
48. Wick H, Baumgarten R: Thiamine dependent pyruvate dehydrogenase deficiency. Pediatr Res 14:167, 1980
49. Mudd SH, Levy HL, Abeles RH: A derangement in the metabolism of B_{12} leading to homocystinemia, cystathioninemia, and methylmalonic aciduria. Biochem Biophys Res Commun 35:121, 1969
50. Levy HL, Mudd SH, Schulman JD, et al: A derangement in B_{12} metabolism associated with homocystinemia, cystathioninemia, hypomethioninemia and methylmalonic aciduria. Am J Med 48:390, 1970
51. Matsui SM, Mahoney MJ, Rosenberg LE: The natural history of the inherited methylmalonic acidemias. N Engl J Med 308:1857, 1983
52. Fenton WA, Rosenberg LE: Inherited disorders of cobalamin transport and metabolism. p. 2065. In Scriver CR, Beaudet AL, Sly, WS, Valle D (eds): The Metabolic Basis of Inherited Disease. 6th Ed. McGraw-Hill, New York, 1989
53. Shih VE, Axel SM, Tewksbury JC, et al: Defective lysosomal release of vitamin B_{12} (cblF): A hereditary cobalamin metabolic disorder associated with sudden death. Am J Med Genet 33:555, 1989
54. Schuh S, Rosenblatt DS, Cooper BA, et al: Homocystinuria and megaloblastic anemia responsive to vitamin B_{12} therapy: An inborn error of metabolism due to a defect in cobalamin metabolism. N Engl J Med 31:686, 1984
55. Watkins D, Rosenblatt DS: Genetic heterogeneity among patients with methylcobalamin deficiency: Definition of two complementary groups, cblE and cblG. J Clin Invest 81:1690, 1988
56. Wynshaw-Boris A, Korson MS, Levy HL: Methylcobalamin (MeCbl) therapy in disorders of cobalamin metabolism. Am J Hum Genet 47:A162, 1990
57. Erbe RW: Inborn errors of folate metabolism. p. 413. In Blakley RL (ed): Folates and Pterins: Nutritional, Pharmacological and Physiological Aspects. Vol. 3. John Wiley & Sons, New York, 1986
58. Wolf B, Heard GS, Weissbecker KA, et al: Biotinidase deficiency: Initial clinical features and rapid diagnosis. Ann Neurol 18:614, 1985

59. Wolf B, Heard GS, Jefferson LG, et al: Clinical findings in four children with biotinidase deficiency detected through a statewide neonatal screening program. N Engl J Med 313:16, 1985
60. Wolf B, Heard GS: Screening for biotinidase deficiency in newborns: Worldwide experience. Pediatrics 85:512, 1990
61. Gregersen N, Wintzensen H, Kølvraa S, et al: C6-C10-dicarboxylic aciduria: Investigations of a patient with riboflavin responsive multiple acyl-CoA dehydrogenation defects. Pediatr Res 16:861, 1982
62. Scriver CR, Kaufman S, Woo SLC: The hyperphenylalaninemias. p. 495. In Scriver CR, Beaudet AL, Sly WS, Valle D (eds): The Metabolic Basis of Inherited Disease. 6th Ed. McGraw-Hill, New York, 1989
63. Smolin LA, Benevenga NJ, Berlow S: The use of betaine for the treatment of homocystinuria. J Pediatr 99:467, 1981
64. Wilcken DEL, Wilcken B, Dudman NPB, Tyrrell PA: Homocystinuria—The effects of betaine in the treatment of patients not responsive to pyridoxine. N Engl J Med 309:448, 1983
65. Gahl WA, Bernardini I, Chen S, et al: The effect of oral betaine on vertebral body bone density in pyridoxine-nonresponsive homocystinuria. J Inher Metab Dis 11:291, 1988
66. Tada K: Nonketotic hyperglycinemia: Clinical and metabolic aspects. Enzyme 38:27, 1987
67. Gitzelmann R, Steinmann B, Otten A, et al: Nonketotic hyperglycinemia treated with strychnine, a glycine receptor antagonist. Helv Paediatr Acta 32:517, 1977
68. McDonald JW, Johnston MV: Nonketotic hyperglycinemia: Pathophysiological role of NMDA-type excitatory amino acid receptors. Ann Neurol 27:449, 1990
69. Kølvraa S, Brandt NJ, Christensen E: Nonketotic hyperglycinemia. Clinical, biochemical and therapeutic aspects. Acta Paediatr Scand 68:629, 1979
70. Wolff JA, Kulovich S, Yu AL, et al: The effectiveness of benzoate in the management of seizures in nonketotic hyperglycinemia. Am J Dis Child 140:596, 1986
71. Budd MA, Tanaka K, Holmes LB, et al: Isovaleric acidemia—Clinical features of a new genetic defect of leucine metabolism. N Engl J Med 277:321, 1967
72. Tanaka K, Ikeda Y, Matsubara Y, Hyman DB: Molecular basis of isovaleric acidemia and medium-chain acyl-CoA dehydrogenase deficiency. Enzyme 38:91, 1987
73. Levy HL, Erickson AM, Lott IT, Kurtz DJ: Isovaleric acidemia: Results of family study and dietary treatment. Pediatrics 52:83, 1973
74. Krieger I, Tanaka K: Therapeutic effects of glycine in isovaleric acidemia. Pediatr Res 10:25, 1976
75. Naglak M, Salvo R, Madsen K, et al: The treatment of isovaleric acidemia with glycine supplement. Pediatr Res 24:9, 1988
76. Roe CR, Millington DS, Maltby DA, et al: L-Carnitine therapy in isovaleric acidemia. J Clin Invest 74:2290, 1984
77. Roe CR, Millington DS, Maltby DA, Bohan TP: L-Carnitine enhances excretion of propionyl coenzyme A as propionylcarnitine in propionic acidemia. J Clin Invest 73:1785, 1984
78. Roe CR, Hoppel CL, Stacey TE, et al: Metabolic response to carnitine in methylmalonic aciduria. Arch Dis Child 58:916, 1983
79. Wolff JA, Carroll JE, Thuy LP, et al: Carnitine reduces ketogenesis in patients with disorders of propionate metabolism. Lancet 1:289, 1986
80. Allison F, Bennett MJ, Variend S, Engel PC: Acylcoenzyme A dehydrogenase defi-

ciency in heart tissue from infants who died unexpectedly with fatty change in the liver. Br Med J 296:11, 1988
81. Roe CR, Millington DS, Maltby, DA: Diagnostic and therapeutic implications of acylcarnitine profiling in organic acidurias associated with carnitine deficiency. p. 97. In Borum PR (ed): Clinical Aspects of Human Carnitine Deficiency. Pergamon, New York, 1986
82. Rizzo WB, Phillips MW, Dammann A, et al: Adrenoleukodystrophy: Dietary oleic acid lowers hexacosanoate levels. Ann Neurol 21:232, 1987
83. Moser AB, Borel J, Odone A, et al: A new dietary therapy for adrenoleukodystrophy. Biochemical and preliminary clinical results in 36 patients. Ann Neurol 21:240, 1987
84. Efron ML, Kang ES, Visakorpi J, Fellers FX: Effect of elevated plasma phenylalanine levels on other amino acids in phenylketonuric and normal subjects. J Pediatr 74:399, 1969
85. de Cespedes C, Thoene JG, Lowler K, Christensen HN: Evidence for inhibition of exodus of small neutral amino acids from non-brain tissues in hyperphenylalaninaemic rats. J Inher Metab Dis 12:166, 1989
86. Gerdes A-M, Nielsen JB, Lou H, Güttler F: Plasma amino acids in term neonates and infants with phenylketonuria before and after institution of the diet. Acta Paediatr Scand 79:64, 1990
87. Acosta PB, Fernhoff PM, Warshaw HS, et al: Zinc and copper status of treated children with phenylketonuria. J Parenter Enter Nutr 5:406, 1981
88. Lombeck I, Kaspareck K, Harbish HD, et al: The selenium state of children. II. Selenium content of serum, whole blood, hair and the activity of erythrocyte glutathionine peroxidase in dietetically treated patients with phenylketonuria and maple syrup urine disease. Eur J Pediatr 128:213, 1978
89. MacCready RA: Admissions of phenylketonuric patients to residential institutions before and after screening programs of the newborn infant. J Pediatr 85:383, 1974
90. Levy HL, Sepe SJ, Shih VE, et al: Sepsis due to *Escherichia coli* in neonates with galactosemia. N Engl J Med 297:823, 1977
91. Menkes JH, Hurst PL, Craig JM: A new syndrome: Progressive familial infantile cerebral dysfunction associated with an unusual urinary substance. Pediatrics 14:462, 1954
92. Waisbren SE, Mahon BE, Schnell RR, et al: Predictors of intelligence quotient and intelligence quotient change in persons treated for phenylketonuria early in life. Pediatrics 79:351, 1987
93. Fishler K, Azen CG, Henderson R, et al: Psychoeducational findings among children treated for phenylketonuria. Am J Ment Defic 92:65, 1987
94. Koff E, Boyle P, Pueschel S: Perceptual-motor functioning in children with phenylketonuria. Am J Dis Child 131:1084, 1977
95. Smith I, Beasley MG, Wolff OH, et al: Behavior disturbances in 8-year old children with early treated phenylketonuria. J Pediatr 112:403, 1988
96. Bickel H, Grubel-Kaiser S: Inborn errors of metabolism—Consequences of long-term treatment for the individual, as derived from observations in phenylketonuria. p. 211. In Cockburn F, Gitzelmann R (eds): Inborn Errors of Metabolism in Humans. Alan R Liss, New York, 1982
97. Realmuto GM, Garfinkel BD, Tuchman M, et al: Psychiatric diagnosis and behavioral characteristics in phenylketonuric children. J Nerv Ment Dis 174:536, 1986
98. Schmidt H, Mahle M, Michel U, Pietz J: Continuation vs discontinuation of low-phenylalanine diet in PKU adolescents. Eur J Pediatr 146(suppl 1):A17, 1987

99. Naughten ER, Kiely B, Saul I, Murphy D: Phenylketonuria: Outcome and problems in a "diet-for-life" clinic. Eur J Pediatr 146(suppl 1):A23, 1987
100. Owada M, Abe M, Tanimoto M, et al: Dietary treatment of PKU using a low-phenylalanine peptide milk. Acta Paediatr Jpn 30:405, 1988
101. Clow CL, Reade TM, Scriver CR: Outcome of early and long-term management of classical maple syrup urine disease. Pediatrics 68:856, 1981
102. Maestri NE, Hauser ER, Brusilow SW: Long-term survival of patients with neonatal onset of urea cycle defects (UCD). Am J Hum Genet 47:A163, 1990
103. Hayes A, Costa T, Scriver CR, et al: The effect of Mendelian disease on human health. II. Response to treatment. Am J Hum Genet 21:243, 1985
104. Levy HL: Inborn errors of metabolism. p. 527. In Avery ME, Taeusch HW Jr (eds): Diseases of the Newborn. WB Saunders, Philadelphia, 1984
105. Avery ME, First LR (eds): Pediatric Medicine. p. 934. Williams & Wilkins, Baltimore, 1989

2

APPROACHES TO THE DIETARY MANAGEMENT OF HYPERPHENYLALANINEMIA

Neil R.M. Buist · Annie P. Prince ·
Kathleen L. Huntington · Judith M. Tuerck ·
Berkley R. Powell · Diane D. Waggoner

INTRODUCTION

In this chapter we discuss only the forms of hyperphenylalaninemia that are caused by mutations of the phenylalanine hydroxylase locus, whether they result in a major or minor metabolic abnormality. Variant (or malignant) forms of hyperphenylalaninemia, caused by defects of folate or pteridine metabolism, are not considered. Detailed guidance for the design and use of phenylalanine-restricted diets is not given. Rather, the historical events that led to the current protocols for treating hyperphenylalaninemia are reviewed and our reasons for thinking the medical foods currently available are not optimally formulated are discussed. Also, some suggestions as to how these products could be redesigned are provided.

Failure to achieve long-term adherence to the diet in phenylketonuria (PKU) can be attributed to a number of different causes, including poor family coping skills; the increasing independence of growing children; the highly restrictive nature of the diet, which severely limits the food choices both for the parents and for the patients; and the unpalatable characteristics of currently available medical foods for the treatment of PKU. PKU diets are usually designed so that medical foods meet the essential nutrient needs of patients, with the exception of the energy requirements. Energy is intended to be derived from foods that are naturally low in phenylalanine (natural foods) or foods specially made to be low in total protein (low-protein foods). We propose that this approach is misguided and perforce leads to the ingestion of too much protein from natural foods in an attempt to provide adequate energy.

We have examined the taste characteristics of the constituents of current medical foods and have identified the components that contribute predominantly to their poor organoleptic properties (*organoleptic* relates to taste, smell, and oral sensory perceptions). We have also compared the nutrient content of a recommended dietary prescription for a group of children with PKU with their actual intakes and with the nutrient intakes of their non-PKU siblings. Few patients were consuming the quantity of medical food prescribed; as a result, these products provided only 75 percent of the protein requirements and 26 percent of the energy needs. Natural foods provided the remaining energy needs, which totaled close to 100 percent of calculated requirements. The result of the unprescribed intake of calories from natural foods was satiety, but with excessive intake of protein, and therefore of phenylalanine.

The approach we have taken significantly changes the role of the medical foods in the diet. Currently, medical foods are prescribed to meet approximately 120 percent of the protein requirements and to provide 80 to 90 percent of the total protein intake. Current products provide only about 30 percent of the energy requirements. Thus, when current diet protocols are followed, the protein intake is significantly in excess of the Recommended Daily Allowance (RDA). We propose the use of an increased energy-to-protein ratio of medical foods for older patients, which should result in a lower protein intake. The consequence of this change is greater emphasis on the contribution of natural foods to the total diet.

Elemental medical foods have poor organoleptic qualities, largely attributable to several of the free L-amino acids. If the level of the worst-tasting compounds were reduced as far as possible, consistent with safe nutritional practice, the products would have far better organoleptic characteristics. We believe that the medical foods should be designed to complement the nutritional contribution of the regular foods, rather than be viewed as the central component in the diet prescription.

No matter what improvements are made in the medical foods intended for PKU or other metabolic disorders, they will count for little if the patients cannot adhere to the diet prescription. The role of education in acceptance of these regimens cannot be overemphasized; most 5- to 6-year-old children are ready to start learning about their diet. A well-balanced educational program should involve them as well as their parents.

EARLY HISTORICAL BACKGROUND

Between 1934 and 1950, PKU became recognized as an unusual metabolic disorder, in which profound central nervous system (CNS) damage was associated with elevated levels of phenylalanine in the plasma and with excretion of this amino acid, and several of its metabolites, in the urine. It was postulated that if it were possible to reduce the dietary intake of phenylalanine, the plasma level would decrease and the neurologic damage might be alleviated or even prevented.[1] Such an approach, using either an elemental diet of free L-amino acids or a casein hydrolysate from which phenylalanine was largely removed by charcoal filtration was proposed by L.I. Woolf (personal communication). This approach became feasible during the early 1950s. It was tried by Dent (unpublished data) and was soon reported by Armstrong et al.[2] and by Bickel et al.[3] It became clear that the diet could be taken with apparent safety by patients with PKU, at least over relatively short periods of time. The result was some improvement in the abnormal biochemical findings. Unfortunately, this approach was prohibitively expensive, and it was not known whether it would be safe or effective for long-term management or whether it could ameliorate the neurologic damage that was the hallmark of PKU. Several manufacturers began to make special nutritional products for the long-term treatment of patients with PKU and, for several years, often provided them at no charge.

It soon became apparent that treatment, when started after several months of age, did little to prevent brain damage, although it often ameliorated the behavioral problems to some extent. It became obvious that treatment had to begin as early as possible, particularly as methods to detect the disease within a few days of birth became available. Several problems that emerged during the early years of use of medical foods led to fierce controversy over their safety and efficacy. The result was prolonged clinical trials and large, long-term, collaborative studies, some of which still continue.

CONTROVERSIES

Our understanding of PKU and how it should be treated has been achieved after much research and considerable heated debate, elegantly reviewed by Scriver et al.[4] and briefly summarized below.

Definition of PKU and Hyperphenylalaninemia

As methods to measure blood phenylalanine became more generally available, it became obvious that many patients had varying degrees of hyperphenylalaninemia and that some of them rarely, if ever, exhibited plasma levels comparable to those of patients with classic PKU syndrome. It was also apparent that some of these patients were completely normal, whereas others had evidence of brain damage. How were such cases to be classified? Did they have "mild" PKU? Was there some magic value of blood phenylalanine above which brain damage was inevitable and below which brain damage did not occur? If so, did the former indicate the diagnosis of PKU and the latter only hyperphenylalaninemia? Studies indicated that different patients could metabolize (or tolerate) different amounts of phenylalanine in the diet without having comparable levels of the amino acid in the blood. This led to the development of a number of different protocols designed to measure phenylalanine tolerance in order to assign a definitive diagnosis of either PKU or some form of milder hyperphenylalaninemia to the patient.[5,6]

There were only three really critical questions in this debate, however: Which children are at risk of developing brain damage? Which children should be treated and for how long? What is an acceptable, or desirable, therapeutic level of plasma phenylalanine? These arguments continued unabated until the advent of molecular biologic studies of the phenylalanine hydroxylase locus. It is now clear that there are many different mutations of this gene and that these are responsible for differing degrees of phenylalanine intolerance.[4] Nonetheless, it is a reflection on the vagaries of metabolic diseases that uniform agreement on the clinical classification of these cases or on the details of treatment has not yet been achieved.

In this chapter, the terms phenylketonuria (PKU) and hyperphenylalaninemia are used interchangeably. The former implies a more severe biochemical defect requiring greater dietary phenylalanine restriction than the latter. We do not subscribe to the notion that PKU and hyperphenylalaninemia should be considered as fundamentally different chemical diagnoses. (Scriver et al.[4] use the terms PKU and non-PKU hyperphenylalaninemia to distinguish between the severe and milder forms of phenylalanine hydroxylase deficiency.)

Degree of Dietary Phenylalanine Restriction

Before the development of the low phenylalanine medical foods, there was no clinical experience with the use of such contrived products for the treatment of metabolic diseases. It was believed that an effective therapy could be designed to lower the blood phenylalanine close to, or within, the normal range. The problem was that some children on the new therapy developed signs of phenylalanine deficiency at intakes that, in other children, were still too high to control the blood levels. Indeed, reports of overtreatment causing protein-calorie malnutrition, failure to thrive, skin rashes, megaloblastic anemia, prolonged seizures, osteolytic lesions, diarrhea, and even death were quite frequent.[7] As these dysnutritional states progressed, the blood phenylalanine level generally increased. In retrospect, this problem was caused by insufficient quantities of the amino acid (or possibly other nutrients) to permit normal endogenous protein synthesis and turnover.[8] It was clear that the symptoms could be rapidly reversed by increasing the phenylalanine intake to a level sufficient to promote protein synthesis. In the wake of this unexpected syndrome, nutritionists became anxious over the possibility that high blood phenylalanine levels, in the absence of any other findings, could be caused by deficiency of the amino acid. This possibility has continued to influence diet management practices.[6] There are still frequent cases in which the phenylalanine intake is inappropriately increased for patients who are clearly eating plenty of everything and consuming excess quantities of the amino acid. In fact, hyperphenylalaninemia caused by phenylalanine deficiency is now extremely rare. It usually requires prolonged, obsessional adherence to an inappropriate diet prescribed without cognizance of standard metabolic nutritional principles. In retrospect, given the state of knowledge at the time, it is not surprising that the syndrome of phenylalanine deficiency occurred.

It was over the syndrome of phenylalanine deficiency and the evidence that the treatment could be deleterious that a major controversy erupted. On the plus side, a properly managed diet was found to lessen the biochemical abnormalities; rapidly accumulating data indicated that the CNS damage could be ameliorated or even prevented. Unfortunately, the diet could cause severe malnutrition, and some children were adversely affected as a result of the treatment. Was there evidence that the diet was truly effective and what parameters could be measured to ensure efficacy and safety? Bessman[9] developed a well-argued (but erroneous) proposition, which he termed the *Justification Theory*. This term was derived from printing terminology, in which lines of text on a printed page are typeset so that they are lined up neatly (i.e., justified) at both sides of the page. He pointed out that phenylalanine and tyrosine, and other aromatic metabolites, are connected through the action of phenylalanine hydroxylase, just as words at the beginning and end of a line are connected by intervening text. Thus, changes in the level of one amino acid, either pre- or postnatally, could have effects, deleterious or beneficial, on other metabolites, and therefore on growth and cellular function. Where was the proof that phenylalanine restriction (already shown to be potentially dangerous) was therapeutically indicated? Might the brain damage not originate prenatally, and might the correct treatment be tyrosine supplements rather than phenylalanine restriction? How could contemporary theories explain the increasing number of patients with significant hyperphenylalaninemia who were completely normal? The brouhaha gradually subsided as experience grew from the burgeoning results of several clinical trials in both the United States and Europe. It is now clear that variable biologic responses to similar biochemical or genetic abnormalities are a frequent, indeed usual, finding in metabolic disorders.

These arguments were extremely valuable because they demonstrated the necessity of prospective studies on all aspects of treatment of the disorder. The idea that there were many degrees of hyperphenylalaninemia and that the tolerance or the requirements of phenylalanine may vary widely was underscored. Moreover, it appeared that sensitivity of the brain to damage from hyperphenylalaninemia varied at different ages.

Duration of Dietary Treatment

The evidence that phenylalanine restriction could alter the prognosis of PKU led to immediate attempts to define when the diet should be started and how long it should be continued. It appeared probable that the diet should be started as soon after birth as feasible. This approach fueled the idea of routine screening of newborn infants for hyperphenylalaninemia (first started in Oregon and Massachusetts in April 1961), and later for other, unrelated, metabolic diseases. In the United States, most infants are tested before 48 hours of age; in some regions newborns are tested within 24 hours of birth. Because the blood phenylalanine level only begins to rise after birth, the earlier most infants are tested, the greater the need for retesting of all infants. This is now done routinely in many screening programs.[10] In other countries, the test is often postponed until 5 to 7 days after birth, obviating the need for routine retesting but possibly delaying the diagnosis of PKU or other diseases with acute manifestations included in the screening panel. The original idea of Guthrie to use filter paper samples for this purpose has permitted the development of screening tests for more than 50 different conditions, including infectious diseases such as toxoplasmosis, and even for human immunodeficiency virus (HIV) infection.[11] It is now accepted that all infants who require therapy for hyperphenylalaninemia should be receiving treatment by 3 weeks of age.

The question of how long the diet should be continued was contentious and remains disputed to this day. At the outset, it was maintained that brain maturation was largely completed by the time a child entered school. It was proposed that children be given a normal diet in school, partly so that they would not be considered different by their peers, and partly because it was more difficult to keep school-age children on a sufficiently restricted diet to maintain blood phenylalanine levels within an "acceptable" range, however "**acceptable**" was defined. Thus, the cohort of children born during the 1960s and 1970s were often deliberately removed from the diet, either because it was believed to be safe or as a result of increasing nonadherence to the diet prescription.[12]

At the same time, many of these children in the United States and other countries were enrolled in collaborative studies designed to answer these questions and to monitor the developmental outcome of treated cases.[13] Through the use of sequential psychomotor testing, it was realized that many (but not all) of the children who were "off diet" or in poor biochemical control displayed deteriorating school performance and behavior.[12] It is now realized that two different types of neurologic problems can occur in PKU. Irreversible, permanent neuroanatomic changes, which occur during normal brain growth and maturation and cannot be reversed by subsequent treatment. The longer an infant's brain is exposed to toxic levels of phenylalanine (either pre- or postnatally), and the higher the phenylalanine levels, the more severe this brain damage is likely to be. Potentially reversible neurologic effects occur later and appear to interfere with brain function, rather than producing detectable anatomic changes.[14] It is presumed that phenylalanine or its metabolites reversibly disrupt normal biochemical brain function much as many drugs can. These neurologic problems can be at least partially reversed by successful reintroduction of a strict diet with lowering of the plasma phenylalanine to an "acceptable" range.[15] Not all children are equally susceptible to this chemical disruption, but the likelihood that it will happen has resulted in major efforts to prolong strict dietary control—certainly during the learning years, if possible.

Policies regarding continuation of strict diet vary in different clinics. In most centers, the expectation is that the diet should be continued indefinitely, into adult life, by both males and females. Some success has been reported with this approach. On the other hand, in the many centers in the United States, including our own program, by the time most patients reach adolescence, only about 30 percent are still considered to be "on diet." Adherence to the diet is so poor and the blood phenylalanine so high, that the patients are, de facto, "off diet." It seems inappropriate to insist on continued ingestion of a medical food when patients cannot be deterred from consuming a regular diet with high-protein foods and when the blood phenylalanine level never falls below 1250 μmol (20 mg/dl), in spite of the best efforts of the metabolic team and the family. When this occurs, it is necessary to make every effort to improve diet adherence; however, this may result in such rebellion and severe psychodynamic problems for the family that it is more appropriate to accept a therapeutic failure, to stop the use of the medical foods, and to encourage a low-protein vegetarian lifestyle. The diet should be continued as long as there is evidence that it is producing a beneficial effect and for as long as it is socially accepted by the patient and the family. Regrettably, these times do not always coincide.

These statements beg the question of how to approach the problem of the maternal PKU syndrome. It seems clear that the blood phenylalanine level must be controlled before a pregnancy is started. However, whether all women should be (1) maintained on an elemental medical food through their childbearing years, (2) only be treated before a planned pregnancy, or (3) counseled only toward alternative family planning options has not been determined. How and when the diet should be relaxed or discontinued remain unanswered questions in the management of hyperphenylalaninemia.

Degree of Dietary Restriction of Phenylalanine

Concomitant with the debate regarding duration of treatment, there were also discussions about how much to restrict the phenylalanine to prevent any brain damage. It was clear, very early, that the goal was not to maintain the blood phenylalanine level within the normal range, that is, 40 to 120 μmol/L; 0.6 to 2 mg/dl (1 mg/dl = 62.5 μmol/L; 100 μmol/L = 1.6 mg/dl); deficiency had to be avoided. Also, both children and adults with normal brain function who had never been treated, yet who had blood phenylalanine values 2 to 10 times normal, were detected.[16] Even rare cases of classic PKU with normal brain function were known.[17] As a result of a great deal of work in collaborative studies, it gradually emerged that the blood phenylalanine level could be safely maintained within the first few years of life at 125 to 500 μmol/L (2 to 8 mg/dl). Consistent values of greater than 750 to 900 μmol/L (12 to 14 mg/dl) in early childhood were more likely to result in a poorer intellectual outcome. After school entry, values were commonly found within this range but, while perhaps not "optimal," were not generally associated with permanent developmental delay. However, when the blood phenylalanine level consistently increased much beyond 625 to 750 μmol/L (10 to 12 mg/dl), there was an increased incidence of psychomotor and behavioral problems.

Appropriate values for treating hyperphenylalaninemia during pregnancy are still not clearly delineated. The fetus may be more sensitive to damage from phenylalanine at certain times of the pregnancy, and fetal damage can occur even when the maternal blood phenylalanine levels are only mildly elevated. There may be a linear relationship between increased maternal blood phenylalanine and decreased head size and weight of the fetus, which may be evident even with only mild degrees of hyperphenylalaninemia in the mother.[18]

Problems With Assays for Amino Acids

Another problem was that amino acids were difficult to quantitate in plasma. Even if they could be measured, the methods were costly, cumbersome, and time consuming. The original method of column chromatography for measuring amino acids took 72 hours for a complete assay. Within 10 years, the time was reduced to 12 to 24 hours, whereas today a complete plasma aminogram can be provided within 60 to 90 minutes. Cost remains an issue however. In the United States such assays can cost more than $200. These vicissitudes led to the development of rapid, inexpensive, and specific assays for phenylalanine and tyrosine that remain in common use today.

CURRENT MANAGEMENT PRINCIPLES

The current approach to the treatment of hyperphenylalaninemia embraces the following principles:

1. All cases should be detected as soon as possible after birth and should be evaluated in a competent metabolic program.
2. All infants in whom the blood phenylalanine level rises above a threshold value, usually 500 to 625 µmol/L (8 to 10 mg/dl) on an unrestricted diet, should have their diet modified to reduce the blood phenylalanine level to 125 to 500 µmol/L (2 to 8 mg/dl), by the use of either breast milk, modified infant formulas, or special elemental medical foods in which the phenylalanine content is intentionally reduced.
3. All infants requiring diet changes should be on treatment by 14 to 21 days of age. The regimen should be continued for as long as practical.
4. The use of special diets in hyperphenylalaninemia and other metabolic disorders requires supervision by a metabolic team experienced in the nutritional management of these disorders.
5. Regular biochemical monitoring of the blood phenylalanine, tyrosine, and other compounds is essential.
6. Optimum values of blood phenylalanine should be defined. They may vary at different ages. For infants and preschool children, postprandial values within the range of 125 to 500 µmol/L (2 to 8 mg/dl) are considered safe. For older children, this degree of control, while desirable, is often not possible, and it may be necessary to accept values as high as 750 to 875 µmol/L (12 to 14 mg/dl). By adolescence, values within the range of 500 to 1250 µmol/L (8 to 20 mg/dl) may be the best that can be expected using currently available medical foods and nutrition protocols. Management of pregnancy in women with PKU/HPA remains extremely difficult. Optimum plasma phenylalanine values remain to be defined. In our opinion, the blood phenylalanine should optimally be 125 to 375 µmol/L (2 to 6 mg/dl) throughout the pregnancy, provided that all other nutritional intakes and assessments are within normal limits.
7. Regardless of the severity of the underlying metabolic defect, the quantity

of phenylalanine required or tolerated varies with age. The following values for patients with relatively low phenylalanine tolerances are given as guidelines only.

Age	mg/kg/d
Newborn and during first months of life	50–80
Later infancy to year 2	30–60
2–4 years	15–40
>4 years	7–25

8. After infancy, all the phenylalanine in the diet should come from natural (low-protein) foods or from special low-protein food products. (Hereinafter, these kinds of foods are classed together as natural or low-protein foods.)
9. Natural foods should be chosen not only for their low phenylalanine content but for high calorie content as well.
10. Natural foods should add texture, flavor, satiety, and variety to the diet.
11. Since natural foods cannot by themselves provide adequate nutrition, they must be supplemented by medical foods low in phenylalanine.
12. The elemental medical food should be designed to provide sufficient quantities of all the nutrients, *such that, when used as a supplement to the natural foods, the total diet provides a balanced and acceptable long-term nutritional regimen.*

BREAST-FEEDING

In many centers, a diet of medical foods, combined with a regular infant formula, constitutes the method of choice for treating hyperphenylalaninemia during infancy. We would like to emphasize our view that the substitution of regular breast-feeding for the infant formula presents an easy and preferable way to manage the diet for these infants.

We have found that breast-feeding results in better acceptance by the parents and that good biochemical control is easy to achieve and maintain. Some investigators[19] have advocated breast-feeding for PKU but require that test weighing should be a regular part of the protocol. We disagree. Test weighing is not very accurate unless very carefully done. It also diminishes the spontaneity and pleasure that breast-feeding should bring and often raises anxiety on the part of the mother. As with other situations in which breast milk and formula feedings are combined, test weighing should only be done when there is concern over the child's weight, growth, biochemical control, or clinical progress.

While other investigators have advocated the use of different formulas, including Analog-XP, PKU-1, and Lofenalac, combined with breast-feeding, we have most experience using Lofenalac. We now prefer Phenyl-Free because it contains no phenylalanine; thus, relatively more breast milk can be consumed than if Lofenalac is used. We use the following approach:

1. A phenylalanine-free product is used at regular dilution (1 scoop (10 g)/2 oz water = 20 kcal/oz) in 30 to 50 ml/kg/d. We usually start with the higher figure to speed the reduction in blood phenylalanine and then reduce the quantity as the blood values fall to the therapeutic range.
2. This formula is then given as 30- to 50-ml supplements before five feedings per day. The supplements must be finished before each breast-feeding is allowed. There is no need to give a supplement during the night.
3. Following the supplements and during the night, breast-feeding is allowed ad lib.
4. Blood tests for phenylalanine are taken three times per week, according to the regular protocol in our clinic, until the blood phenylalanine level is stable. The intervals are gradually lengthened to weekly samples as indicated.
5. Adjustment in the volume of formula is used as the only variable in controlling the blood phenylalanine.
6. Test weighing is not done and is unnecessary.
7. We are experimenting with the possibility of using all the formula in one or two feedings per day and then allowing ad lib breast-feeding thereafter. There are theoretical advantages and disadvantages to doing this.

CURRENT PRODUCTS

Although it is clear that long-term control of hyperphenylalaninemia is necessary, there have been relatively few attempts to manufacture more products in which the phenylalanine content is reduced or eliminated. To be fair, it should be pointed out that, until 1972, the products were considered drugs by the U.S. Food and Drug Administration (FDA). This definition limited the ability of industry to develop and test new formulations. After 1972, however, the FDA reclassified them as *medical foods*. (On June 6, 1973, S.D. Fine, Associate Commissioner for Compliance of the FDA, sent a letter to the manufacturers.) This designation removed the restrictions that apply to new investigational drugs and was expected to encourage manufacturers to develop new products. During the past few years, some new products have appeared, spurred partly by the fact that diet therapy is being prolonged into adolescence and also by the evidence that current treatment for the maternal PKU syndrome is horrendously difficult.

Three kinds of products can be identified:

1. Medical foods primarily designed for infants
2. Medical foods designed for older children and adults
3. Low protein "natural" foods, such as starches, baking mixes, dairy substitutes, and special pastas, which can be used for the treatment of a variety of disorders requiring a low-protein diet

Groups 1 and 2 can be considered to be somewhere between a food and a drug and are termed *medical foods*, usually only available on prescription. In some states in the United States, they are now covered by insurance as regular prescription items. This concession by the insurance industry was not won easily and has often been vigorously contested. In several states, they are still not covered by any private insurance companies, although the same insurers may accept them as medical prescription items in other states.

In the United States, some of the food products in group 3 are available in groceries; others can usually be obtained on request. They are not covered by insurance companies as necessary medical supplies. In other countries, however, such as Britain, they can be obtained on prescription through the National Health Service.

The first group are exemplified in the United States by Lofenalac (Mead Johnson) and in the United Kingdom by Minafen (Cow and Gate), both of which are based on Woolf's early concept of modification of an enzymatic hydrolysate of casein. Neither product has been changed substantially during the past 25 years. Albumaid XP (Scientific Hospital Supplies) is also based on a casein hydrolysate and PKU-1 (Milupa) and Analog-XP (Scientific Hospital Supplies) are elemental formulas; both are designed for infants. (In the United States, Scientific Hospital Supplies products are marketed by Ross Laboratories and products made by Milupa are marketed by Mead Johnson.)

The second group includes Phenyl-Free (Mead Johnson), PKU-2 (Milupa), Aminogran (Cow and Gate), and Maxamaid X-P (Scientific Hospital Supplies). Two products, designed more specifically for the treatment of maternal PKU, include PKU-3 (Milupa) and Maxamum XP (Scientific Hospital Supplies). In all these products, the protein equivalent is derived from elemental L-amino acids. (The formation of a peptide bond between each amino acid in proteins eliminates a molecule of water, thereby reducing the molecular weight of each amino acid by 18. To account for the difference in molecular weight, the protein equivalent of a mixture of free amino acids is calculated as 20 percent lower than if equimolar quantities were all incorporated into a protein.) They are considered elemental medical foods. Their calorie content, protein equivalent, and amino acid profile are summarized in Table 2-1.

In most of these products, the amino acids are combined with vitamins and minerals to provide all or most of the RDA for such compounds. Some provide significant quantities of calories in the form of dextrins and oils. Others are little more than protein substitutes and are almost devoid of carbohydrate and fat. These elemental foods are expected to be used in combination with low-protein, natural foods; with foods from group 3; and with high-energy, protein-free foods, such as sugars, dextrins, or starches for carbohydrates and cooking oils for fats. Product 80056 (Mead Johnson) and Periflex (Scientific Hospital Supplies) are protein-free mixtures of dextrins, oils, minerals, and vitamins that can provide a reasonably palatable source of calories and nonprotein nutrients if used in sufficient quantities.

All the elemental medical foods currently available have some undesirable organoleptic properties, which are largely attributable to certain L-amino acids. In the early years, children usually accept these products quite well, and some children come to like them. For a few, such a taste preference lasts for years,

Table 2-1. Medical Foods for Treatment of Hyperphenylalaninemia[a]

Amino Acids (g)	Amino-gran (Allen & Hanburys)	Minafen (Cow and Gate)	Lofenalac (Mead Johnson)	Milupa (PKU-1) (Mead Johnson)	Milupa (PKU-2 & -3) (Mead Johnson)	Phenyl-Free (Mead Johnson)	Albumaid XP (SHS/Ross)	Analog XP (SHS/Ross)	Maxamaid XP (SHS/Ross)	Maxamum XP (SHS/Ross)	PK Aid 1 (SHS/Ross)
Alanine	2.40	N/S	0.64	2.40	3.10	0	2.10	0.58	1.03	1.60	8.00
Arginine	4.20	0.45	0.34	2.00	2.70	0.69	1.70	1.03	2.21	3.02	6.00
Aspartic acid	6.10	N/S	1.34	5.70	7.60	5.30	4.20	0.96	1.85	2.81	14.00
Cystine	1.50	N/S	0.025	1.40	1.80	0.35	1.70	0.38	0.71	1.11	5.00
Glutamic acid	12.20	N/S	3.78	12.00	16.00	1.90	4.80	1.17	2.40	4.56	16.00
Glutamine	0	0	0	0	0	4.80		0.11	0.22	0.34	0
Glycine	6.10	N/S	0.35	1.40	1.80	3.30	1.60	0.90	1.77	2.82	8.00
Histidine	3.40	0.26	0.26	1.40	1.80	0.47	1.00	0.59	1.27	1.71	3.00
Isoleucine	8.50	1.12	0.75	3.40	4.50	1.10	2.00	0.90	1.70	2.66	3.00
Leucine	10.00	1.36	1.41	5.70	7.60	1.73	3.40	1.55	2.91	4.56	5.00
Lysine	6.60	1.67	1.57	4.00	5.40	1.89	3.30	1.06	2.22	3.49	9.50
Methionine	3.00	0.21	0.45	1.40	1.80	0.63	1.00	0.25	0.48	0.73	2.00
Phenylalanine	0	<0.02	0.08	0	0	0	<0.01	Trace	Trace	Trace	0
Proline	4.30	N/S	1.15	5.40	7.10	0	1.90	1.10	2.05	3.23	2.50
Serine	8.40	N/S	1.02	3.00	4.00	0	2.60	0.68	1.26	2.00	2.50
Taurine	Nil	N/S	N/S	N/S	N/S	0	N/S	0	0	0	Nil
Threonine	5.70	0.88	0.77	2.70	3.60	0.94	2.60	0.76	1.42	2.23	4.00
Tryptophan	1.50	0.21	0.19	1.00	1.40	0.28	0.40	0.30	0.57	0.89	1.50
Tyrosine	6.40	0.81	0.81	3.40	4.50	0.37	3.20	1.37	2.56	4.03	6.00
Valine	7.00	0.96	1.20	4.00	5.40	1.26	2.50	0.99	1.85	2.92	4.00
Protein equivalent (g)	86	12.5	15	50	67	20	33.8	13	25	39	88
Energy (kcal)	400	509	460	272	295	410	324	475	350	340	400

[a] Amino acid, protein equivalent, and energy content per 100 g powder.

but most come to dislike or to reject their "milk," as they realize how vile the products taste and smell. We propose that improving these characteristics would prolong dietary adherence, particularly if the product could be incorporated into other foods, such as sauces, soups, breads, pastas, and puddings. The selection of such low-protein foods is still pitifully small in the United States. In Europe, there is a wider variety, which certainly makes the long-term management of PKU more feasible than it was 10 to 15 years ago. Schuett[20] describes more than 300 low-phenylalanine recipes using low-protein products and natural low-protein foods. All parents should be encouraged to try out recipes from this or other sources. Increased use of low-phenylalanine foods increases the contribution of calories and many nutrients from these sources. The result is that relatively less nutrition needs to be provided from an elemental medical food. Thus, it is timely to consider whether the formulation of existing elemental medical foods can be modified to improve either their nutritional or organoleptic characteristics.

HOW COULD THE DIET BE MODIFIED?

From the patient's point of view, the overriding problems with the diet are its monotony and the organoleptic properties of the medical foods. There is a relatively low but steady demand for specialty foods low in natural protein for patients with renal or hepatic failure as well as for different kinds of metabolic diseases. It is to be hoped that the food industry will increase the number and variety of such products. It is clearly the responsibility of metabolic teams to encourage the families of patients with these disorders to experiment with low-protein cooking either by following Schuett's lead or by designing new recipes that can be shared through parent organizations and professional publications. Our major concern relates to the formulation of the present elemental medical foods and to the possible design and use of new products to improve the organoleptic properties and to recognize the changing role of elemental medical foods in the overall management of hyperphenylalaninemia.

ENERGY/PROTEIN RATIOS OF MEDICAL FOODS

The protein intake of a normal average American 6-year-old child is about 50 to 60 g/d, which compares with the RDA of 24 g protein for the same child. Most young children consume about the RDA for energy (90 kcal/d), but unnecessarily high energy intakes are common in older children and in adults. Thus, the average American diet provides excess of both protein and energy above the RDA.[21] (We do not wish to enter a discussion as to whether the RDA are optimal for a normal child, but it is accepted that, given a normal amino acid profile of the protein or protein equivalent, chronic ingestion of the RDA of the protein in an otherwise balanced diet is not associated with nutritional disequilibrium.)

Most dietary proteins, whether of animal or vegetable origin, contain 4 to 6 percent of the amino acids as phenylalanine, but the range is 2 to 9 percent. Thus, it is impossible to devise a natural protein intake that is specifically restricted in this amino acid. Even low-protein foods, such as vegetables, fruits, and special low-protein foods, all provide about 40 to 50 mg of phenylalanine per gram of protein (Table 2-2). Thus, in designing the natural food portion of the diet, it is necessary either to rely on foods that have a naturally low-protein content, such as most fruits and some vegetables, or to use the special low-phenylalanine foods. An excellent reference source is provided by Acosta.[22] Since school-age children with PKU are likely to tolerate only 7 to 20 mg/kg/d of phenylalanine, this restricts a 6-year-old child with PKU to no more than 15 g of natural protein per day. Thus, it is important to increase the bulk and variety and, above all, the energy content of the natural foods. Using some ingenuity and the available special low-protein products, the nonmedical food intake can be designed to provide 100 to 120 kcal/g protein. We have used the figure of 115 kcal/g protein for this source as determined from a study of a group of 6- to 12-year-old PKU children in our clinic.[23]

The bulk and the energy-to-protein ratio of foods differs enormously (Table 2-2). Protein-dense foods such as eggs or meat provide only 8 to 10 kcal/g of protein. They should clearly be avoided in these diets, not only for their high phenylalanine content, but also because of the low energy-to-protein ratio. Surprisingly, many vegetables, including broccoli, cabbage, cauliflower, and even lettuce, also have energy-to-protein ratios of around 10 to 20 kcal/g protein. These foods are nonetheless critical to a low-protein diet because large volumes can be ingested to provide bulk, fiber, taste, texture, and satiety. They become more valuable still as energy sources, when cooked with oil or prepared with a salad dressing. For some vegetables and most fruits, the energy-to-protein ratio is about 40 to 80 kcal/g protein.[20]

From the above figures we can now calculate the probable contribution of natural foods to the overall diet plan for a 6-year-old child with PKU. In Table 2-3, we have deliberately chosen the example of a child with quite severe PKU

Table 2-2. Energy-to-Protein Ratio of Different Foods for the Treatment of Hyperphenylalaninemia[a]

	Protein (kcal/g)		Protein (kcal/g)
Fruits	60–120	"Lo-pro" breads, pasta	200–500
Starches	25–60	Phenyl-Free	20
Vegetables	10–40	Maxamaid XP	14
Meat	8–10	PKU-2 & -3	4.4
Whole cow's milk	20	"Ideal" EMF	23
"Average" U.S. diet	22	Non-EMF PKU diet	100–120

[a] Average energy per gram protein: 1 g natural protein = approximately 40–60 mg phenylalanine.
Abbreviations: EMF, elemental medical food; PKU, phenylketonuria.

whose phenylalanine tolerance is 12.5 mg/kg/d. The natural low-protein foods should provide all the phenylalanine (250 mg), limiting the total protein intake from these sources to about 5 g/d. Given an average of 115 kcal/g protein, low-protein foods can be expected to provide around 600 kcal/d. The balance of the protein, energy, and other nutrients must come from an elemental medical food. Table 2-3 indicates the protein and energy that a medical food should contain to complement the natural foods. These figures are compared with those of Phenyl-Free. It can be seen that the use of Phenyl-Free to obtain sufficient calories raises the protein intake to an unnecessarily high level.

By reference to Table 2-1, it can be seen that the existing formulas do not have an energy-to-protein ratio that comes close to complementing the natural foods according to the requirements shown in Table 2-3. When they are prescribed to meet the protein needs, energy requirements are not met. Conversely, if sufficient calories are ingested, the protein intake is unnecessarily high. In our clinic, Prince et al.[23] have shown that PKU children consume very close to 100 percent of the predicted energy requirement. If this is provided by the elemental medical food, the protein intake is excessive. The one exception to this is Analog-XP, which has a energy-to-protein density of 28 kcal/g protein. This product, designed for use during an infant's first 2 years of life, comes closest to an energy-to-protein ratio similar to our specifications, but its poor organoleptic properties limit its use in older patients. Since it is the L-amino acids that contribute most to these poor organoleptic properties, the current medical foods condemn these patients to unwarranted ingestion of the most unpalatable portion of their already dreary diet. It is small wonder that, as these patients grow and become more independent and discriminating, they begin to consume less of the medical foods. The consequence is that their hunger drives them to increase their intake of natural foods.

Our dissatisfaction with the existing medical foods for older patients has led us to examine their characteristics to see whether we could design a theoretical product with improved nutritional and organoleptic characteristics. We use Phenyl-Free as the primary comparator because it is the product commonly used in our program. This does not imply that we consider Phenyl-Free to be

Table 2-3. Standard Diet Prescription for a 6-Year-Old Child With Phenylketonuria[a]

	RDA	Natural Foods	"Ideal" EMF	Phenyl-Free
Phenylalanine (mg) 12.5 mg/kg/d	250	250	0	0
Protein equivalent (g) + 20%	28	5	23	60
Energy (kcal) (kJ/4.2)	1800	600	1200	1200
kcal/g protein	(64)[b]	120	52	20

[a] This diet is constructed for a 20-kg 6-year-old child with hyperphenylalaninemia in whom the phenylalanine tolerance is 12.5 mg/kg/d.
[b] Calculated from the RDA for protein and energy.

optimally formulated, nor that we promote this product over others that are available. Rather, we use it in this context because it is the product with which we are most familiar.

STUDIES ON THE ORGANOLEPTIC PROPERTIES OF ELEMENTAL MEDICAL FOODS

It is uniformly agreed that elemental medical foods smell and taste unpleasant. This is largely because of the poor organoleptic properties of several of the free L-amino acids. This perception is not necessarily held by smaller children who have been fed on these products since infancy. However, as these children grow and their discriminatory abilities increase, they become more aware of the taste of their "milk" and of the stringent limitations of their "nonmilk" foods. It seems likely, but has not been proved, that a better-tasting product, one preferably compatible with some cooking techniques, would be better accepted. We have therefore examined the taste discrimination for individual L-amino acids in a group of 24 volunteers, a PKU patient, and his mother.[23] Each amino acid and several elemental medical foods were prepared as a 0.5 percent (w/v) aqueous solution, as suggested by Solms et al.[24] Subjects were asked to describe what primary tastes they experienced and to rate taste and smell on a semantic differential scale of 0 (pleasant) to 10 (most unpleasant). The results from these tests are summarized in Figure 2-1. The more unpleasant solutions appear as taller columns on the right. In this experiment, the amino acids rated as neutral or pleasant tasting and those considered unpleasant are similar to those reported by others.[25] It can be seen that the most unpleasant solutions tested were the products Phenyl-Free, Maxamaid, and PKU-2 or -3 and the amino acids glutamate, methionine, and aspartate.

AMINO ACID PROFILE OF EXISTING ELEMENTAL MEDICAL FOODS

It is not clear how the amino acid profile and quantity of amino acids in current elemental medical foods were decided. Why, for example, do all the products contain relatively large quantities of L-glutamate and L-aspartate, when both are dispensable amino acids that contribute so largely to the unpleasant organoleptic qualities? Either their amino acid content is based on an erroneous assessment of the normal profile of casein or some other biologically important protein or the cost of individual components has been a factor in the design. Alternatively, they may have been chosen for their stability in a complex mixture.

If we consider the state of science during the 1950s and 1960s, acid hydrolysis was the only method on which to base the amino acid profile of proteins. It was not unreasonable to design elemental food products on the basis of this kind of information. As was known even in the 1950s, however, all the glutamine and asparagine residues are converted to their dicarboxylic equivalents by acid hydrolysis, whereas approximately 50 percent of the glutamate and aspartate in the hydrolysate actually represents the amidated amino acids. The complete amino acid sequence of bovine caseins has not yet been described. In other high-quality proteins, such as lactalbumin and egg white proteins,[26] aspartate and asparagine are present in about equal quantities, as is true for glutamate and glutamine. Thus, there is a biologic reason to incorporate the amidated amino acids in a medical food. In addition, both glutamine and asparagine have infinitely better taste qualities than do glutamate and aspartate, respectively (Fig. 2-1).

Similarly, when medical foods are consumed according to most standard diet prescriptions, several of the other unpleasant amino acids are ingested in quantities considerably in excess of the RDA. Therefore, it should be possible to reduce these and to add increased quantities of some of the pleasant or neutral-tasting compounds. In addition, it should be possible to increase the energy-to-protein ratio of the products. The amino acid content of most of the existing medical foods bears little relationship to the profile of cow's milk or other first-class proteins. In Phenyl-Free, for example, there is no alanine, proline, serine, or asparagine, yet this product is clearly capable of sustaining health in PKU patients over many years. There are no RDAs for dispensible amino acids. Indeed, if they are truly dispensible, there may be no need for any of them, provided enough nitrogen is provided in a bioavailable form. It would seem entirely inappropriate to design a product devoid of all dispensible amino acids for routine long-term use. It is possible, however, to eliminate some of

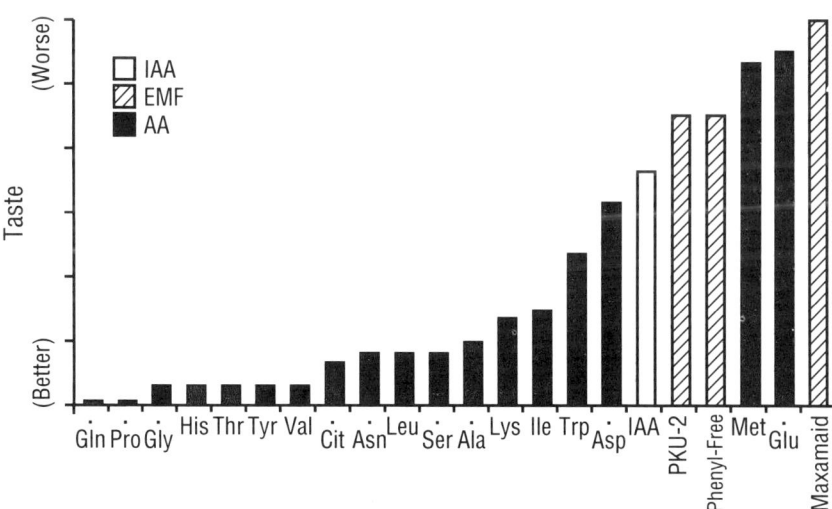

Fig. 2-1. Taste characteristics of L-amino acids and selected medical foods. A low score represents neutral or pleasant organoleptic characteristics. A high score indicates unpleasant properties. The score scale was 0 to 48 with 48 representing maximal unpleasant characteristics of smell, taste, and aftertaste. AA, amino acid; EMF, elemental medical food; IAA, indispensable amino acids.

them. Therefore, it should be possible to eliminate some of the unpleasant-tasting amino acids and to substitute a nonprotein amino acid for compounds such as L-arginine, which also have highly distinctive tastes. Finally, despite some efforts to find alternative forms of specific amino acids, such as oligopeptides, none of these has been tried, as far as we know, in a commercial product.

How, then, should an appropriate quantity and profile of amino acids be determined? Through the use of an expanded nutritional database that contains the complete amino acid profile of 1800 foods, one of us has determined the complete amino acid intake patterns in a cohort of children with PKU and in their control siblings[23] (Nutrition and Diet Services, 927 SE Rimrock Lane, Portland, OR 97267). We have used this information and the ideas presented in this chapter to design the amino acid profile and energy-to-protein ratio of a new or "ideal" medical food and are in the process of assessing its nutritional effects and its acceptance by our patients.

We propose that the taste qualities of an elemental medical foods can be radically altered by increasing the energy-to-protein ratio and by altering the amino acid profile while continuing to use free L-amino acids as the protein equivalent. Further refinement, using novel amino acid complexes, is to be hoped for. The nutritional quality of a medical food can also be improved by designing it to complement the nutritional contribution of the natural low-protein foods.

Should Tyrosine Be Added to Existing Products?

Over the past few years, it has been realized that fasting plasma tyrosine values in treated patients are often below the normal range, and our studies confirm these values. While some products have more tyrosine than Phenyl-Free, this latter product contains only 18 mg/g protein (9 mg/100 kcal). It seems clear that this is often not enough to maintain biochemical normality. It is not known in these cases, however, whether the low plasma tyrosine levels result in other biochemical abnormalities, such as reduced tissue levels of catecholamines, nor is it evident that they are associated with any clinical consequences. Nonetheless, it seems prudent to increase the daily tyrosine intake sufficiently to normalize the plasma levels.

The quantity of tyrosine required to achieve this is unknown. It probably varies for different ages and patients. It is unlikely that there is a stoichiometric relationship between increases in the dietary content of free L-tyrosine and its absorption, as limited solubility of the free amino acid under physiologic conditions (500 mg/L) limits its bioavailability. Since an additional source of tyrosine is needed in Phenyl-Free, and possibly in the other elemental medical foods as well, we currently add tyrosine to the dietary plan for our patients who use over-the-counter 500 mg capsules of L-tyrosine. Such bolus administration is likely to be less well absorbed than a similar quantity distributed throughout the day. It also presents a less balanced mixture of amino acids for absorption that could contribute to suboptimal use of limiting nutrients for physiologic purposes.

For these reasons, it is appropriate to consider alternative forms of tyrosine, such as oligopeptides or N-acctyl tyrosine, which is more soluble than the free amino acid (50,000 mg/L). The latter is already used in intravenous hyperalimentation solutions, although it is not currently an approved nutritional additive.

Should Other Amino Acids Be Increased?

Recently, Berry et al.[27] suggested that the branched-chain amino acids, valine, isoleucine, and leucine, can inhibit the entry of phenylalanine into the brain. These workers suggest that a mixture of these compounds could improve the reversible neuropsychiatric disorders of poorly treated PKU. They are currently embarked on a clinical trial of its efficacy.

ALTERING THE ROLE OF MEDICAL FOODS IN THE DIET

When the medical diet is eaten according to current prescription, it meets all nutritional requirements. Most medical foods provide excess quantities of protein and most nutrients. The nonformula foods are the "side dishes" of lesser nutritional importance. However, nonformula foods can be made to provide as much as 40 to 50 percent of the energy needs but less than 30 percent of the RDA for protein for most cases. With careful planning and perhaps new low-protein foods, it should be possible to increase the energy contribution from these sources further, although the tolerance for regular protein will remain unchanged. In reality, medical foods are often consumed in less than prescribed quantities, thereby reducing the energy intake, even when the intake of protein and most other nutrients is still adequate or even excessive. Moreover, diminishing the intake of medical food results in a compensatory increase of protein from other sources.

The importance of low-protein foods in preparing attractive and exciting meals cannot be overestimated. Without them, a diet consisting predominantly of formula and elementary foods, such as pure starches, sugars, and oils, is draconian. Therefore, every effort should be made to maximize the nutritional contribution from these low-protein products and natural foods first. Only the balance of the nutrients need be derived from a medical food. This approach demotes the medical foods from the status of the predominant source of essential nutrients to that of a nutritional supplement.

This switch represents a major philosophic change, altering the focus of nutritional concerns for the metabolic team. It entails careful analysis of the nutritional characteristics of the natural foods and according them sufficient importance in nutritional homeostasis. Conversely, since the formula provides less nutrition, there is an increased onus on the metabolic team to ensure the nutritional adequacy and balance of the total diet.

Although this approach makes sense in humanistic terms, it remains unproven whether any restrictive dietary regimen, no matter how palatable and varied, will be consistently adhered to by a young person to whom the widening scope of choices and self-determination are becoming increasingly apparent. What is clear is that the present system leaves much to be desired and that most school-age children and adolescents are not able to comply with current diet recommendations.

THE ROLE OF EDUCATION

There is extensive experience in treating children and adolescents with juvenile diabetes. Such patients are expected to learn about their diet and, by the age of 8 to 9 years, are expected to be able to perform their own injections and blood tests. While there are daily anxieties over the treatment of diabetes, the diet does not need to be monotonous or very restrictive. Treatment of PKU presents obvious contrasts to this, but it is our experience that from the age of 5 to 6 years, PKU patients are able to start learning about their diet and, indeed, are often eager to do so. In our clinic, it is customary to begin with simple games that have a nutritional message and to progress to more sophisticated programs. If we do not succeed in convincing the children that their diet really must be different, it will not matter how varied or palatable we make the medical foods or the recipes.

SUMMARY

Currently available nutritional products and protocols for treating hyperphenylalaninemia are far from optimal. All elemental medical foods have unpleasant organoleptic properties mostly attributable to certain of the L-amino acids. The energy-to-protein ratio of these products should be increased to provide a better balance of nutrients. Reduction in the total amino acids in medical foods should improve the organoleptic properties of these products, which may therefore be taken more willingly. This in turn will reduce the likelihood that patients will consume excess quantities of other foods to meet energy needs. In addition, the amino acid profile of these products should be altered to reduce the intake of the most obnoxious ones as far as feasible and possibly to increase the nitrogen intake by providing more of the neutral or pleasant-tasting ones. Some products appear to warrant an increase in their tyrosine content. The quantity and form that this should take remain uncertain. Consideration should be given to alternative compounds for the most unpleasant-tasting amino acids. We have designed, and are testing, a new product based on the principles outlined in this chapter.

Pressure should be put on industry to encourage the development of new low-protein and low-phenylalanine foods; low-protein recipe development should be continued. Low-protein products, combined with the low-protein natural foods, should be considered the primary source of nutrition for these patients. The elemental medical foods should then be designed to supplement these to provide a well-balanced diet close to the RDA. This change in emphasis places an increased onus on metabolic teams to ensure that the proportions of the diet from each source are properly balanced. The importance of education of patients regarding their dietary limitations and opportunities is emphasized.

ACKNOWLEDGMENT

This work was supported in part by grant HD26360-01 from the National Institute of Child Health and Human Development.

REFERENCES

1. Woolf LI, Vulliamy DG: Phenylketonuria with a study of the effect upon it of glutamic acid. Arch Dis Child 26:487, 1951
2. Armstrong MD, Tyler FH: Studies on phenylketonuria. I. Restricted phenylalanine intake in phenylketonuria. J Clin Invest 34:565, 1955
3. Bickel H, Gerrard I, Hickmans EM: Influence of phenylalanine intake on phenylketonuria. Lancet 2:12, 1953
4. Scriver CR, Kaufman S, Woo SLC: The hyperphenylalaninemias. p. 495. In Scriver CR, Beaudet AL, Sly WS, Valle D (eds): The Metabolic Basis of Inherited Disease. 6th Ed. McGraw Hill, New York, 1989
5. Blaskovics ME, Schaeffler GE, Hack S: Phenylalaninaemia: Differential diagnosis. Arch Dis Childh 49:835, 1974
6. Francis DEM: Phenylketonuria. p. 224. In Diets for Sick Children. 4th Ed. Blackwell Scientific Publications, Oxford, 1987
7. Rouse BM: Phenylalanine deficiency syndrome. J Pediatr 69:246, 1966
8. Smith I, Francis DEM: Amino acid disorders. p. 295. In McLaren D, Burman D (eds): Textbook of Paediatric Nutrition. 2nd Ed. Churchill Livingstone, Edinburgh, 1982
9. Bessman SP: Genetic failure of fetal amino acid "justification": A common basis for many forms of metabolic, nutritional and "non-specific" mental retardation. J Pediatr 81:834, 1972
10. Andrews L: State Laws and Regulations Governing Newborn Screening. American Bar Foundation, Chicago, 1985
11. Buist NRM: Laboratory aspects of newborn screening for metabolic disorders. Lab Med 19:145, 1988
12. Smith I, Lobascher ME, Stevenson JE, et al: Effect of stopping low-phenylalanine diet on intellectual progress of children with phenylketonuria. Br Med J 2:723, 1978
13. Kock R, Azen C, Freidman EG, Williamso ML: Paired comparisons between early treated PKU children and their matched sibling controls on intelligence and school achievement test results at eight years of age. J Inher Metab Dis 7:86, 1984
14. Lou HC, Guttler F, Lykkelund C, et al: Decreased vigilance and neurotransmitter synthesis after discontinuation of dietary treatment of phenylketonuria in adolescents. Eur J Pediatr 17:17, 1985
15. Clarke JTR, Gates RD, Hogan SE, et al: Neuropsychological studies on adolescents with phenylketonuria returned to phenylalanine-restricted diets. Am J Ment Retard 92:255, 1987

16. Waisbren SE, Schnell R, Levy HL: Intelligence and personalty characteristics in adults with untreated atypical phenylketonuria and mild hyperphenylalaninemia. J Pediatr 105:955, 1984
17. Primrose DA: Phenylketonuria with normal intelligence. J Ment Defic Res 27:239, 1983
18. Smith I, Glossop J, Beasley M: Fetal damage due to maternal phenylketonuria: Effects of dietary treatment and maternal phenylalanine concentrations around the time of conception. (An interim report from the U.K. Phenylketonuria Register.) J Inher Metab Dis 13:651, 1990
19. Yannicelli S, Ernest AE, Neifert MR, McCabe ERB: Guide to breastfeeding the infant with PKU. U.S. Department of Health and Human Services, Public Health Service, Bureau of Maternal and Child Health and Resources Development, Washington, DC, 1988
20. Schuett VE: Low Protein Cookery for PKU. 2nd Ed. University of Wisconsin Press, Madison, WI, 1988
21. National Research Council (U.S.): Recommended Daily Allowances. 10th Ed. National Academy of Sciences, Washington, DC, 1989
22. Acosta PB: The Ross Metabolic Formula System. Nutrition Support Protocols. Ross Laboratories, Columbus, OH, 1989
23. Prince A, Lecklem J, Buist NRM: Evaluation of protein and energy in elemental medical foods for use in phenylketonuria. (submitted)
24. Solms J, Vuataz L, Egli RH: The taste of L- and D-amino acids. Experientia 21:692, 1965
25. Schiffman SS, Gagnon J: Comparison of taste qualities and thresholds of D- and L-amino acids. Physiol Behav 27:51, 1981
26. Brew K, Vanaman TC, Hill RL: Comparison of the amino acid sequence of bovine alpha lactalbumin and hen's egg-white lysosome. J Biol Chem 242:3747, 1967
27. Berry HK, Branner RL, Hunt ML, White PP: Valine, isoleucine and leucine. A new treatment for phenylketonuria. Am J Dis Child 144:539, 1990

3

ADVANCES IN THE TREATMENT OF AMINO ACID AND ORGANIC ACID DISORDERS

William L. Nyhan · Marylynne Rice-Asaro · Phyllis Acosta

INTRODUCTION

Treatment of the inborn errors of amino acid metabolism, and particularly the organic acidemias, is demanding. Some notable successes in bringing some of these patients to adulthood provide ample evidence that these diseases are treatable. Yet most patients with certain diseases are still dying in infancy and childhood. Even simple things are known with little security, and even among experts with considerable experience there is very little consensus. Nevertheless, advances continue to be made in the understanding of these diseases and in their management. We can expect that the future will bring further progress in our quest for optimal therapy.

CLASSIC NUTRITIONAL MANAGEMENT

What is now considered classic treatment for the inborn errors of metabolism is represented by the nutritional restriction of precursors of the compound whose metabolism is blocked as a consequence of deficiency of an enzyme. This approach was ushered in by the development of successful therapy for phenylketonuria (PKU).[1] The defective enzyme in this disorder is phenylalanine hydroxylase. Deficiency in its activity leads to accumulation of phenylalanine. The amino acid is then converted to phenylpyruvic acid, the phenylketone that gave the disease its name. Additional products include phenylacetic acid, which is responsible for the musty or animal-like odor of the patient with untreated PKU. Accumulation of phenylalanine and its products results in metabolic milieu that is unfriendly for the developing central nervous system (CNS). Untreated patients generally develop intelligent quotient (IQ) levels of less than 50, virtually all within the range of severe mental retardation.

Fundamental to treatment is a diet so restricted in phenylalanine that the concentrations of phenylalanine in the plasma are maintained at an acceptable level, and all the secondary metabolites are absent from the urine. Acceptable levels have generally been 180 to 900 µmol/L. Smith et al.[2] have recommended a smaller window of 120 to 300 µmol/L. Their data appear to show a linear relationship between IQ and mean concentration during therapy of more than 300 µmol/L, but the differences were not clear until levels exceeded 800 µmol/L. Setting the lower level at 120 µmol/L is less secure; patients enduring long periods below this level appeared to have low IQ levels, only in the early cohort born before 1971. No other lower limit was assessed. The combination of early diagnosis through the neonatal screening programs and early effective therapy results regularly in a normal IQ.

Phenylalanine as an essential amino acid must be provided in quantities sufficient for growth. Requirements change progressively in infancy and childhood. Optimal management requires repeated nutritional assessment of dietary adequacy and the measurement of optimal growth in height, weight, and head circumference, as well as the quantitative determination of concentrations of phenylalalanine in plasma. Preparations are available for use in infant formulas, such as Analog XP (Ross) and Lofenelac (Mead Johnson), that facilitate dietary management. Lists are available for the phenylalanine contents of foods.[3]

Nutritional management is much more demanding in most of the organic acidemias in which the catabolism of more than one amino acid is altered by the defective enzyme. Management of patients with disorders of propionate exemplifies these issues. Propionic acidemia and methylmalonic acidemia present in the classic organic acidemia fashion with vomiting, ketonuria, and metabolic acidosis leading to dehydration, coma, apnea, and death. The enzymatic defects in the pathway are in propionyl CoA carboxylase, a biotin-requiring enzyme, and methylmalonyl CoA mutase, an enzyme with a vitamin B_{12} cofactor, deoxyadenosylcobalamin. In patients with methylmalonic acidemia as a result of defects in the pathway from dietary vitamin B_{12} to deoxyadenosylcobalamin, the mainstay of management is the administration of pharmacologic doses of vitamin B_{12}. Therefore, it is critical to determine early whether a patient is B_{12} responsive. Although we test all patients with propionic acidemia for biotin responsiveness and currently treat one patient with biotin on the basis of some chemical evidence of responsiveness,[4] we have yet to encounter clinical evidence that any patient with propionic acidemia benefited from treatment with biotin.

In patients with apoenzyme defects of either enzyme, nutritional therapy is required. Our guiding principle is to restrict the intake of the four essential amino acid precursors, isoleucine, valine, threonine, and methionine, to the quantities necessary to meet the anabolic needs of a growing patient. We assume that intake in excess of requirement leads to accumulation of toxic intermediates. We monitor the urinary excretion of organic acids: methylmalonic acid in methylmalonic acidemia, and methylcitric and hydroxypropionic acids in propionic acidemia. We also monitor growth, especially in weight, nitrogen balance, and the concentrations of albumin and amino acids in the plasma. Requirements for essential amino acids vary among patients and change with age. Thus, they are determined individually and redetermined after the passage of time. We have conceptualized theoretical curves for this process, in which a patient's plateau level of organic acid in the urine is appreciably above zero, reflecting the constant availability of substrates to catabolic enzymes.[5] This should actually be a horizontal line, even during anabolism, in which excretion is plotted against the intake of protein or precursor amino acid. When the intakes of precursor amino acids required exceed requirements for growth, we expect a change in the slope of the line and thereafter a linear relationship between precursor provided and product excreted. The point at which the slope changes provides our ceiling for intake, and we do not exceed these amounts of precursor. The floor is provided by the level of protein intake at which growth in weight occurs and nitrogen balance reaches a satisfactory positive level. In our view, the optimum dietary quantities of precursor amino acids should be between the floor and the ceiling.

This approach is most readily pursued in the management of patients with methylmalonic acidemia, as the excretion of methylmalonate so closely mirrors the metabolic condition of the patient[6] (Fig. 3-1). The infants illustrated were 3 to 8 months and 13 to 16 months of age when these data were obtained. Their sizes were virtually identical during the study. The points marking departure of the slope from the plateau level were at 0.75 and 1.17 g/kg of protein. Thereafter, there was a linear relationship in which 1 mole of methylmalonate was excreted for each mole of precursor ingested. In the older patient, in whom the point was at 1.17 g/kg of protein, nitrogen retention was 51 mg/kg at 0.75 g/kg of protein and 80 mg/kg at 1.0 g/kg. The mean daily gain in weight was 15 g at 0.75 g/kg and 70 g at 0.95 g/kg. His requirements were therefore considered to be 0.75 to 1.17 g/kg, and a diet containing 1.0 g/kg was employed. When this patient ingested 1.5 g/kg of protein, a negative nitrogen balance developed, indicating the toxic effect of accumulated metabolites on protein synthesis, and consistent with the failure to thrive characteristic of this disease. The younger infant could not tolerate a protein intake of 1.5 g/kg without ketoacidosis and clinical illness. Her excretion point occurred at 0.75 g/kg of protein. Nitrogen balance was satisfactory at 0.70 g/kg, as was weight gain. Therefore, her requirement was considered to be 0.70 to 0.75 g/kg.

These levels of protein intake are just short of protein malnutrition. Concentrations of albumin in plasma tend to be low. In order to avoid nutritional edema and interference with growth, we avoid levels lower than 2.5 g/dl. Concentrations of amino acids in plasma are also lower than those generally accepted as normal. However, occasionally the level of a single amino acid is so low as to limit growth. In such cases, we tend to supplement with that amino

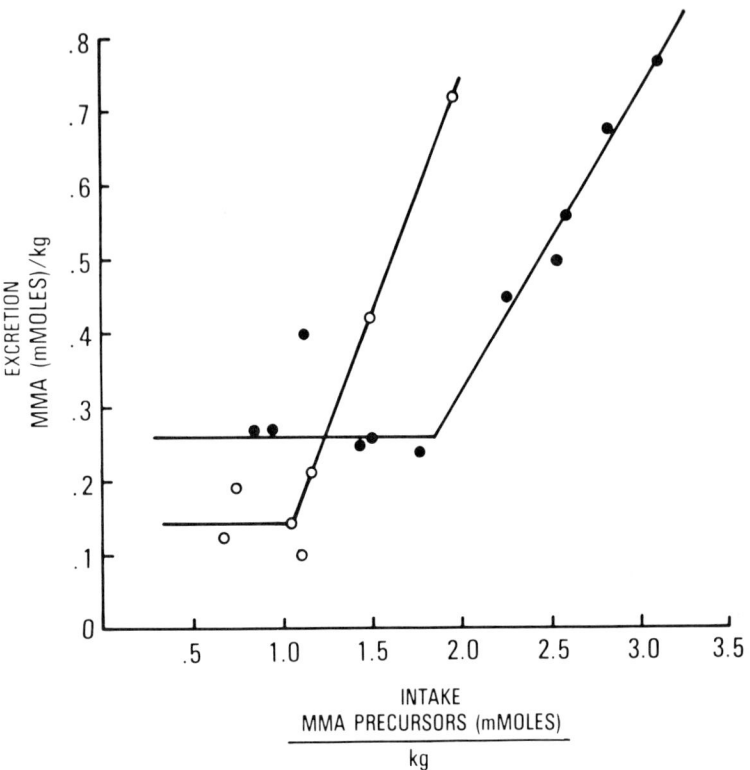

Fig. 3-1. Urinary methylmalonate excretion and its relationship to the intake of amino acid precursors of methylmalonate provided by whole protein in two infants with methylmalonic acidemia. (From Ney et al.,[6] with permission.)

acid. Otherwise we supply the amino acids as whole protein and the calories as carbohydrate and fat. In some instances, it is appropriate to use special mixtures from which the precursor amino acids have been excluded, such as Analog and Maxamaid XMET, THRE, VAL, ISOLEU (Ross) and OS1 and OS2 (Mead Johnson).

ANABOLISM AS AN ADJUVANT TO THERAPY

The provision of mixtures of amino acids lacking those whose metabolism is altered permits a unique use of the forces of anabolism in therapy. This principle is best illustrated in the management of the acute crisis in maple syrup urine disease.[7] This condition may be the initial neonatal episode of coma or a later episode. It usually follows an intercurrent infection but may also be a consequence of dietary indiscretion. This disease remains potentially lethal whenever an acute episode supervenes. Catabolic forces are so powerful that a patient who is unable to eat, even if receiving parenteral 10 percent glucose, will experience progressive elevation of plasma levels of leucine.

The fundamental defect in maple syrup urine disease is in the branched-chain ketoacid decarboxylase that catalyzes the conversion of the ketoacid analogues of leucine, isoleucine, and valine to their respective CoA ester products. The activity of the enzyme is conveniently measured in leukocytes, fibroblasts, or amniocytes by measuring the conversion of ^{14}C-leucine to $^{14}CO_2$. In patients with classic maple syrup urine disease, the activity approximates zero. This defect leads to the accumulation of leucine, isoleucine, and valine in plasma, as well as alloisoleucine. Concentrations of alanine are decreased. The concentration of leucine is the best index of the patient's clinical condition. Levels of this amino acid may be elevated in urine; however, the renal conservation of branched-chain amino acids is so efficient that the condition is assessed by assay of plasma, not of urine.

Autopsy studies in this disease provided early evidence of delayed myelination in the central nervous system (CNS). This may now be followed by magnetic resonance imaging (MRI). Hypodensity of the white matter is readily observed in the untreated patient. The success of treatment may be reflected in a substantial increase in the density of the white matter. We are also embarked on sequential studies of the development of language and cognition in these patients. Slow development or abnormalities in language may be seen despite normal motor development.

The usual approach to management of the acute episode has been exchange transfusion or peritoneal dialysis, or both, but these measures are not very effective. Hemodialysis would probably be a better modality, but there is little experience with this form of therapy, and its logistics are formidable. We began considering the potential of anabolic forces some years ago. Initial experience was published by Saudubray et al.[8] Their calculations indicated that the anabolic effect of a gain in body weight of even as little as 50 g could be associated with sufficient incorporation of leucine into protein to decrease the plasma concentration of leucine by 3000 µmol/L. Conceptually, one could approach the power of anabolism in a patient with maple syrup urine disease by providing the standard MSUD formula, which contains no leucine, isoleucine, or valine. Together, the amino acids in this mixture might be expected to lay down accumulated leucine, isoleucine, and valine into protein, clearing body fluids. Each of the patients reported was also treated with peritoneal dialysis, so that it was not really clear which was responsible for the improvement.[8] However, measurement of the leucine in the peritoneal dialysate indicated that it was so small that it was likely that anabolism had brought the leucine down.

We treated a patient in the initial episode without the use of dialysis or exchange transfusion.[9] This patient was not in a deep coma, but he was virtually flaccid. Amino acids and calories were provided by nasogastric drip. There was a very prompt linear fall in the plasma concentration of leucine from a level of 3540 µmol/L (Fig. 3-2). When the leucine reached reasonable levels, the patient's state of consciousness improved, so much so that protein-contain-

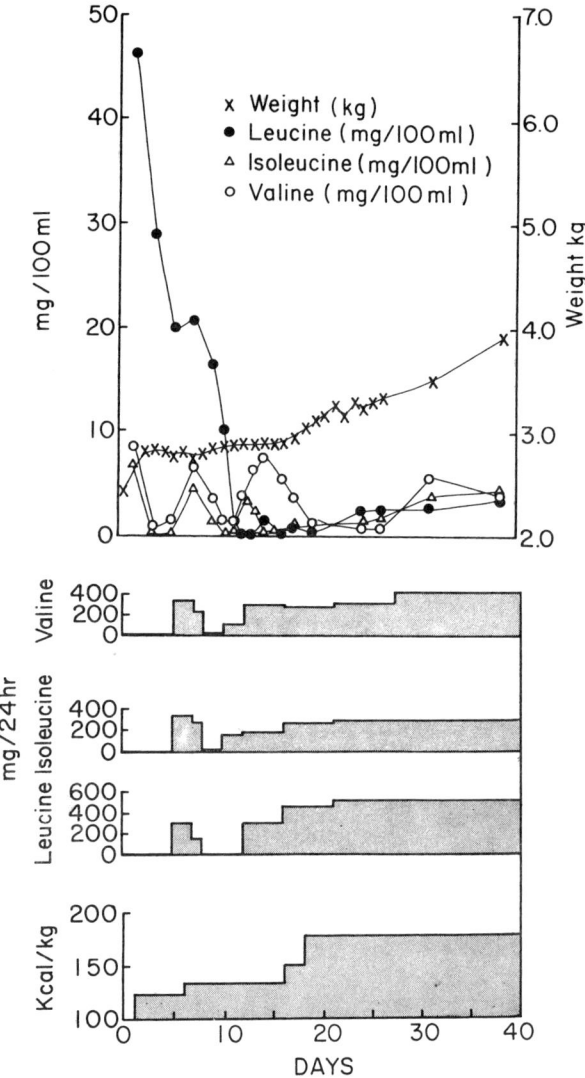

Fig. 3-2. Effect of anabolism in therapy. Management of the initial episode in maple syrup urine disease by provision of calories and amino acids, excluding the branched-chain amino acids by nasogastric tube. (From Nyhan and Rice,[9] with permission.)

ing feedings were initiated. Examination of leucine concentration indicated, however, that it decreased to only 1500; it also began to rise. The protein was removed again until levels reached the normal range. This approach has now been used successfully on a number of occasions.

Such an approach to therapy has some major conceptual problems. In many intensive care units, one would not be permitted to employ nasogastric feedings in comatose patients. Furthermore, these patients frequently vomit, even in health. In an acute episode, they are sometimes unable to retain even slowly introduced nasogastric feedings. Aspiration is a realistic risk in such a patient. Another issue relates to the anorexia that characterizes the patient with maple syrup urine disease. Once committed to nasogastric feeding, it may be exceedingly difficult to convince these patients to eat and drink on their own again. For these purposes, an intravenous formulation was required. Experience with such a solution is illustrated in Figure 3-3. The patient was acutely ill after the development of intercurrent illness. Vomiting was inclusive of virtually everything. On admission he was treated with parenteral fluid containing large amounts of water and glucose. On this regimen, this serum concentration of leucine rose from approximately 400 µmol/L to almost 1200, as the patient became ataxic and increasingly lethargic. He was then treated with our first mixture of amino acids designed for this disease. In this acute emergency, we employed the Trophamine mixture from which the isoleucine, leucine, and valine had been removed and replaced with alanine. The test for ketones in the urine became negative for the first time in 34 hours. The fall in the concentrations of leucine, isoleucine, and valine was dramatic, and the patient appeared well. Again this led to the early introduction of feedings by mouth containing protein, which were progressively increased from day 2 to 0.5 g/kg whole protein, this patient's normal daily prescription. The plasma concentration of leucine rose, and the amounts of protein of the plasma were decreased, followed by a decrease to normal limits by day 8 of the plasma concentration of leucine. Caloric intake during this period amounted to one-third to one-half the patient's usual requirement, but this did not inhibit the effectiveness of the solution. Alanine in a concentration of more than 1 g/dl was effective in maintaining a normal concentration of alanine during inadequate caloric intake. The gluconeogenic properties of this amino acid may have provided substrate for energy production and promotion of anabolism. When the metabolic status was improved and energy needs met, the plasma concentration of alanine rose. By that time, the intake of the amino acid mixture had essentially doubled, so the alanine concentration in plasma may have reflected this major excess.

The composition of the most recent formulation is shown in Table 3-1. A tendency of the solution to turn yellow was associated on analysis with a disappearance of tyrosine. This effect and problems of solubility led to the substitution of *N*-acetyltyrosine. The solution has been employed with success in two episodes in a patient with maple syrup urine disease (Fig. 3-4, left). The three episodes illustrate the similarity in the response of the plasma concentrations of amino acids to treatment intravenously or by nasogastric tube as in Figure 3-2. The parenteral solution has been prepared fresh before each use and, with experience, the time interval for shipment has been shortened. Nasogastric administration was started while awaiting the arrival of the parenteral solutions (Fig. 3-4, left). The other two episodes were treated by nasogastric administration.

In general, we have provided the parenteral mixture at a rate of 0.5 g/kg/d of amino acids. We recommend withholding whole protein until the concentration of leucine is below 400 µmol/L. In the episode at the left (shown in Fig. 3-2), he initially received 0.55 g/kg of MSUD mixture nasogastrically and

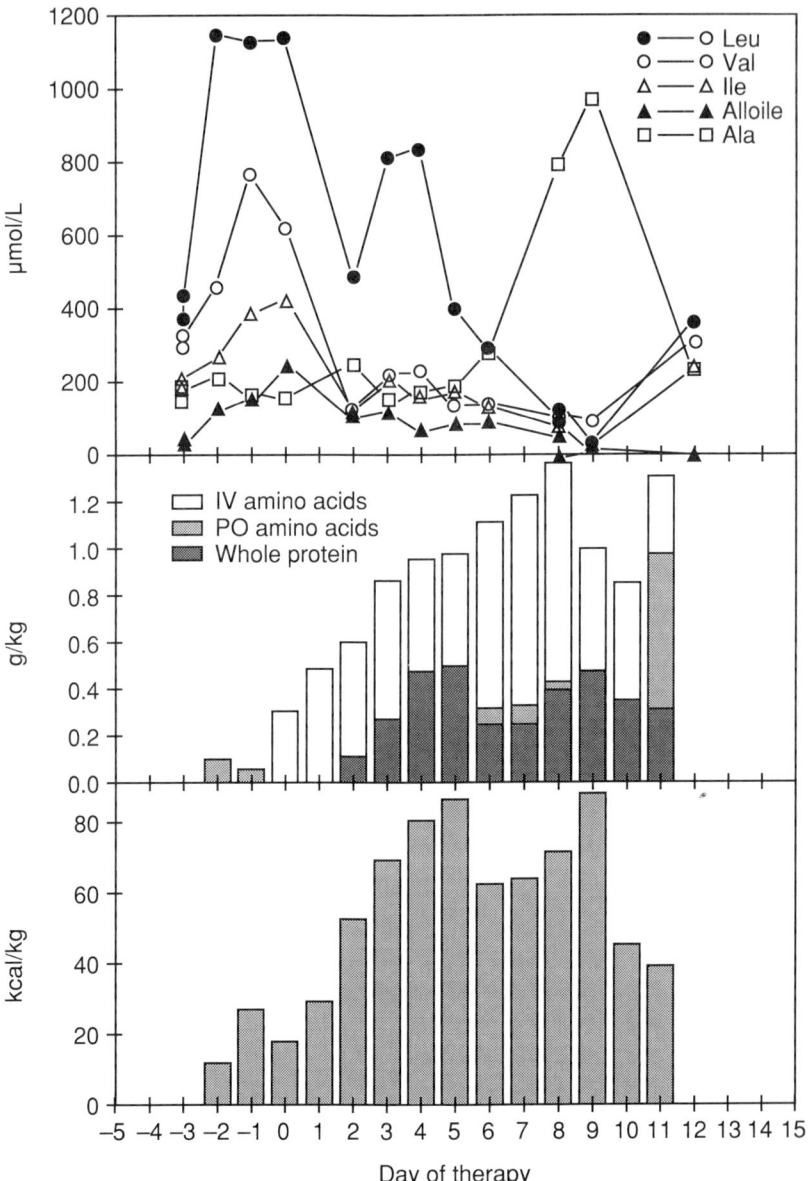

Fig. 3-3. Further effects of anabolism. Use of a parenteral mixture of amino acids excluding branched-chain amino acids in an acute relapse in maple syrup urine disease. (From Nyhan and Rice,[9] with permission.)

on day 3, 0.34 g/kg of the solution orally and 0.2 g/kg nasogastrically. Overall the concentration of leucine fell from 877 to 177 on day 4 and to 113 on day 5. The modest rise on the last day shown reflected efforts to convert him to oral intake before discharge, by deletion of the nasogastric feeding.

Table 3-1. Parenteral Solution for Acute Intervention in Maple Syrup Urine Disease[a]

Amino Acid	mg/2.5 g
Alanine	600
Arginine	300
Aspartic acid	40
Cysteine	100
Glutamic acid	50
Glycine	50
Histidine	150
Isoleucine	0
Leucine	0
Lysine	300
Methionine	150
Phenylalanine	200
Proline	150
Serine	100
Taurine	10
Threonine	200
Tryptophan	50
N-Acetyl tyrosine	50
Valine	0

[a] Solution prepared by Pharmathera, 1785 Nonconnah Blvd., Memphis, TN 88132.

SOLUTIONS FOR PARENTERAL NUTRITION IN INBORN ERRORS OF METABOLISM

In addition to the anabolic approach to therapy outlined above, there are a variety of more general indications for special mixtures for use in patients with metabolic disease. Among the first we encountered before the development of parenteral solutions is illustrated in Figure 3-5. A 6-year-old boy with maple syrup urine disease developed obstructive apnea because of enormous tonsils, for which he underwent a tonsillectomy. A variety of complications made it necessary to manage this patient without the benefit of his usual alimentation, for a period of 5 days. A period this long, especially along with surgery, would doubtless have led to major elevations of concentrations of leucine and other branched-chain amino acids. A parenteral maintenance solution would have been ideal. Failing that, we inserted a nasogastric tube at surgery, and the MSUD mixture was dripped in slowly. Calories were provided by parenteral glucose, and small amounts of whole protein were added to the nasogastric mixture. Levels of branched-chain amino acids in plasma were maintained satisfactorily. The rise on the last day reflected his continued anorexia and the cessation of the intravenous solution, to encourage return to oral intake. He took in considerably less fluid and calories than optimal that day. Nevertheless, he continued to improve in this respect as the day went on, remained well, and was discharged.

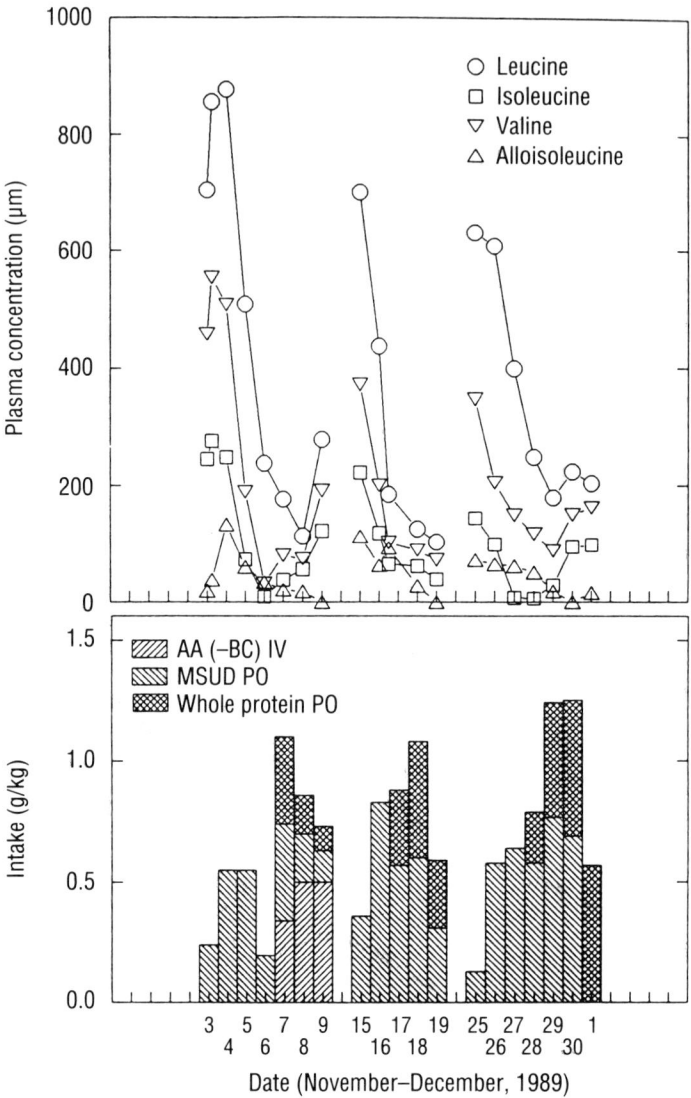

Fig. 3-4. Treatment of three mild episodes in a patient with maple syrup urine disease by a combination of intravenous and nasogastric administration. PO, almost exclusively nasogastric feeding.

These considerations indicated that there was a need not only for therapeutic solutions but also for solutions that could be used for maintenance in a patient unable to tolerate enteral feedings for a period. A solution devised for this purpose in maple syrup urine disease is shown in Table 3-2. We tested this solution in two patients with maple syrup urine disease while they were in good health. In each case we first conducted a 48-hour nitrogen balance at the Clinical Research Center, while the patient received the usual regimen. The

Advances in the Treatment of Amino Acid and Organic Acid Disorders   55

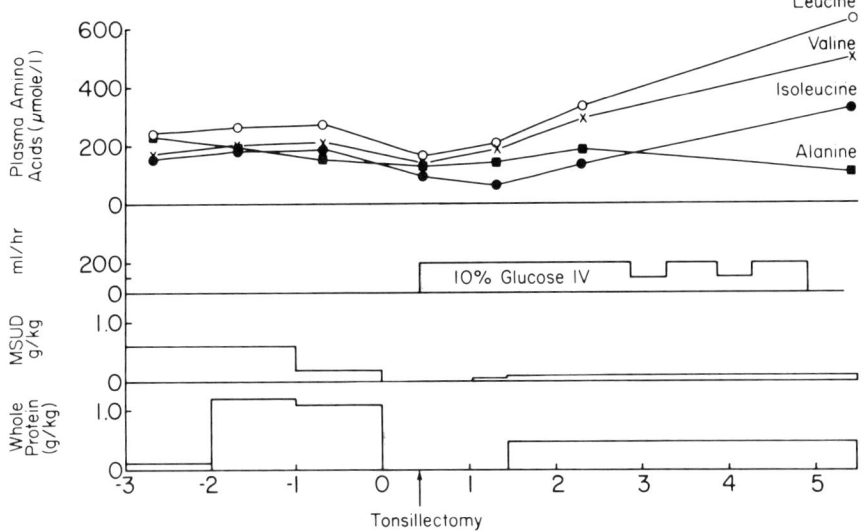

Fig. 3-5. Maintenance of a patient with maple syrup urine disease (MSUD) during recovery from tonsillectomy. MSUD (ordinate) represents provision of standard oral MSUD mixture by continuous nasogastric drip postoperatively and by mouth preoperatively.

Table 3-2. Parenteral Solution Designed for Maintenance of Patients With Maple Syrup Urine Disease

Amino Acid	mg/2.5 g
Alanine	500
Arginine	320
Aspartic acid	40
Cysteine	100
Glutamic acid	50
Glycine	50
Histidine	150
Isoleucine	40
Leucine	50
Lysine	300
Methionine	150
Phenylalanine	150
Proline	150
Serine	100
Taurine	10
Threonine	200
Tryptophan	50
N-Acetyl tyrosine	50
Valine	40

plan was to give a high-calorie mix containing no protein for 12 to 16 hours, after which the patient received all calories and amino acids intravenously, and the nitrogen balance was repeated. Plasma concentrations of amino acids were determined.

In the first patient (Table 3-3), in his usual state of health and usual dietary regimen, amino acid concentrations are generally either normal or only slightly elevated. The prescription was for 0.66 g/kg of whole protein and 0.65 g/kg of the amino acids from MSUD Diet Powder. Within 2 days on this regimen, he was shown to be in nitrogen balance. His serum concentration of leucine on admission was 555 µmol/L. The corresponding values for isoleucine and valine were 204 and 371 µmol/L. The alanine was 112 µmol/L, somewhat reduced. Two days after admission, he was given his usual quantity of calories in the mixture containing no protein at all, but this was carried out for a period of 23 hours. During that time, his plasma leucine rose to 788 µmol/L. In response to 12 hours of administration of the parenteral formula, leucine fell to 601, isoleucine to 142, and valine to 271 µmol/L. By 24 hours, the corresponding values were 494, 137, and 234, respectively, and by 48 hours, 328, 110, and 157 µmol/L, respectively. Alanine concomitantly rose, the values ranging from 485 at 12 hours to 224 µmol/L at 48 hours, from 129 µmol/L before treatment..

We concluded from these data that the parenteral mixture is certainly capable of maintaining exemplary levels of the branched-chain amino acids in a short trial such as this. However, the 23-hour period without protein was clearly catabolic, even though he maintained his usual quantity of calories. This was evident particularly in the increased concentration of leucine in the plasma. The forces of anabolism were evident by the fact that the provision of the parenteral mixture led to a substantial decrease, not just in the concentration of leucine, but in the other branched-chain amino acids as well. These data indicate that, under certain conditions, this mixture might well be therapeutic.

The patient was in positive nitrogen balance during the 2 days of the parenteral alimentation. No stool was collected during this period, so this finding may be somewhat artificial, as compared with an average of 0.57 g of nitrogen for 24 hours in his stool on the oral regimen. It is of interest that the urine nitrogen excretion also decreased on the parenteral mixture, although the numbers were not very great. On the oral regimen, it was 1.92 and 2.60 g; on the parenteral regimen, it was 1.57 and 1.87 g. The 48-hour total urinary nitrogen was 4.52 g of nitrogen on the oral regimen and 3.44 g on the parenteral regimen. In our other patient with maple syrup urine disease, the nitrogen balance was appreciably more positive with the parenteral mixture than on his oral usual regimen. It was not surprising that in 48 hours he wasted 0.96 g of stool nitrogen during the oral regimen, decreasing to 0.22 on the parenteral regimen. Why he wasted 8.53 g of urinary nitrogen during the 48 hours of oral and only less than one-half that much (3.67 g), when he was receiving the parenteral mixture is unclear (Table 3-4). Concentrations of amino acids in plasma in this patient did not increase just before the parenteral solution—he received the no-protein diet for less than 12 hours. However, concentrations of

Table 3-3. Parenteral Maintenance For Maple Syrup Urine Disease in Patient A

	Home Diet		Protein-Free Formula	IV Amino Acids		
	\multicolumn{6}{c}{Day}					
	1	2	3	4	5	6
				Maintenance Mixture		
Diet						
g/kg	0.65 MSUD 0.66 whole protein	0.64 0.65		1.5 IV	1.5 IV	—
cal/kg	71	62		71 PO 6 IV	72 PO 6 IV	—
Weight (kg)	19.3	19.4	19.4	19.4	19.4	19.6
Nitrogen balance Output/24 h						
Urine	1.92	2.60		1.57	1.87	
Stool	0.57	0.57		0	0	
Intake/24 h	4.08	3.98		4.8	4.8	
Balance (g)	+1.59	+0.81		+3.23	+2.93	
Plasma concentrations of amino acids (μmol/L)						
Leucine	555		788	601	494	328
Isoleucine	204		197	142	137	110
Valine	371		357	271	234	157
Alanine	112		129	485	384	224
Alloisoleucine	96		72	11	93	83
Tyrosine	45		13	26	32	24

Abbreviation: MSUD, maple syrup urine disease mixture.

leucine and the other amino acids in this patient also decreased, while that of alanine increased.

Although the entire source of tyrosine during parenteral maintenance was *N*-acetyltyrosine, no acetyltyrosine was found in the urine. On this regimen, plasma concentrations of tyrosine were maintained. In fact, in each patient the plasma concentration of tyrosine fell during the baseline period without protein, and the level rose on administration of the *N*-acetyltyrosine. The first patient received 10.8 mmol of *N*-acetyltyrosine each day (0.55 mmol/kg). The second patient received 10.8 to 10.6 mmol of *N*-acetyltyrosine (0.3 mmol/kg).

Table 3-4. Parenteral Maintenance For Maple Syrup Urine Disease in Patient B

	Home Diet			IV Amino Acids		
	Day					
	1	2	3	4	5	6
				Maintenance Mixture (0.86 g/kg/amino acid/d)		
Diet						
g/kg	0.25 MSUD 0.56 whole protein	0.25 MSUD 0.51 whole protein				
cal/kg	63.2	72.6		75 PO 3.6 IV	75 PO 3.2 IV	
Weight (kg)	32.0	31.75	31.75	32.0	32.5	32.25
Nitrogen balance						
Output/24 h (48 h)[a]						
Urine	5.01	3.52	(8.53)	1.59	2.08	(3.67)
Stool	0.48	0.48	(0.96)	0.11	0.11	(0.22)
Intake/24 h	4.13	4.19		4.75	4.20	
Balance (g)	−1.36	+0.19		+3.05	+2.01	
Concentration of amino acids (μmol/L)						
Leucine	281			181	134	94
Isoeucine	122			88	70	70
Valine	257			167	128	103
Alanine	166			226	365	512
Alloisoleucine	28			25	25	24
Tyrosine	48			22	28	39

[a] The number in parentheses represents the 48-hour nitrogen totals.
Abbreviation: MSUD, maple syrup urine disease mixture.

The absence of N-acetyltyrosine in the urine was unexpected. We are accustomed to finding this compound in the urine of patients receiving Trophamine. In healthy adult volunteers, as much as 60 percent of infused N-acetyltyrosine was excreted in the urine within 8 hours.[10] Trophamine contains 40 mg/2.5 g of N-acetyltyrosine and 17 mg/2.5 g of tyrosine, while our solutions contain 50 mg/2.5 g of N-acetyltyrosine. A diet of 1 g/kg of protein provides 20 mg (0.36 mmol)/kg.

Solutions have also been designed for use in PKU, propionic acidemia, methylmalonic acidemia, tyrosinemia, and isovaleric acidemia (Table 3-5). For the organic acidemias, we have followed the pattern of experience with maple syrup urine disease and designed acute intervention formulations as well as

Table 3-5. Parenteral Amino Acid Solutions for Patients With a Variety of Disorders of Amino Acid Metabolism (in mg/2.5 g)

Amino Acid	PA/MMA			IVA			
	Acute Intervention	Maintenance #1	Maintenance #2	Acute Intervention	Maintenance	PKU	TYR
L-Alanine	500	500	425	400	400	200	200
L-Arginine	300	250	150	290	240	200	300
L-Aspartic	40	40	40	40	40	40	80
L-Carnitine[a]	150	150	150	150	150	10	20
L-Cysteine	100	80	80	65	65	65	65
L-Glutamic	50	25	25	50	50	50	50
L-Glutamine	150	150	150	150	150	150	150
Glycine	25	25	25	50	50	50	50
L-Histidine	150	150	125	130	130	150	150
L-Isoleucine	0	40	80	150	150	150	150
L-Leucine	200	300	300	0	50	300	300
L-Lysine	300	300	300	300	300	300	300
L-Methionine	0	25	50	60	60	60	30
L-Phenylalanine	150	150	125	130	130	30	30
L-Proline	175	50	40	190	190	190	150
L-Serine	90	40	40	90	90	90	90
Taurine	20	20	20	20	20	20	10
L-Threonine	0	40	80	100	100	100	110
L-Tryptophan	50	50	50	55	55	55	55
L-Tyrosine	50	50	50	80	80	100	30
L-Valine	0	45	80	150	150	200	200

[a] Carnitine was not included in the calculation of the 2.5-g total.
Abbreviations: PA, propionic acidemia; MMA, methylmalonic acidemia; IVA, isovaleric acidemia; PKU, phenylketonuria; TYR, tyrosinemia.

maintenance formulas. For PKU and tyrosinemia, we have designed only maintenance formulas. For the disorders of propionate metabolism, the two maintenance formulas contain different amounts of the precursor amino acids, while the acute intervention formulas did not include precursor amino acids.

Patients with inborn errors of metabolism experience periods of anorexia, catabolism, or inanition, during which parenteral administration of a standard total parenteral nutrition (TPN) formula is useful, as reported in two patients by Kahler et al.[11] We have used this approach particularly in the case of intercurrent diarrhea, in which a patient receiving a severely protein-restricted diet with abundant calories can rapidly develop kwashiorkor.

AMINO ACID SUPPLEMENTATION: ALANINE EFFECT

The use of special mixtures of amino acids leads logically to consideration of more specific supplementation with single amino acids for specific purposes. We began working with alanine[12] on the basis of considerations of the management of disorders of propionate metabolism. In these infants, catabolism of

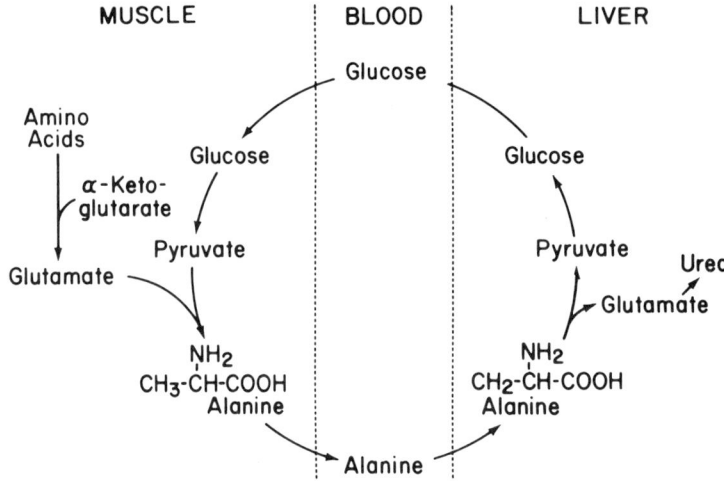

Fig. 3-6. Glucose–alanine cycle. Amino acids are catabolized to provide nitrogen for the net synthesis of alanine in muscle. It is transported to the liver, where it provides substrate for gluconeogenesis. (From Kelts et al.,[12] with permission.)

certain branched-chain and other amino acids led to the accumulation of propionate manifested as clinical illness. Our hypothesis was that some of the obligatory catabolism of these amino acids was to feed the alanine glucose cycle (Fig. 3-6) and serve the needs of glyconeogenesis. Also, a large supply of exogenous alanine might have a sparing effect in this process. Net synthesis of alanine takes place in muscle and net uptake of this gluconeogenic amino acid occurs in the liver. In muscle, amino acids and, particularly, branched-chain compounds are transaminated with 2-oxoglutarate to form glutamate, which in turn serves the transamination of pyruvate to form alanine. We discovered that alanine was quite specifically anabolic and that the effect was not at all limited to patients with disorders of propionate metabolism.

In the most stringent test of the hypothesis, patients with propionic acidemia and methymalonic acidemia had protein intakes that were sufficiently reduced to prevent growth from taking place. The addition of 1.0 g/d of alanine was followed by satisfactory growth.

In these infants, alanine was found capable of promoting growth on a variety limited intakes of protein (Fig. 3-7) not simply at the extreme at which control growth was absent. In each instance, the addition of alanine resulted in an increase in the rate of weight gain. Nitrogen retention was also significantly increased when the alanine supplement was given. The anabolic effect of alanine was not limited to patients with disorders of propionate metabolism. We have reported its use in patients with isovaleric acidemia and ornithine transcarbamylase(OTC) deficiency.[13] We employed it with success more recently in patients with malnutrition (e.g., the malabsorption of sucrase deficiency). This effect was evident not only with the very low intakes of protein employed in disorders of propionate metabolism, but also with the relatively high intakes of

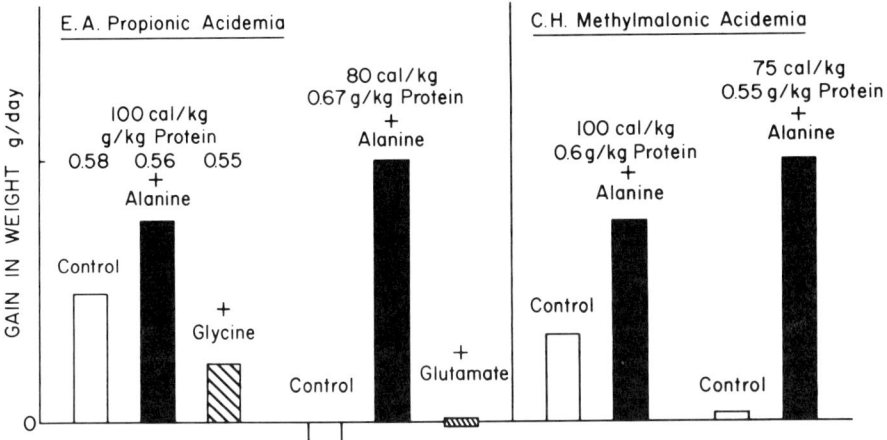

Fig. 3-7. Effects of supplementation with alanine. (From Nyhan,[5] with permission.)

protein in patients with isovaleric acidemia and OTC deficiency. Differences were significant statistically. The specificity of the effect for alanine was indicated by a failure to duplicate it by supplementation with glutamic acid or glycine, excluding the nutritional role of so-called nonessential nitrogen.[14]

ROLE OF LYSINE IN HYPERORNITHINEMIA HYPERAMMONEMIA HOMOCITRULLINURIA SYNDROME

The hyperornithinemia hyperammonemia homocitrullinuria (HHH) syndrome is a consequence of defective transport of ornithine into the mitochondria (Fig. 3-8). It is characterized by episodic hyperammonemia not unlike that of the urea cycle defects. Most patients have exhibited delayed mental development and neurologic abnormalities, such as ataxia and spastic paraparesis.

It has not been emphasized in the literature, but failure to thrive appears to be a feature of this disease. Our single patient has consistently been below the 5th percentile for both height and weight.

Metabolic abnormality is usually suggested first by elevated levels of ammonia in the blood during an attack of vomiting, stupor, or coma. Concentrations of glutamine and alanine may be elevated in plasma, and the urinary excretion of orotic acid may be increased. Plasma concentrations of ornithine are markedly elevated. In addition, excretion of homocitrulline in the urine is increased. Defective transport of ornithine may be demonstrated in cultured fibroblasts. The disorder has generally been treated by restricted intake of protein, which should prevent most attacks of hyperammonemia. Supplementation with ornithine and arginine has been reported to lead to decreased postprandial concentrations of ammonia and urinary homocitrulline.[15,16]

Our interest in lysine supplementation was prompted by the failure to thrive in our patient, in association with very low levels of lysine in plasma (Table 3-6). Concentrations of lysine were elevated during the acute attack of hyperam-

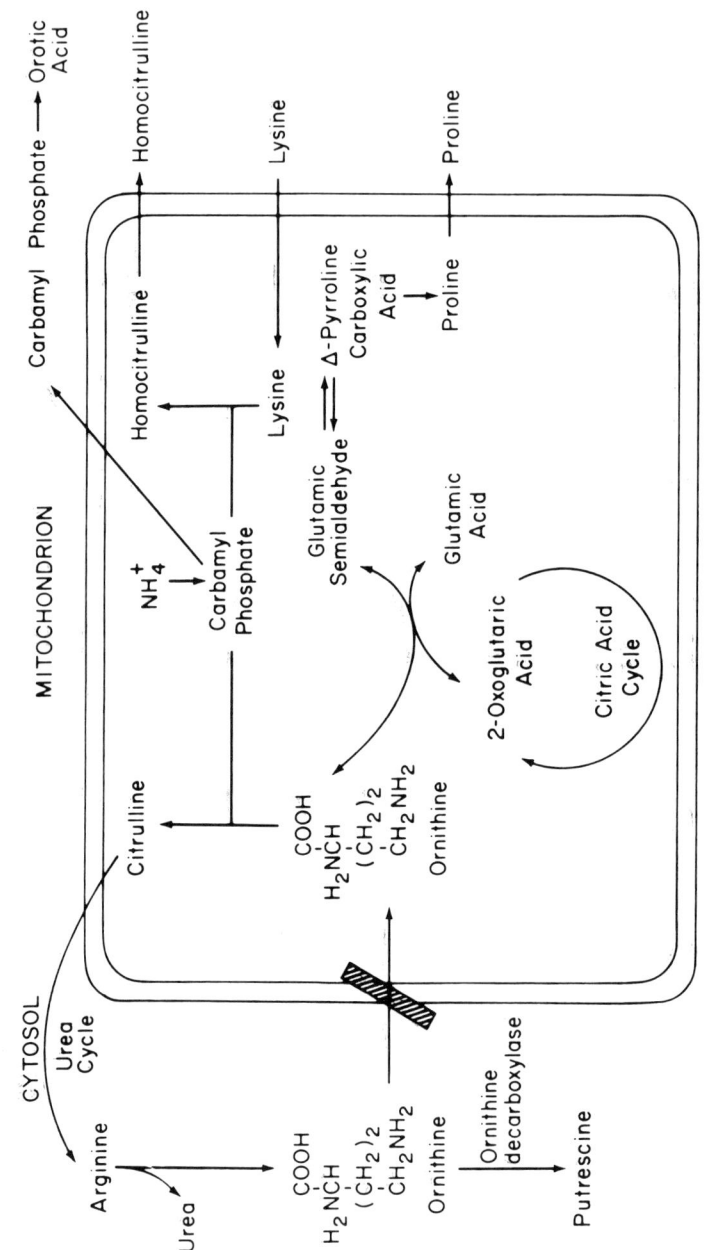

Fig. 3-8. Metabolic defect and relevant pathways in hyperornithemia hyperammonemia homocitrullinuria (HHH) syndrome.

Table 3-6. Effects of Lysine in a Patient With HHH Syndrome

Age (yr/mo)	Weight (kg)	Height (cm)	Protein (g/kg)	Cal (kg)	Lysine (kg)	Plasma Lysine (μmol/L)
5/10	13.6	98	3.7	85	0	15
6/1	14.5	99	4.7	96	0	—
6/8	15.6	101	5.4	129	0	34
7/3	15.7	106	6.5	171	0	—
7/7	16.5	106	4.5	111	0	64
7/11	17.2	109	3.1	94	0^a	38
8/4	17	112	3.8	112	0.2^b	24
8/6	17.2	112	3.5	95	0	67
8/9	17.8	112	—	—	0^c	—
9	18.3	113	3.6	86	0	24
9	18.3	114	3.8	92	0.2	172
9/4	18.4	115	3.5	70	0.23	209
9/6	18.6	115.4	4.0	87	0.23	—

^{a} Started lysine supplement. 4.4 g lysine monohydrochloride = 3.45 g lysine (0.2 g/kg).
^{b} Reported minimal compliance during previous 6 weeks.
^{c} Changed to lysine orotate 7.19 g = 3.45 g lysine (0.2 g/kg).
Abbreviation: HHH, hyperornithinemia hyperammonemia homocitrullinuria.

monemia, but this nonspecific effect, like those of glutamine and alanine, simply reflects the overproduction of ammonia. During steady-state conditions, concentrations are usually low.[17]

It would be logical for lysine to be limiting in this disease (Fig. 3-8). The failure of ornithine to enter the mitochondria would lead to amination of lysine from carbamylphosphate to form homocitrulline as a major route for the removal of waste nitrogen. We attempted supplementation with lysine hydrochloride without success, because its bad taste led to almost zero compliance. However, the use of lysine orotate has led to a major increase in plasma concentrations of lysine. Growth had virtually ceased when this was begun and has been rewarding during the 6 months of follow-up evaluation to date.

CARNITINE

A major development in the management of inherited diseases of metabolism has been the recognition of an expanded role for carnitine. For some time, there has been a certain literature on what was thought to be primary carnitine deficiency. It is now recognized that virtually all deficiencies of carnitine are secondary, that such deficiency is common in a wide variety of inherited metabolic diseases, and that most are improved by supplementation with carnitine.

Carnitine is a natural product, normally abundant in muscle and other tissues. It is biosynthesized from lysine. Carnitine is an essential component for the oxidation of long-chain fatty acids. After lipolysis, free fatty acids are released. Fasting leads to a rise in their concentration in serum. Fatty acids are then converted to CoA esters, which react with carnitine to form acylcarnitine esters. The reactions are catalyzed by a group of carnitine acyltransferases. The acylcarnitines are effectively transported across the mitochondrial membrane, whereas the free acids or CoA esters of long-chain fatty acids are not. Once inside the mitochondria, the carnitine esters are hydrolyzed, releasing the fatty acids for β-oxidation. The ultimate products are acetoacetate and 3-hydroxybutyrate, indicative of effective ketogenesis and the normal operation of the entire process of lipid catabolism.

A disorder has recently been described, sometimes referred to as primary carnitine deficiency,[18] in which the fundamental deficiency is an abnormality in the transport of carnitine into a variety of cells, including muscle, renal tubule, and cultured fibroblasts. Affected patients have increased urinary excretion of carnitine,[19] as well as microvesicular myopathy and cardiomyopathy. Clinically, they may present with hypotonia, weakness, or cardiac failure. Another form of carnitine deficiency in which there is primary renal loss of carnitine is cystinosis and the renal Fanconi syndrome.[20] The other typical presentation for carnitine deficiency is a syndrome of hypoketotic hypoglycemia. The presentation may suggest a diagnosis of Reye syndrome. Recurrent episodes that may lead to coma often follow acute intercurrent infectious illness. Dicarboxylic aciduria is found on analysis of the organic acids of the urine. This is a consequence of the ω-oxidation that comes into play when β-oxidation is defective. These syndromes of clinical carnitine deficiency may occur in a variety of patients with inherited metabolic disease, particularly disorders of fatty acid oxidation. The most common of these is medium-chain acyl-CoA dehydrogenase deficiency.

Secondary deficiency of carnitine also occurs in patients with the classic organic acidemias, such as propionic acidemia, methylmalonic acidemia, isovaleric acidemia, and glutaric acidemia. The mechanism for production of carnitine deficiency in these disorders, as well as in the disorders of fatty acid oxidation, is that accumulated CoA esters (e.g., propionyl CoA) form carnitine esters that are excreted in the urine. In these disorders, we have found that despite severe carnitine deficiency, hypoketosis does not occur. The cardinal characteristic of these patients is the development of massive ketosis. Testing their ketogenesis with a 19-hour fast demonstrated that the same type of excessive ketosis occurred as in their episodes of clinical illness (Fig. 3-9). Repetition of the test after therapy with carnitine replacement resulted in a modulation of fasting ketogenesis, as measured by quantification of acetoacetate and 3-hydroxybutyrate.[21] Dose-response data indicated further modulation with increased dosage. A diminution in the tendency of these patients to develop ketoacidosis should be therapeutic. Conceptually this is a different role for carnitine in therapy. In addition to correcting carnitine deficiency, carnitine can effect a detoxification by the formation of carnitine esters of accumulated CoA esters in patients with organic acidemia. These esters of carnitine are excreted efficiently in the urine. For both reasons, treatment with carnitine is a useful adjunct to therapy.

The compound is virtually nontoxic. The dosage tolerated is limited only by diarrhea. The usual initial oral dose of L-carnitine is 50 to 100 mg/kg. Many patients tolerate 300 mg/kg.

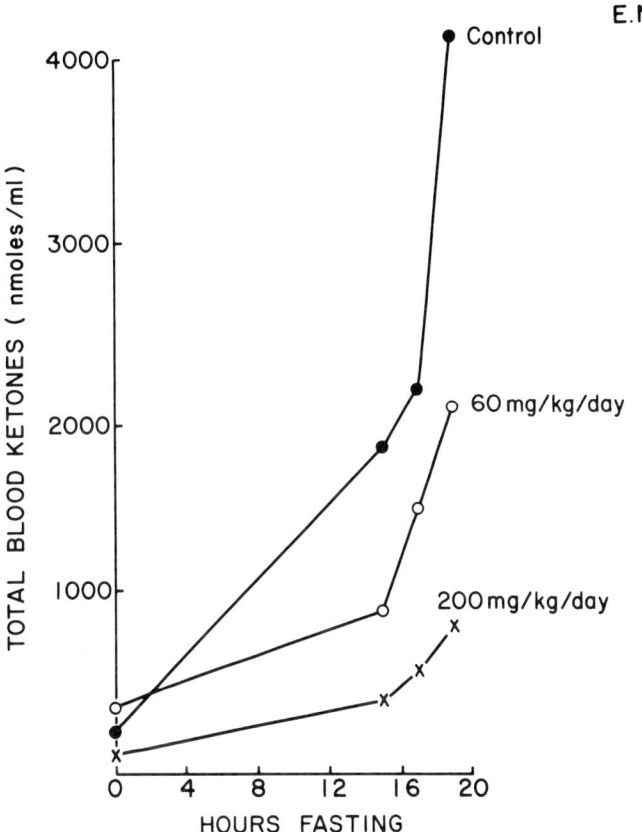

Fig. 3-9. Effects of carnitine on fasting ketogenesis in a patient with propionic acidemia. (From Wolff et al.,[21] with permission.)

Optimal therapy requires monitoring the concentrations of carnitine in blood and urine.[22] It is relatively simple to give enough carnitine to normalize free carnitine levels in blood. In patients with the transport defect or renal losses, it may take more carnitine and a longer period to increase levels in muscle. In patients in whom the aim is detoxification, we seek a dose at which urinary ester excretion is maximal. In propionic acidemia, loose stools may limit dosage before plateau levels are achieved. In other conditions, a plateau in ester excretion indicates the achievement of optimal dosage of carnitine.

ACKNOWLEDGMENTS

This work was aided by U.S. Public Health Service grant HD04608 from the National Institute of Child Health and Human Development; by General Research Center grant M01RR00827, from the Division of Research Resources,

National Institutes of Health, Bethesda; and by a grant from the William W. Allen Foundation in Midland, Michigan.

REFERENCES

1. Nyhan WL, Sakati NO: Phenylketonuria (PKU). p. 100. In Diagnostic Recognition of Genetic Disease, Lea & Febiger, Philadelphia 1987
2. Smith I, Beasley MG, Ades AE: Intelligence and quality treatment in phenylketonuria. Arch Dis Child 65:472, 1990
3. Acosta PB, Schaeffler GE, Wenz E, Koch R: Phenylketonuria (PKU)—A Guide to Management. California State Department of Health, Berkeley, CA, 1972
4. Barshop BA, Yoshida I, Ajami A, et al: Metabolism of 1-^{13}C-propionate in vivo in patients with disorders of propionate metabolism. Pediatric Res (in press)
5. Nyhan WL: Disorders of propionate metabolism. p. 363. In Bickel H, Wachtel U (eds): Inherited Diseases of Amino Acid Metabolism. Recent Progress in the Understanding, Recognition and Management, International Symposium in Heidelberg, 1984. Georg Thiem Verlag, Stuttgart, 1984
6. Ney DN, Bay C, Saudubray JM, et al: An evaluation of protein requirements in methylmalonic acidaemia. J Inher Metab Dis 8:132, 1985
7. Nyhan WL, Sakati NO: Maple syrup urine disease. p. 31. In Diagnostic Recognition of Genetic Disease. Lea & Febiger, Philadelphia, 1987
8. Saudubray JM, Ogier H, Charpentier C, et al: Neonatal management of organic acidurias. J Inher Metab Dis 7(suppl 1):2, 1984
9. Nyhan WL, Rice ML: Aminoazidopathien und organoazidopathien. In Proceedings of the Gesellschaft fur Neuropadiatre, Göttingen, West Germany, October 27–29, 1989. Springer-Verlag, Stuttgart, 1990
10. Magnusson I, Ekman L, Wangdahl M, Wahren J: N-Acetyl-1-cysteine as tyrosine and cysteine precursors during intravenous infusion in humans. Metabolism 38:957, 1989
11. Kahler SG, Millington DS, Cederbaum SD, et al: Parenteral nutrition in propionic and methylmalonic acidemia. J Pediatr 115:235, 1989
12. Kelts DG, Ney D, Bay C, et al: Studies on requirements for amino acids in infants with disorders of amino acid metabolism. I. Effect of alanine. Pediatr Res 19:86, 1985
13. Wolff JA, Kelts DG, Algert S, et al: Alanine decreases the protein requirements of infants with inborn errors of amino acid metabolism. J Neurogenet 2:31, 1985
14. Snyderman SE, Holt LE, Dancis J, et al: "Unessential" nitrogen: A limiting factor for human growth. J Nutr 78:57, 1962
15. Shih VE, Effron ML, Moser HW: Hyperornithinemia, hyperammonemia, and homocitrullinuria. A new disorder of amino acid metabolism associated with myoclonic seizures and mental retardation. Am J Dis Child 117:83, 1969
16. Winter HS, Perez-Staude AR, Levy HL, Shih VE: Unique hepatic ultrastructural changes in a patient with hyperammonemia (HAM), hyperornithinemia (HOR) and homocitrullinuria (HC). Pediatr Res 14:583, 1980
17. Rodes B, Ribes A, Poneda M, et al: A new family affected by the syndrome of hyperornithinemia, hyperammonemia and homocitrullinuria. J Inher Metab Dis 10:73, 1987

18. Stanley C, Treem WR, Coates PM, et al: Primary carnitine deficiency due to a defect in carnitine transport. Pediatr Res 23:397A, 1988
19. Waber LJ, Valle D, Neill C, et al: Carnitine deficiency presenting as familial cardiomyopathy: A treatable defect in carnitine transport. J Pediatr 101:700 1982
20. Gahl WA, Bernardini I, Dalakas M, et al: Oral carnitine therapy in children with cystinosis and renal Fanconi syndrome. J Clin Invest 81:549, 1988
21. Wolff JA, Thuy LP, Haas R, et al: Carnitine reduces fasting ketogenesis in patients with disorders of propionate metabolism. Lancet 1:289, 1986
22. McGarry JD, Foster DW: An improved and simplified radiologic assay for the determination of free and esterified carnitine. J Lipid Res 17:277, 1976

4

THERAPEUTIC APPLICATIONS OF L-CARNITINE IN METABOLIC DISORDERS

Charles R. Roe · David S. Millington ·
Stephen G. Kahler · Naoki Kodo ·
Daniel L. Norwood

SYNTHESIS AND METABOLISM

The synthesis of L-carnitine (β-hydroxy-γ-N-trimethylammonium butyrate) has been well described.[1] Lysine residues incorporated into protein are methylated to trimethyllysyl (TML) residues by the enzyme methylase III.[2] Proteins with TML residues are degraded in lysosomes.[3] The released TML is then hydroxylated in the mitochondria to hydroxy-TML (HTML). HTML is then acted on by 3-hydroxy-6-N-trimethyllysine aldolase, which appears to be the same as the serine hydroxymethyltransferase enzyme involved in folate metabolism.[4] The products of the reaction with HTML as substrate are γ-butyrobetaine aldehyde (γ-BBA) and glycine. γ-BBA is then dehydrogenated to γ-butyrobetaine (γ-BB). The final step to L-carnitine involves γ-butyrobetaine, 2-oxoglutarate dioxygenase (γ-butyrobetaine hydroxylase) (Fig. 4-1).

The liver and kidney are capable of the complete synthesis of carnitine in humans, whereas skeletal and cardiac muscle are able to synthesize γ-BB but must export this intermediate for final hydroxylation to L-carnitine by the kidney or liver. In addition to endogenous synthesis, tissue carnitine stores can be replenished by dietary intake. Meat products (especially red meat) and dairy products are rich in carnitine.

The only known metabolic fate for L-carnitine is the formation of acylcarnitines effected by several carnitine acyltransferases. Unlike D-carnitine, the L-isomer is not degraded to smaller molecular forms; however, bacterial degradation in the gastrointestinal tract does result in the removal of the qua-

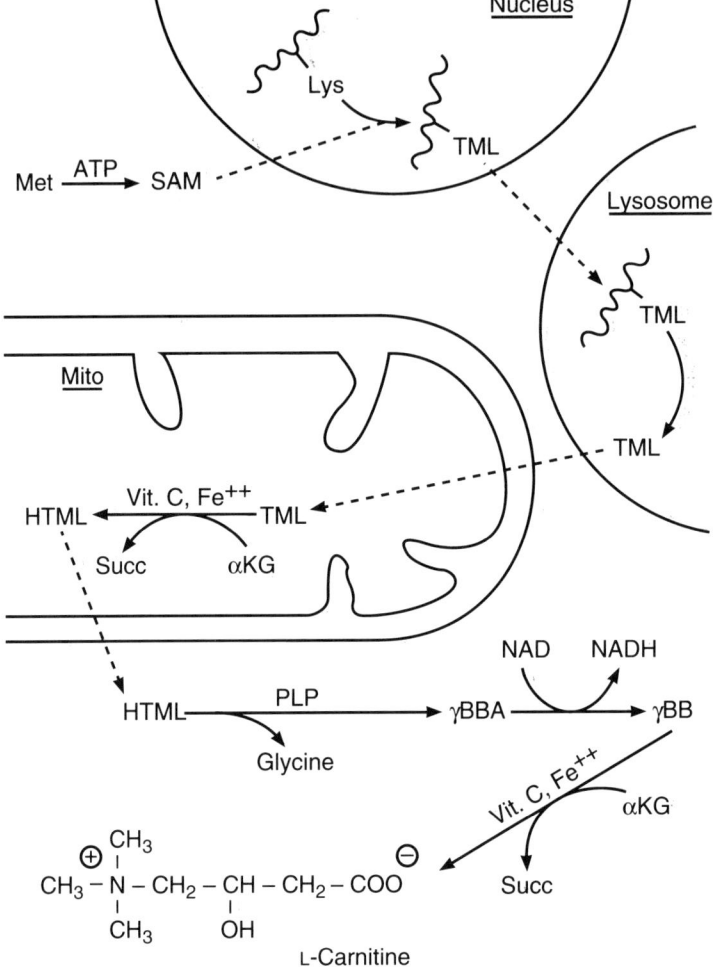

Fig. 4-1. Carnitine biosynthesis. α-KG, ketoglutarate (2-oxoglutarate); Lys, lysine; Met, methionine; PLP, pyridoxal phosphate; SAM, S-adenosylmethionine; Succ, succinate. See text for details.

ternary end of the molecule producing trimethylamine. This compound is normally oxidized by hepatic trimethylamine oxidase and is excreted in the urine.[1] Children receiving large oral doses of L-carnitine (more than 300 mg/kg/24 h) may produce trimethylamine in excess of the capacity of the oxidase. This produces a body odor much like rotten fish, but no toxicity. The odor disappears as the dose is reduced.

DIAGNOSTIC APPLICATIONS

The use of fast-atom bombardment mass spectrometry (FAB-MS) has led to the detection and characterization of specific acylcarnitines.[5] In particular, this technique identified new acylcarnitines, whose structures were subsequently demonstrated by auxiliary techniques.[6] Patients with inherited disorders of branched-chain amino acid and fatty acid metabolism were found to excrete disease-specific acylcarnitines that reflected abnormal acyl-coenzyme A (CoA) thioesters accumulating at or near the site of the enzyme deficiency.[7] In situations in which these unusual acyl-CoA thioesters do not accumulate, such as maple syrup urine disease (MSUD) or phenylketonuria (PKU), unusual acylcarnitines are not detected. After oral loading with L-carnitine, acylcarnitines are much more easily detected, since excretion of the abnormal diagnostic species in patients with metabolic disorders is then enhanced.

In some disorders, a single diagnostic acylcarnitine is observed, such as isovalerylcarnitine in isovaleric acidemia[8] or propionylcarnitine in propionic acidemia.[9] In disorders of fat metabolism, including medium-chain acyl-CoA dehydrogenase (MCAD) deficiency or multiple acyl-CoA dehydrogenase deficiency (MADD), several different acylcarnitines are excreted. These acylcarnitine profiles are extremely useful for diagnosis. Since the levels of diagnostic organic acids vary considerably with clinical state[10] in disorders of fat metabolism, the diagnosis of these disorders is greatly improved by including an analysis of both organic acids and acylcarnitines.

In an older child, or in an infant who has not been breast-fed, a significant carnitine deficiency can develop in MCAD deficiency, making the FAB-MS profile unclear or uninterpretable. The safe and simple expedient of collecting urine after an oral carnitine load of 100 mg/kg enhances the diagnostic acylcarnitine profile and removes ambiguity.[11]

When a child with MCAD deficiency is given an oral or intravenous bolus of methyl-^{2}H$_3$-L-carnitine, a stable isotope-labeled form of carnitine, each acylcarnitine signal is accompanied by a "satellite" with a mass increment of +3 from the deuterium atoms in the labeled carnitine. This has also been demonstrated in a patient with propionic acidemia.[12] These results indicate that exogenous carnitine equilibrates rapidly with the intramitochondrial acyl-CoA pool, revealing the general specificity of the carnitine acyltransferases to form short- and medium-chain length acylcarnitines in vivo. It also represents a single definitive method of confirming the identities of acylcarnitine signals observed in the FAB-MS spectrum, which is especially important in identifying unknown acylcarnitines. These experiments with stable isotopes also illustrate and document the role of carnitine for detoxification in several of the organic acidurias. Carnitine can rapidly transport toxic acyl groups out of the cell as nontoxic acylcarnitines, which are then excreted. These findings support the biochemical rationale for L-carnitine therapy in the organic acidurias.

Successful detection of acylcarnitines at low concentration in urine, liver tissue,[13] and blood plasma by FAB-MS has been possible but usually requires extensive sample purification. A more selective method has been developed using a triple quadrupole, an example of a tandem mass spectrometer that incorporates two stages of mass analysis (MS/MS) in a single instrument. As with other combined techniques, such as gas chromatography-mass spectrometry (GC/MS) and liquid chromatography-mass spectrometry (LC/MS), MS/MS offers the potential for improving specificity and enabling the direct analysis of mixtures.[14] FAB-MS/MS has provided additional specificity and sensitivity and permits quantitative analysis of individual acylcarnitines in urine, plasma, and tissue by isotope dilution assay.[15]

The FAB mass spectra of acylcarnitine methyl esters and the daughter ion spectra of the M$^+$ ions generated by tandem MS show a prominent common fragment at m/z 99.[7,16] This ion is derived from the loss of both the acyl moiety, as the corresponding acid, and the quaternary ammonium function as trimethylamine. The structure of this highly characteristic fragment can be formally represented as $^+CH_2$—CH=CH—CO_2CH_3 and represents the backbone of the acylcarnitine molecule. Isotopically labeled forms, having ^2H or ^{13}C in either the acyl or trimethylamine group, also exhibit the m/z 99 fragment. Because the precursors of m/z 99 are predominantly acylcarnitine molecular cations, the new scan function, when applied to a biologic sample, generates a metabolic profile of acylcarnitines in the sample.[15,16]

The most immediately obvious differences between acylcarnitine profiles obtained by FAB-MS/MS and those performed by the standard FAB-MS[7] procedure is a large reduction in chemical noise. This results from the increased selectivity (specificity) of the analytic procedure, which has improved the detection limit for individual acylcarnitine methyl esters in urine from 50 nmol/ml to less than 1 nmol/ml.[15] It is now possible to observe diagnostic acylcarnitine profiles in urine, plasma, or tissue samples even from neonatal MCAD deficient patients (with carnitine deficiency) that would have been obscured by chemical interference with the older method.

Most of the even-mass ions above m/z 200 in the m/z 99 precursor ion spectra of urine samples are derived from acylcarnitines. This is based on the profiles of patients with well-defined metabolic defects whose diagnostic acylcarnitines have been previously characterized and reported.[5,10,17] Thus, as observed in these earlier studies,[7] the normal profile is dominated by acetylcarnitine (m/z 218), with lesser amounts of C3, C4, C5 and C8:1 (m/z 300) acylcarnitines. In isovaleric acidemia (IVA), isovalerylcarnitine (m/z 260) is the dominant species[8]; in propionic acidemia (PA), propionylcarnitine (m/z 232) is the major species.[9] In methylmalonic aciduria (MMA), propionylcarnitine is also very prominent and is accompanied by a prominent signal for acetylcarnitine.[5] In many cases, a signal corresponding to methylmalonylcarnitine (m/z 290) is also observed. Similarly, the profiles of glutaric aciduria type I (glutaryl-CoA dehydrogenase deficiency [GAI]), β-ketothiolase deficiency, and 3-hydroxy-3-methylglutaryl-CoA lyase deficiency reveal the expected dominant molecular cations corresponding to glutarylcarnitine (m/z 304), tiglylcarnitine (m/z 258), and 3-methylglutarylcarnitine (m/z 318), respectively.[17,18] The profile from a patient with MCAD deficiency is characterized by medium-chain acylcarnitines[10,11]: hexanoyl (m/z 274), octanoyl (m/z 302), octenoyl (m/z 300), and 4-cis-decenoyl (m/z 328). Patients with MADD show increased excretion of isobutyryl- and butyrylcarnitine (m/z 246) as well as C5, C6, C8 acylcarnitines and glutarylcarnitine (m/z 304).

Using the precursors of m/z 99 scan (MS/MS), individual acylcarnitines can be detected in plasma at concentrations of less than 0.5 nmol/ml.[15] This compares with typical physiologic concentrations for total acylcarnitines of 6 to 10 nmol/ml, which can increase two- or threefold in patients with metabolic disorders.[19]

Detection of acylcarnitines in blood is very important, since it permits the postmortem recognition and diagnosis of inherited metabolic diseases. The overall composition of blood is much less variable than that of urine, and the ranges of acylcarnitine concentrations are much narrower. The profile of acylcarnitines in blood is a better index of the carnitine status in tissue than the profile in urine. Initial results with blood have been very encouraging. For example, diagnostic profiles representative of the metabolic diseases have been achieved, permitting postmortem diagnosis of propionic acidemia and MCAD deficiency.

Detection of organic acidurias from plasma by tandem MS has now led to diagnosis from Guthrie cards by simple extraction and analysis of acylcarnitine profiles. This has been successful with each of the disorders in which diagnostic species occur. Automation of this methodology is under development in a pilot screening project for North Carolina. Early recognition of disorders such as MCAD deficiency should reduce the risk of sudden infant death syndrome in that population.

TREATMENT WITH L-CARNITINE

Many of the organic acidurias have an associated secondary carnitine deficiency.[19] In some disorders, such as cystinosis, an inability to conserve carnitine at the renal level is also a reasonable explanation.[20] In several other disorders, the total quantities of free carnitine and acylcarnitines excreted per kg body weight per 24 hours are much less than normal. Untreated patients with IVA[8] and MCAD deficiency,[10,11] for example, have very low levels of plasma free carnitine associated with decreased excretion of total carnitine. Renal loss, therefore, does not appear to be an adequate explanation for their carnitine deficiency. Carnitine synthesis may be inhibited in some of these disorders.

In PA and MMA, normal levels of total carnitine in the plasma and normal quantities excreted in the urine are usually observed; however, most of the carnitine is esterified. Although not a true deficiency state, it represents a relative insufficiency of carnitine to meet metabolic needs. Oral supplementation with L-carnitine in PA results in increased excretion of propionylcarnitine as well as an increase in free carnitine, suggesting that additional carnitine is needed to handle the large quantities of propionyl-CoA being produced in that disease.

Despite the realities of carnitine deficiency or insufficiency and the biochemical evidence for its role in detoxification, there is considerable controversy about its use as treatment of the organic acidurias. This is surprising, in view of the acceptance of glycine therapy in IVA,[21] which is also used to enhance the removal of a toxic metabolite by conjugation. Carnitine supplementation in PA, MMA, and IVA is analogous and, in the latter case, at least, has the advantage of correcting a known deficiency. In MCAD deficiency, carnitine supplementation would also seem to be justified on the basis of correcting a true deficiency state and for conjugation and excretion of toxic medium-chain acyl-CoA derivatives. The range of disorders treated includes PA, IVA, MMA, HMG lyase, β-ketothiolase deficiency (KT), MCAD deficiency, and MADD. Typically, daily oral doses are 200 mg/kg for PA and MMA and 100 mg/kg for the others. Only about 15 percent of the oral dose is actually absorbed. The daily supplement is divided into four doses because of the speed of the biochemical response and its rapid excretion.

In most disorders, except for MCAD deficiency, parents report increased social interaction and awareness of the environment and an apparent subjective overall clinical improvement. Most patients with IVA, KT, HMG lyase, and MADD have experienced no further hospitalizations while on carnitine supplement. In IVA, carnitine supplementation has been used successfully in place of glycine. When carnitine treatment was stopped in several patients with IVA, KT, and MADD, serious illness requiring hospitalization occurred.

Although carnitine supplementation in PA, MMA, and MCAD deficiency has not eliminated recurrent illness, it has reduced the severity of illness and the number of hospitalizations. Another encouraging observation has been that patients with MCAD deficiency have had chickenpox without requiring hospitalization. Because the clinical course is so variable in this disorder, it is difficult to determine the significance of these observations. There are, however, compelling reasons to support chronic carnitine therapy for MCAD deficiency. One-third of children with MCAD deficiency die with the first episode of illness. Systemic carnitine deficiency in an untreated child could seriously limit mobilization of sufficient carnitine to conjugate the rapidly accumulating toxic medium-chain fatty acyl-CoA compounds during illness. Observations in a family with four affected children, two of whom had died reportedly with a Reye-like episode and sudden infant death syndrome[11] before a diagnosis was made, tend to support these concepts. MCAD deficiency was diagnosed in the surviving sibling at 2 months of age before clinical symptoms by analysis of organic acids and acylcarnitines. The diagnosis was subsequently confirmed by enzyme assay in cultured fibroblasts. Before carnitine therapy, this patient, who was breast-fed at the time, excreted 2 μmol/mg creatinine total carnitine, of which 43 percent was acylated. Quantitative analysis by isotope-dilution FAB-MS[5] showed that octanoylcarnitine exceeded the acetylcarnitine concentration in this urine by a factor of 3. While asymptomatic and receiving oral L-carnitine supplement at 100 mg/kg/d, the output of total carnitine was 7 μmol/mg creatinine, of which only 0.5 μmol was esterified. During severe illness, she received IV carnitine (30-mg/kg bolus followed by 30 mg/kg over the next 24 hours) and recovered rapidly over the next 5 hours. Total carnitine excreted during IV therapy was 61 μmol/mg creatinine, 48 percent of which was acylated. Octanoylcarnitine and acetylcarnitine concentrations were 13 and 16 μmol/mg, respectively. A repeat of the same IV carnitine regimen when the patient was clinically well revealed a much lower acylcarnitine level as a percentage of the total, with octanoylcarnitine representing about 25 percent of the acylated fraction. Analysis of postmortem urine from the deceased untreated sibling indicated a total carnitine excretion of only 3 µmol/mg, of which about 90 percent was esterified. These data suggest that untreated MCAD deficiency patients are at increased risk because of the decreased availability of carnitine during clinical episodes. Supplementation provides additional carnitine for detoxification during acute illness. Intravenous carnitine has been used during acute illnesses with excellent clinical results and no toxicity in patients with MCAD deficiency, MADD, IVA, and PA.

CARNITINE TRANSPORT

Plasma and tissue carnitine levels reflect the balance between intake, production, and loss and the ability of various tissues to maintain a concentration of carnitine higher than in the extracellular fluid. Acetylcarnitine and the various other acylcarnitines also play a role. A kinetic three-compartment model, representing the extracellular space and low- and high-affinity tissues, fits the observed distribution data well. High-affinity organs include cardiac and skeletal muscle. Clinical and laboratory data suggest that they share a transport system with fibroblasts, renal tubule epithelium, and the gut.[22,23]

Plasma carnitine concentrations are typically about 50 µmol/L (0.05 µmol/ml), while liver may contain 2 to 3 µm/g wet weight (wet wt.), heart contains 1 to 5 µm/g wet wt., and skeletal muscle contains 2 to 9 µm/g wet wt.[1,24] The fractional excretion of L-carnitine, at usual plasma concentrations, is less than 0.10; acylcarnitines may be lost at a greater rate.[25] This is apparently true for octanoylcarnitine as well, produced in MCAD deficiency.[26] The most dramatic example of preferential loss is pivampicillin degradation. As it is not resorbed at all, it can seriously deplete body carnitine stores.[27]

Several families with primary systemic carnitine deficiency have been reported. This is a disorder of impaired carnitine transport in renal tubule and other cells. Dilated or hypertrophic cardiomyopathy, beginning in infancy or childhood, is the usual presentation. Dysrhythmias and sudden death may occur. Skeletal muscle weakness is common. Lipid myopathy may be found on biopsy. The heart may show abnormalities in mitochondrial number or morphology. Hypoglycemia, encephalopathy, apnea, or coma may be precipitated by fasting. The transaminases and muscle enzymes may be elevated. Plasma carnitine levels are typically less than 5 µm; skeletal and cardiac muscle carnitine levels are severely depressed to 1 to 5 percent of normal, while the liver carnitine level may be one-third of normal. Hypoketosis may be noticed during a fast, and there may be dicarboxylic aciduria.[1,22,28–30] Carnitine absorption in the gastrointestinal tract is impaired, as is renal reabsorption of filtered carnitine. Uptake in cultured fibroblasts and skeletal muscle is impaired, suggesting a common transport mechanism.[22]

Therapy with high-dose oral carnitine will lead to a gradual resolution of cardiac symptoms, restoration of normal cardiac function, and tolerance of fasting. Doses used have ranged from 100 to 1000 mg/kg/d PO, given in divided doses. Strength and coordination also have improved. The carnitine level in skeletal muscle may rise to about one-half of normal, and the lipid vacuoles may disappear. The surviving patient of Tripp et al.,[28] treated with L-carnitine since 8 years of age, had persistent dysrhythmias in adolescence. A permanent pacemaker was placed at age 13; at age 17, she has normal cardiac function and strength but significantly impaired aerobic exercise capacity (Tripp ME: personal communication).

SUMMARY

The detection of acylcarnitines by FAB-MS in human physiologic fluids has added another valuable diagnostic tool for the recognition of specific metabolic diseases. The newly developed technique of FAB-MS/MS, which embodies the principles of tandem mass spectrometry, affords a quantum leap in specificity. FAB-MS/MS permits the detection and quantification of acylcarnitines in concentrations well within the physiologic ranges in urine, blood, and tissue. As a therapeutic agent, L-carnitine appears to be useful in providing protection against the harmful consequences of catabolism in those organic acid disorders associated with secondary carnitine deficiency. Definitive clinical improvements have been shown in disorders of carnitine transport. It has also been shown that there is no discernable toxicity associated with carnitine therapy, even when administered intravenously at high doses.

ACKNOWLEDGMENTS

These studies were supported by FDA Orphaned Products Division grant FD-R-000177, Food and Drug Administration, Washington, DC; by NIH grants HD-22704 and HD-24908, National Institutes of Health, Bethesda; and by the RR-30 General Clinical Research Centers Program Division of Research Resources, National Institutes of Health, Bethesda. The expert assistance of Diane Gale and Ann Burrus is gratefully acknowledged.

REFERENCES

1. Bremer J: Carnitine—metabolism and functions. Physiol Rev 63:1420, 1983
2. Paik WK, Kim S: Solubilization and partial purification of protein methylase III from calf thymus nuclei. J Biol Chem 245:6010, 1970
3. Labadie J, Dunn WA, Aronson NN: Hepatic synthesis of carnitine from protein-bound trimethyl-lysine. Lysosomal digestion of methyl-lysine labelled asialo-fetuin. Biochem J 160:85, 1976
4. Hulse JD, Ellis SR, Henderson LM: Carnitine biosynthesis: β-Hydroxylation of trimethyllysine by an α-ketoglutarate-dependent mitochondrial dioxygenase. J Biol Chem 253:1654, 1978
5. Millington DS, Roe CR, Maltby DA: Application of high resolution fast atom bombardment and constant B/E ratio linked scanning to the identification and analysis of acylcarnitines in metabolic disease. Biomed Mass Spectrom 11:236, 1984

6. Millington DS: New methods for the analysis of acylcarnitines and acyl-CoA compounds. p. 97. In Gaskell SJ (ed): Mass Spectrometry in Biomedical Research. John Wiley & Sons, Chichester, 1986
7. Roe CR, Millington DS, Maltby DA: Diagnostic and therapeutic implications of acylcarnitine profiling in organic acidurias associated with carnitine insufficiency. p. 97. In Borum PR (ed): Clinical Aspects of Human Carnitine Deficiency. Pergamon Press, New York, 1986
8. Roe CR, Millington DS, Maltby DA, et al: L-Carnitine therapy in isovaleric acidemia. J Clin Invest 74:2290, 1984
9. Roe CR, Millington DS, Maltby DA, et al: L-Carnitine enhances excretion of propionyl coenzyme A as propionylcarnitine in propionic acidemia. J Clin Invest 73:1785, 1984
10. Roe CR, Millington DS, Maltby DA, et al: Diagnostic and therapeutic implications of medium-chain acylcarnitines in medium-chain acyl-CoA dehydrogenase deficiency. Pediatr Res 19:459, 1985
11. Roe CR, Millington DS, Maltby DA, Kinnebrew P: Recognition of medium-chain acyl-CoA dehydrogenase deficiency in asymptomatic siblings of children dying of sudden infant death or Reye-like syndromes. J Pediatr 108:13, 1986
12. Millington DS, Maltby DA, Gale D, Roe CR: Synthesis and human applications of stable isotope-labelled L-carnitine. p. 189. In Baillie TA, Jones R (eds): Synthesis and Applications of Isotopically Labelled Compounds 1988. Elsevier, Amsterdam, 1989
13. Roe CR, Millington DS, Maltby DA, Wellman RB: Post-mortem recognition of inherited metabolic disorders from specific acylcarnitines in tissue in cases of sudden infant death. Lancet 1:512, 1987
14. Yost RA, Enke CG: Tandem quadrupole mass spectrometry. p.175. In McLafferty FW (ed): Tandem Mass Spectrometry. John Wiley & Sons, New York, 1983
15. Millington DS, Norwood DL, Kodo N, et al: Application of fast atom bombardment with tandem mass spectrometry and liquid chromatography/mass spectrometry to the analysis of acylcarnitine in human urine, blood and tissue. Anal Biochem 180:331, 1989
16. Norwood DL, Kodo N, Millington DS: Application of continuous-flow liquid chromatography/fast atom bombardment mass spectrometry to the analysis of diagnostic acylcarnitines in human urine. Rapid Commun Mass Spectrom 2:269, 1988
17. Millington DS, Roe CR, Maltby DA: Characterization of new diagnostic acylcarnitines in patients with β-ketothiolase deficiency and glutaric aciduria type 1 mass spectrometry. Biomed Environ Mass Spectrom 14:711, 1987
18. Roe CR, Millington DS, Maltby DA: Identification of 3-methylglutaryl-carnitine. A new diagnostic metabolite of 3-hydroxy-3-methylglutaryl-coenzyme-A lyase deficiency. J Clin Invest 77:1391, 1986
19. Chalmers RA, Roe CR, Stacey TE, Hoppel CL: Urinary excretion of L-carnitine and acylcarnitine by patients with disorders of organic acid metabolism: Evidence for secondary insufficiency of L-carnitine. Pediatr Res 18:1325, 1984
20. Bernardini I, Rizzo WB, Dalakas M, et al: Plasma and muscle free carnitine deficiency due to renal Fanconi syndrome. J Clin Invest 75:1124, 1985
21. Yudkoff M, Cohn RM, Ruschak R, et al: Therapeutic effects of glycine in isovaleric acidemia. Pediatr Res 10:25, 1976
22. Treem WR, Stanley CA, Finegold DN, et al: Primary carnitine deficiency due to a failure of carnitine transport in kidney, muscle, and fibroblasts. N Engl J Med 319:1331, 1989

23. Siliprandi N, Sartorelli L, Ciman M, Di Lisa F: Carnitine: Metabolism and clinical chemistry. Clin Chim Acta 183:3, 1989
24. Rebouche CJ: Carnitine metabolism and function in humans. Annu Rev Nutr 6:41, 1986
25. Ohtani Y, Nishiyama S, Matsuda I: Renal handling of free and acylcarnitine in secondary carnitine deficiency. Neurology 34:977, 1984
26. Schmidt-Sommerfeld E, Penn D, Kerner J, et al: Quantitation of urinary carnitine esters in a patient with medium-chain acyl-coenzyme A dehydrogenase deficiency: Effect of metabolic state and carnitine therapy. J Pediatr 115:577, 1989
27. Melegh B, Kerner J, Bieber LL: Pivampicillin-promoted excretion of pivaloylcarnitine in humans. Biochem Pharmacol 36:3405, 1987
28. Tripp ME, Katcher ML, Peters HA, et al: Systemic carnitine deficiency presenting as familial endocardial fibroelastosis: A treatable cardiomyopathy. N Engl J Med 305:385, 1981
29. Vici CD, Bertini E, Bartuli A, Sabetta G: Carnitine in lactic acidosis (letter). J Pediatr 112:678, 1988
30. Rodrigues P, Scholte HR, Luyt-Houwen IE, Vaandrager-Verduin MH: Cardiomyopathy associated with carnitine loss in kidneys and small intestine. Eur J Pediatr 148:193, 1988

5

TREATMENT OF UREA CYCLE DISORDERS

Saul W. Brusilow

INTRODUCTION

The physiologic defect imposed by inborn errors of urea synthesis is the failure to synthesize and excrete waste nitrogen. Waste nitrogen can be defined as dietary nitrogen not used for net protein biosynthesis nor lost from the body by other means (e.g., hair, skin, stool). It follows that treatment should be directed toward two goals: (1) to minimize the requirement for waste nitrogen synthesis and excretion and (2) to discover products that will substitute for urea as a vehicle for waste nitrogen synthesis and excretion.[1,2] Furthermore, because arginine is an indispensable amino acid in patients with urea cycle disorders (apart from arginase deficiency), it is necessary to provide supplementary dietary arginine.[3] For patients with carbamylphosphate synthetase deficiency (CPSD) and ornithine transcarbamylase deficiency (OTCD) this may best be done with citrulline, which offers the advantages of adding only one nitrogen atom to the free amino acid pool per mole administered, as compared with arginine, which adds two nitrogen atoms to the free amino acid pool per mole administered.

It should be recognized that there is great phenotypic and presumed genetic heterogeneity for each urea cycle disorder. This would suggest that some variability is possible in the approach to therapy. Therapy for the most severely affected patients, that is, patients with little or no residual enzymatic activity, is described below. Using this approach, modifications of therapy can be adopted for more mildly affected children.

DIETARY THERAPY

To minimize the requirement for waste nitrogen synthesis, it is recommended that the minimum dietary protein intake be no less than approximately 1.6

g/kg/d from birth to 4 months of age.[4] (Recently we have been able to demonstrate that, during this period, infants with OTCD treated with high-dose phenylbutyrate or phenylacetate therapy can tolerate up to 2 g/kg/d.) From 4 months to 12 months of age, it is recommended that the infant receive approximately 1.4 g/kg/d, and from 12 to 36 months of age, the protein intake should be no less than 1.2 g/kg/d. It is usually necessary in neonatal-onset cases to take advantage of the lower nitrogen density of essential amino acids by prescribing a mixture of essential amino acids and protein. (Protein has a nitrogen density of approximately 0.16 as compared with an essential amino acid mixture, which has a nitrogen density of 0.12.)

Patients with deficiencies of argininosuccinic acid synthetase (ASD) and argininosuccinase (ALD) tolerate higher nitrogen intakes derived entirely from natural protein. It is rarely necessary to substitute essential amino acids for natural protein in these patients.

OTHER PATHWAYS THAT MAY SERVE AS SUBSTITUTES FOR UREA SYNTHESIS

Nitrogen-containing compounds that can substitute for urea are shown in Figures 5-1 to 5-3. For patients with ALD (Fig. 5-1) and ASD, dietary arginine supplementation promotes argininosuccinate (ASA) synthesis and citrulline synthesis (Fig. 5-2), respectively, both of which may serve as waste nitrogen products. ASA is far more effective than citrulline as a waste nitrogen product because, unlike citrulline, which can be reabsorbed by the renal tubule, all filtered ASA is excreted. Furthermore, ASA contains one more waste nitrogen atom than does citrulline.

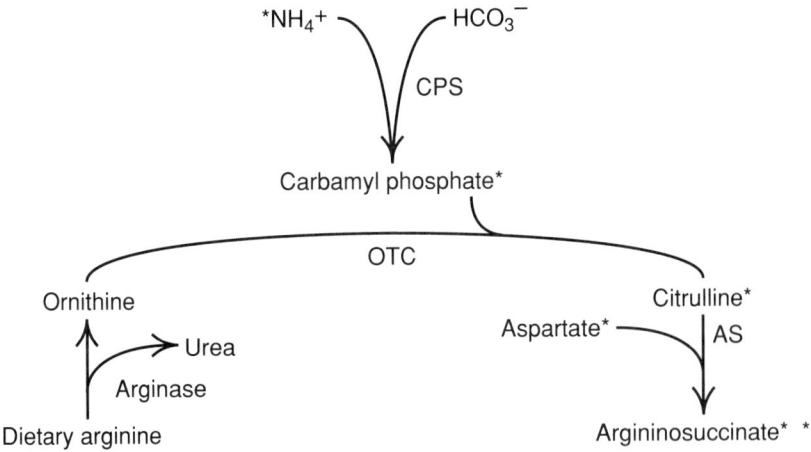

Fig. 5-1. Pathway of waste nitrogen synthesis in patients with argininosuccinase deficiency (ALD). Supplementary dietary arginine supports the continued synthesis of argininosuccinate (ASA), and hence its excretion as a waste nitrogen product. Asterisks (*) denote the number of waste nitrogen atoms contained in various substrates and products. (From Brusilow and Horwich,[2] with permission.)

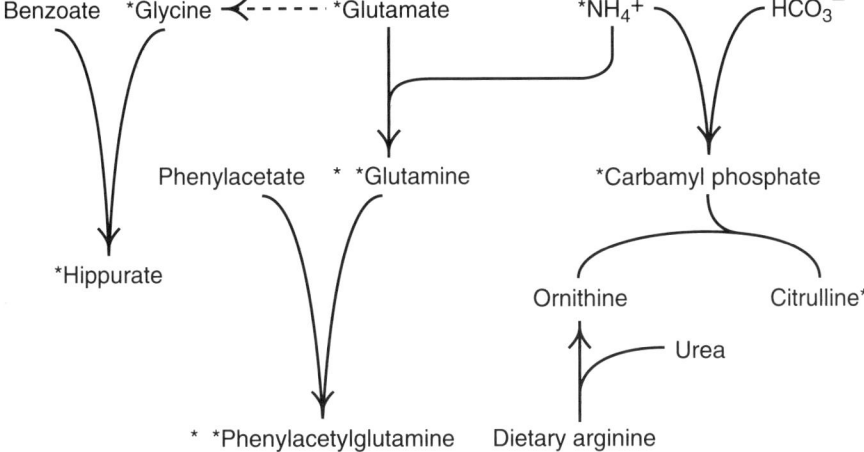

Fig. 5-2. Pathways of waste nitrogen synthesis in patients with arginosuccinic acid synthetase deficiency (ASD). Supplementary dietary arginine supports the continued synthesis of citrulline, hence its excretion as a waste nitrogen product. Benzoate and phenylacetate function as described in Figure 5-3. Dashes (- - -) indicate a series of reactions described in Figure 5-3; asterisks (*) denote nitrogen atoms destined for waste nitrogen excretion in citrulline, hippurate, and phenylacetylglutamine. (From Brusilow and Horwich,[2] with permission.)

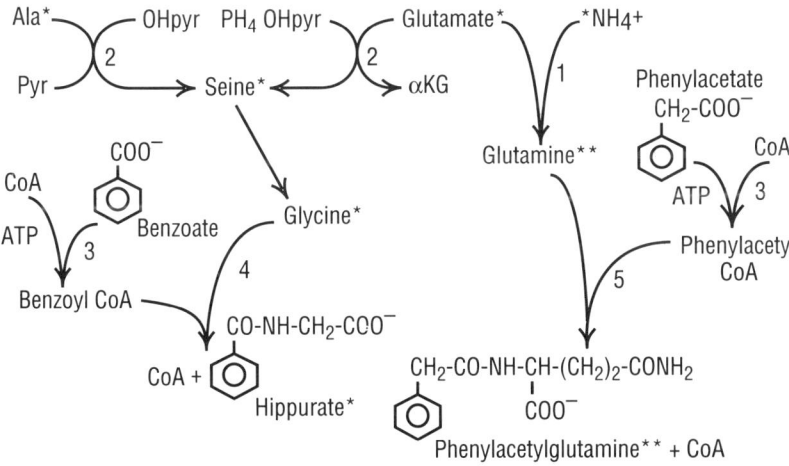

Fig. 5-3. Pathways of waste nitrogen synthesis in patients with carbamylphosphate synthetase deficiency (CPSD) and ornithine transcarbamylase deficiency (OTCD). Asterisks (*) denote nitrogen atoms destined for waste nitrogen excretion in hippurate and phenylacetylglutamine. The enzymatic reactions are numbered: 1, glutamine synthetase; 2, transamination; 3, medium-chain fatty acyl CoA ligase; 4, benzoyl CoA:glycine acyltransferase; 5, phenylacetyl CoA:glutamine acetyltransferase. αKG, α-ketoglutarate; Ala, alanine; OHpyr, hydroxypyruvate; PO₄OHpyr, phosphohydroxypyruvate; Pyr, pyruvate. (From Brusilow and Horwich,[2] with permission.)

Because of the limited ability of citrulline to serve as a waste nitrogen product in patients with ASD and the absence of any excretable accumulated nitrogen product in patients with deficiencies of CPSD or OTCD, it is necessary to exploit latent biochemical pathways. These pathways can be activated to synthesize products that may serve as vehicles for waste nitrogen synthesis and excretion. These include amino acid acylation and acetylation pathways (Fig. 5-3).

STOICHIOMETRY BETWEEN DIETARY NITROGEN INTAKE AND WASTE NITROGEN EXCRETION

In order to prescribe therapy that promotes waste nitrogen excretion in compounds other than urea, the following question must be answered: What is the relationship between urea excretion and nitrogen intake? Once the amount of urea synthesized and excreted by normal infants at different protein intakes has been ascertained, it will be possible to determine the required magnitude of other pathways of waste nitrogen synthesis and excretion to be used in patients with urea cycle disorders.

The relationship between dietary nitrogen intake and urea excretion is most dramatically shown in preterm and full-term infants (Fig. 5-4). Preterm infants excrete only 13.6 percent of their dietary nitrogen as urea nitrogen, whereas infants (aged 12 to 45 days) excrete 19 percent of their dietary nitrogen as urea nitrogen ($P < 0.007$).[5,6] The explanation for these low urea excretions is apparent: much of the dietary nitrogen is incorporated into body protein. This avid nitrogen retention during the first months of life accounts for the protein tolerance of infants who suffer from urea cycle disorders. It also accounts for the reduced protein tolerance that occurs toward the end of the first year of life.

Unfortunately, there are very few descriptions of urea excretion over a wide range of nitrogen intake in patients on isocaloric intakes. Calloway and Margan[7] reported such a study in adults. Figure 5-5 also shows urea nitrogen excretion in an adult receiving 100 g of protein per day.[8]

At high protein intakes (100 g/d), more than 80 percent of dietary nitrogen is excreted as urea nitrogen. At minimum protein intakes (approximately 42 g), 46 percent of dietary nitrogen is excreted as urea nitrogen. It may be concluded from these data that on low-protein intakes, a substitute pathway for waste nitrogen synthesis should be capable of excreting 46 percent of dietary nitrogen.

No studies have been reported in which urine urea nitrogen has been measured in children receiving a wide range of dietary nitrogen. Waterlow,[9] however, reported data from which this relationship can be calculated. Figure 5-6 shows the relationship between dietary nitrogen and urea nitrogen in children aged 6 to 24 months; these derived data are similar to those described by Callaway and Margan. At high-protein intakes, large amounts of urea are synthesized and excreted. At minimum dietary protein intakes (approximately 1.25 g/kg/d), slightly less than a 0.1 g/kg/d (0.093 g/kg/d) of urea nitrogen is excreted. This represents 47 percent of dietary nitrogen.

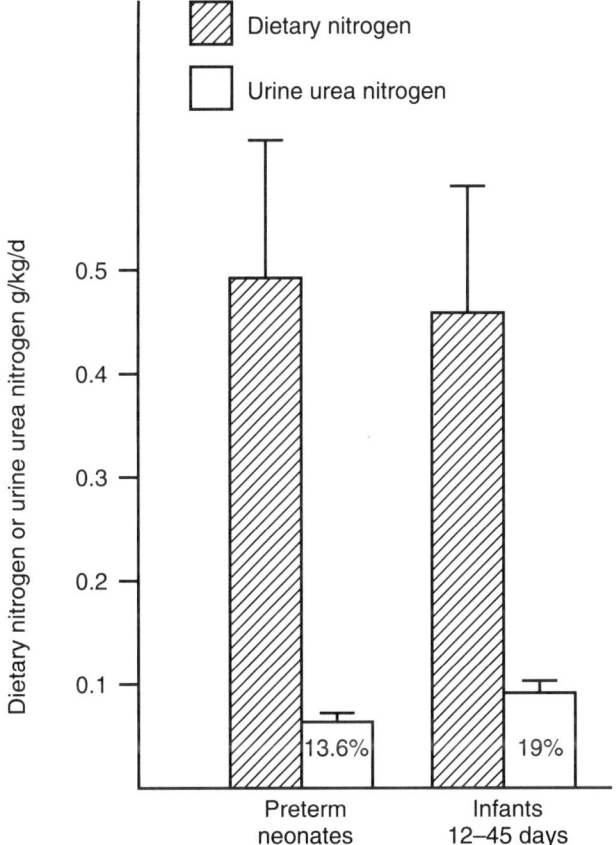

Fig. 5-4. Relationship between protein intake and urine urea nitrogen in preterm and term infants.

Thus, for a patient with little or no urea synthetic capacity, it may be concluded that an alternative pathway of nitrogen excretion should approximate the 0.1 g/kg/d of urea nitrogen excreted by a child on a low-protein (1.25 g/kg/d) diet. Nitrogen-containing compounds that may substitute for urea as a vehicle for waste nitrogen excretion include hippurate, phenylacetylglutamine, citrulline, and argininosuccinic acid.

HIPPURATE AND PHENYLACETYLGLUTAMINE AS WASTE NITROGEN PRODUCTS

Figure 5-3 shows the pathway for the biosynthesis of hippurate and phenylacetylglutamine. Two aspects of this pathway deserve special consideration. It should be noted that hippurate contains one nitrogen atom derived from glycine and phenylacetylglutamine contains two nitrogen atoms, both derived

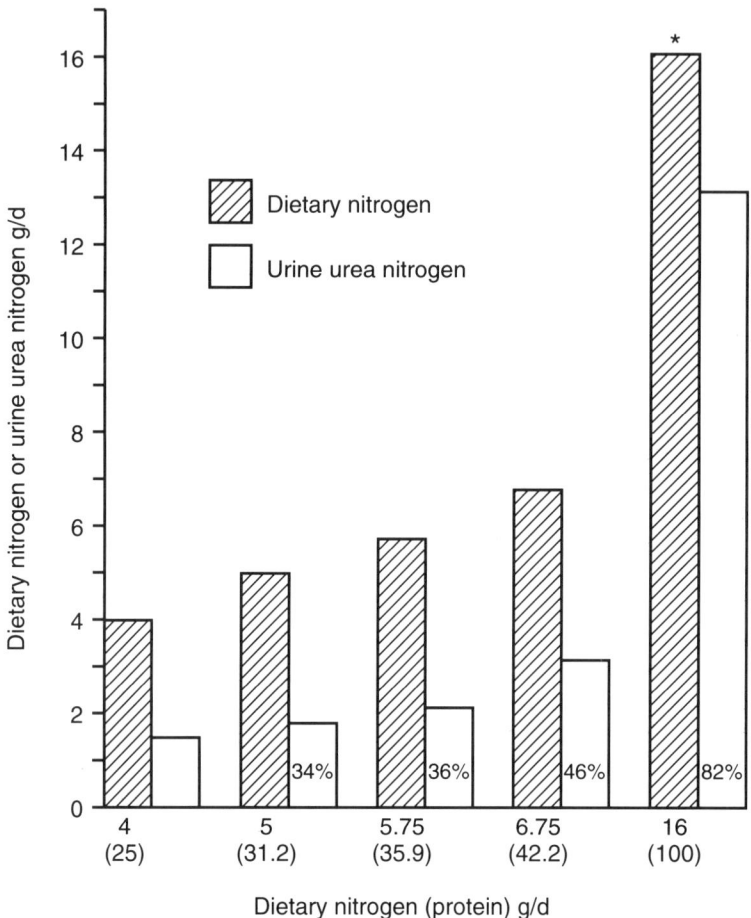

Fig. 5-5. Relationship between variations of dietary nitrogen (protein) and urine urea nitrogen in adults.

from glutamine. Thus, on a molar basis, phenylacetylglutamine is twice as effective as hippurate as a waste nitrogen product. Therefore, for a given dose of phenylacetate or benzoate, the former is approximately twice as effective as the latter. For this reason, and because of the offensive odor of phenylacetate (it is employed as a defensive weapon in the Stinkpot Turtle[10]), we have, whenever possible, substituted sodium phenylbutyrate. Sodium phenylbutyrate is β-oxidized to phenylacetate in vivo.

The theoretical relationships between benzoate and phenylacetate doses and waste nitrogen excretion, as hippurate and phenylacetylglutamine (PAG), is demonstrated in Table 5-1. It is apparent that when benzoate and phenylacetate are administered, each at a dose of 0.25 g/kg/d, the amount of waste nitrogen excreted as hippurate and PAG achieves only 71 percent of the

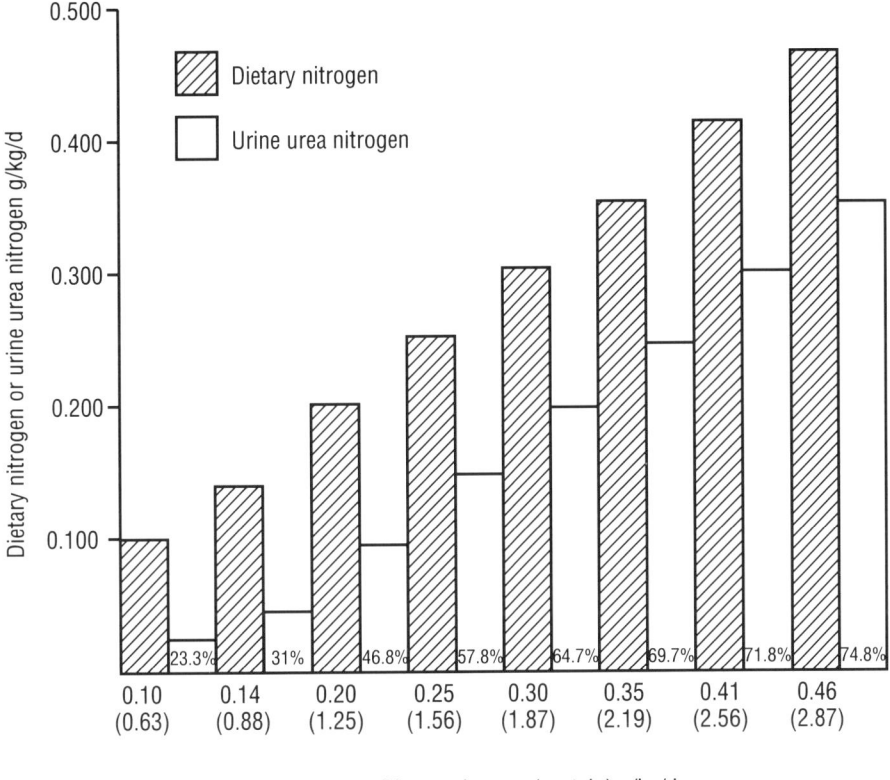

Fig. 5-6. Relationship between variations of dietary nitrogen (protein) and urine urea nitrogen in children aged 6 to 24 months.

Table 5-1. Theoretical Stoichiometry of Benzoate and Phenylacetate Administration and Nitrogen Excretion Afforded by Their Respective Amino Acid Conjugation Products, Hippurate and Phenylacetylglutamine[a]

	Dose (g/kg/d)	Hippurate or Phenylacetylglutamine Nitrogen (g/kg/d)
Sodium benzoate	0.25	0.024
Sodium phenylacetate	0.25	0.044
Total		0.068
Goal		0.093
Percentage of goal		73%

[a] The sum of hippurate and phenylacetylglutamine nitrogen (0.068 g/kg/d) is only 73% of the goal of waste nitrogen excreted as urea nitrogen (0.093 g/kg/d) in patients receiving 1.25 g/kg/d of protein (see Fig. 5-6).

Table 5-2. Theoretical Stoichiometry of Phenylacetate or Phenylbutyrate Administration and Nitrogen Excretion in Phenylacetylglutamine[a]

	Dose (g/kg/d)	Phenylacetylglutamine Nitrogen (g/kg/d)
Sodium phenylacetate	0.5 (3.16)	0.088
Sodium phenylbutyrate	0.6 (3.22)	0.090
Goal		0.093
Percentage of goal		95–97%

[a] Phenylacetylglutamine nitrogen excretion of 0.09 g/kg/d meets the requirement for waste nitrogen excretion (0.093 g/kg/d) in patients receiving 1.25 g/kg/d of protein (see Fig. 5-6).
[b] Numbers in parentheses are in mmol/kg/d.

requirement for waste nitrogen excretion described earlier (0.093 g/kg/d) for patients on a low-protein diet.

Table 5-2 shows the theoretical relationship between high-dose sodium phenylacetate or sodium phenylbutyrate therapy and waste nitrogen excretion (benzoate is not included). The amount of waste nitrogen excreted as PAG fulfills the waste nitrogen excretion requirement for patients on a low-protein diet.

These theoretical considerations were tested in an 8-year-old boy with neonatal-onset CPSD. As shown in Table 5-3,[11] 80 percent or more of administered sodium phenylacetate or sodium phenylbutyrate appeared in the urine as PAG. That phenylacetylglutamine nitrogen (PAG-N) substituted for urea as a vehicle for waste nitrogen excretion may be adduced by the finding that PAG-N accounted for 44 percent of dietary nitrogen.

It may be concluded from these theoretical and experimental observations that high-dose phenylacetate or phenylbutyrate are capable of providing a pathway of waste nitrogen synthesis and excretion that may substitute for urea nitrogen as a vehicle for waste nitrogen excretion.

Table 5-3. Studies of Phenylacetylglutamine Excretion[a]

	Period I Sodium Phenylacetate	Period II Sodium Phenylbutyrate	Period III Sodium Phenylbutyrate
g/3 d	30	36	42
Predicted PAG excretion (mmol)	190	193	225
Measured PAG excretion (mmol)	157	174	181
$\frac{\text{Measured PAG}}{\text{Predicted PAG}} \times 100$	83%	90%	80%
$\frac{\text{PAG-N}}{\text{Dietary N}} \times 100$	38.1%	42%	44%

[a] The urinary excretion of PAG during 3-day periods of treatment of a 7½-year-old boy with CPSD. Treatment consisted of the sodium salts of phenylacetate and phenylbutyrate (g/3 d). Also shown is a calculation of the percentage of dietary nitrogen excreted as PAG-N.

Abbreviations: CPSD, carbamylphosphate synthetase; PAG, phenylacetylglutamine; PAG-N, phenylacetylglutamine nitrogen.

(From Brusilow,[11] with permission.)

Preliminary data from this laboratory suggest that patients with neonatal-onset CPSD or OTCD who are receiving high-dose phenylacetate or phenylbutyrate appear to have an improved survival rate.

Notwithstanding improved survival with therapy as described, the neonatal forms of CPSD and OTCD are very serious diseases, imposing a severe burden on families. All patients are vulnerable to episodes of hyperammonemia, requiring frequent hospitalizations. The therapeutic protocol for management of these hyperammonemic episodes is described in the Appendix. Notwithstanding the value of this protocol in treating intercurrent hyperammonemic episodes, the chief cause of death occurs during a hyperammonemic episode. Improved methods of treating such episodes, as well as novel experimental approaches to this problem, are very much in demand.

For those patients who are rescued from neonatal coma, severe mental retardation, cerebral palsy, and/or seizures are inevitable. For those neonates at risk who are treated prospectively and in whom severe neonatal hyperammonemia is prevented, the neurologic and developmental outcome is much better. But the aforementioned burdens for ostensibly normal children present an even greater problem. Because of these medical burdens, five patients (four with OTCD and one with CPSD) have had orthotopic liver transplants.

CITRULLINE AND ARGININOSUCCINIC ACID AS WASTE NITROGEN PRODUCTS

As shown in Figures 5-1 and 5-2, supplementary dietary arginine will promote the biosynthesis, and hence the excretion, of ASA in patients with ALD and of citrulline in patients with ASD. The relative roles of citrulline, ASA, hippurate, and PGA in serving as vehicles for waste nitrogen synthesis are shown in Table 5-4, which shows the partition of effective urinary nitrogen as measured in three groups of patients: (1) those with CPSD and OTCD treated with benzoate, phenylacetate, and citrulline; (2) those with ASD treated with benzoate, phenylacetate, and arginine; and (3) those with ALD treated with arginine alone. "Effective" urinary waste nitrogen excludes urea from the calculations because urinary urea is derived entirely (or nearly so) from supplementary citrulline or arginine. "Effective" waste nitrogen also excludes 50 percent of the ASA nitrogen and 66 percent of the citrulline nitrogen, because they are also derived from supplementary dietary arginine. It is concluded that citrulline nitrogen contributes significantly to waste nitrogen excretion and that ASA nitrogen by itself serves as a vehicle for waste nitrogen synthesis. Survival of patients with ASD and ALD treated in this manner is now approximately 90 percent. (Several ASD patients are receiving high-dose phenylacetate or phenylbutyrate rather than the benzoate-phenylacetate combination.)

Table 5-5 summarizes the current recommended dietary and drug therapy for urea cycle disorders. Table 5-6 lists the current vendors and sources of drugs and nutritional supplements used in urea cycle disorders.

Not included in these recommendations is the recent suggestion that citrate therapy benefited two patients with ALD.[12] This finding is contrary to prelimi-

Table 5-4. Partition of Urine "Effective" Waste Nitrogen[a]

Diagnosis	Age (mo)	d	NH$_2$-N	Cr-N	αNH$_2$-N	CIT-N	ASA-N	HAN	PAG-N
CPSD	19	3	10	3	3	—	—	27	51
CPSD	4	2	11	4	12	—	—	24	33
OTCD	11	2	15	3	5	—	—	23	40
OTCD	21	3	9	4	5	—	—	18	37
OTCD	9	2	16	2	2	—	—	17	30
ASD	2	3	9	3	—	9	—	18	24
ASD	16	4	13	4	—	19	—	15	21
ASD	48	3	15	3	—	9	—	17	22
ALD	2	2	16	7	—	—	40	—	—
ALD	43	3	16	—	—	—	42	—	—

[a] The partition of urine "effective" waste nitrogen excretion in patients with inborn errors of metabolism treated as described in Table 5-5. "Effective" waste nitrogen excludes urea nitrogen, 66% of citrulline nitrogen, and 50% of ASA nitrogen, because these nitrogen components are derived from supplementary dietary arginine.

Abbreviations: d, collection period in days; CR-N, creatinine nitrogen; α NH$_2$-N, α-amino nitrogen; CIT-N, citrulline nitrogen; ASA-N, argininosuccinate acid nitrogen; HAN, hippurate nitrogen; PAG-N, phenylacetylglutamine nitrogen.

Table 5-5. Recommended Management of Patients With Urea Cycle Disorders

Deficiency	Diet	Medication
Carbamylphosphate synthetase or ornithine transcarbamylase	Essential amino acids[a] 0–0.7 g/kg/d Protein 1.0–2.0 g/kg/d[b] Caloric supplementation with Mead Johnson protein-free diet powder 4.9 cal/g	Sodium phenylbutyrate[c] 0.45–0.60 g/kg/d 9.9–13.0 g/m^2/d Citrulline[d] 0.17 g/kg/d OR 3.8 g/m^2/d
Argininosuccinic acid synthetase	Protein 1.25–2.00 g/kg/d Caloric supplementation with Mead Johnson protein-free diet powder 4.9 cal/g	Sodium phenylbutyrate[c] 0.45–0.60 g/kg/d OR 9.9–13.0 g/m^2/d Arginine (free base) 0.4–0.7 g/kg/d OR 8.8–15.4 g/m^2/d
Argininosuccinase	Protein 1.25–2.00 g/kg/d Caloric supplementation with Mead Johnson protein-free diet powder 4.9 cal/g	Arginine (free base) 0.4–0.7 g/kg/d OR 8.8–15.4 g/m^2/d

[a] An essential amino acid mixture, UCD2, is available from Mead Johnson. UCD2 contains 80 g of an essential amino acid mixture per 100 g. Therefore, it is necessary to prescribe 1.25 g of UCD2 to supply 1 g of essential amino acids.

[b] The goal of therapy is to promote growth and development. To achieve these ends, fasting plasma levels of ammonium, branched-chain amino acids, arginine, and serum plasma protein should be maintained within normal limits and plasma glutamine at levels of <1000 μM. The degree to which nitrogen intake is partitioned into natural protein and essential amino acids is a function of age, residual enzymatic activity, and dose of sodium phenylbutyrate. It has become apparent that infants with neonatally expressed disorders treated as described may tolerate as much as 2 g/kg/d of natural protein during the first few months of life. Protein tolerance will decrease as the infant's growth rate decreases, requiring reduced nitrogen intake. Patients with partial deficiencies, including females heterozygous for OTCD, initially receive a diet containing the age-determined minimal daily protein requirement, which may be increased as tolerated.

[c] The precise dose of sodium phenylbutyrate will depend on clinical circumstances. Lower doses are recommended for infants. The highest dosage is recommended for all other patients, although a lower dose may suffice for patients with significant residual enzymatic activity. Because phenylacetate and phenylbutyrate on a molar basis are twice as effective as benzoate, the use of benzoate is no longer recommended. If only sodium phenylacetate is prescribed, the dose is 0.40–0.50 g/kg/d (10–11 g/m^2/d). If only sodium benzoate and sodium phenylacetate in combination are available, the doses are 0.25 g/kg/d (5.5 g/m^2/d) and 0.25 g/kg/d (5.5 g/m^2/d), respectively. (Sodium benzoate and sodium phenylacetate in combination is supplied in liquid dosage form [Ucephan, Kendall McGaw].) Sodium phenylbutyrate (which is rapidly oxidized to phenylacetate in vivo) has been substituted for phenylacetate because of the unpleasant odor of the latter. As a consequence of its higher molecular weight, the dose of sodium phenylbutyrate is 17% greater than the dose of sodium phenylacetate.

[d] Arginine (free base) may be substituted for citrulline (costing three to four times that of arginine) in some patients with the late-onset form of CPS and OTC deficiency.

Abbreviations: CPS, carbamylphosphate synthetase; OTC, ornithine transcarbamylase; OTCD, ornithine transcarbamylase deficiency.

Table 5-6. Sources of Medicines and Nutritional Supplements

L-Citrulline	L-Arginine
L-Citrulline is available as a powder[a] from the following sources: Seybridge Pharmacy 37 New Haven Road Seymour, Connecticut 06483 (203)888-0073 (ask for Peter Przybylski) Approximate price: $200.00/kg	L-Arginine (free base) is available as a powder[a] from the following sources: Ajinomoto 500 Frank Burr Boulevard Teaneck, New Jersey 07666 (201)488-1212 Approximate price: $50.00/kg Minimum order: 1000 g
Ajinomoto 500 Frank Burr Boulevard Teaneck, New Jersey 07666 (201)488-1212 Approximate price: $200.00/kg Minimum order: 1000 g	Tanabe 7071 Convoy Court San Diego, California (619)571-8410 Approximate price: $50.00/kg Minimum order: 1000 g
Tanabe 7071 Convoy Court San Diego, California 92111 (619)571-8410 Approximate price: $200.00/kg Minimum order: 1000 g	L-arginine (free base) is available as a powder[a] and in capsules from the following sources: Tyson and Associates 1661 Lincoln Boulevard Santa Monica, California 90494 (213)452-7844 Minimum order for powder, 150 g; approximate retail price $28.00 per 150 g (187 kg). L-Arginine (free base) in 700-mg gelatin capsules; approximate retail price $20.00/100 capsules
Triple Crown America, Inc. 13 North 7th Street Perkasie, PA 18944 (215)453-2500 Approximate price: $180.00/kg Minimum order: 1000 g	
Tyson and Associates 1661 Lincoln Boulevard Santa Monica, California 90494 (213)452-7844 Approximate retail price: $90.00/150 g ($600.00/kg) Minimum order: 150 g Tyson also supplies L-citrulline in 600-mg gelatin capsules; approximate retail price $32.50/50 capsules	Intravenous L-arginine hydrochloride is supplied by KabiVitrum Inc. (Franklin, OH) as a sterile 10% solution.

(Continues)

nary data in one similarly treated patient who showed no change in plasma levels of ammonium, glutamine, citrulline, lactate, pyruvate, α-ketoglutarate, glucose, and ASA or its anhydrides while being treated with citrate (unpublished observations). Liver size was unchanged. The hypothesis on which citrate therapy is based states that a depletion of tricarboxylic acid (TCA) substrates occurs secondary to loss of the aspartate moiety of ASA, the aspartate being a transamination product of the TCA substrate oxaloacetate. Ignored in this argument is the ready availability of oxaloacetate via the pyruvate carboxylase reaction. The

Table 5-6. Sources of Medicines and Nutritional Supplements (*Continued*)

Sodium Benzoate/sodium phenylacetate	Essential amino acids
Kendall-McGaw [Irvine, CA (800)854-6851] supplies an oral liquid dosage form of a 10% solution of sodium benzoate and sodium phenylacetate under their tradename Ucephan. It will be unavailable after March 7, 1992. The intravenous dosage form of sodium benzoate and sodium phenylacetate is an Investigational New Drug available to physicians upon application to: Saul Brusilow, MD The Johns Hopkins Hospital 600 N. Wolfe street Baltimore, Maryland 21205 (301)955-0885 **Sodium phenylbutyrate** Sodium phenylbutyrate is an Investigational New Drug undergoing a Phase III clinical trial. It is available as a powder and as 0.5-g tablets to physicians upon application to: Saul Brusilow, MD The Johns Hopkins Hospital 600 N Wolfe Street Baltimore, Maryland 21205 (301)955-0885	Mead Johnson (Evansville, IN) supplies a mixture of essential amino acids (EEA) under their tradenames UCDI (68 g EAA/100 g) and UCDII (81 g EAA/100 g). **Protein-free diet powder** Mead Johnson (Evansville, IN) supplies a protein-free diet powder useful for protein-free caloric supplementation.

[a] Because the powders of L-citrulline and L-arginine (free base) are usually administered as teaspoonsful (or fractions thereof), it will be necessary to determine how many grams of either L-citrulline or L-arginine (free base) are contained in a level teaspoonful or half-teaspoonful of kitchen-type measuring spoons. (Products of different manufacturers have different densities.)

purpose of this homeostatic anapleurotic mechanism is to prevent deficiency of oxaloacetate or other TCA (or Krebs) cycle substrates.[13] It can be calculated that in an ALD patient on a low-protein diet, the loss of oxaloacetic in the aspartate moiety of ASA is quite small and can be replaced by de novo oxaloacetate synthesis via the pyruvate carboxylase reaction at a rate that is 10 percent of the capacity of this pathway when used for gluconeogenesis.

ACKNOWLEDGMENTS

This research was supported by grants HD 11134, HD 26358, and RR 00052 from the National Institutes of Health, by grant FD R 00198 from the Food and Drug Administration, by the Kettering Family Foundation, by the T.A. and M.A. O'Malley Foundation, and by the National Organization for Rare Diseases.

APPENDIX: PROTOCOLS FOR MANAGEMENT OF INTERCURRENT HYPERAMMONIA IN PATIENTS WITH UREA CYCLE DISORDERS

Early Diagnosis and Therapy

Early diagnosis and therapy are the most important aspects of intercurrent hyperammonemia. Delays are disastrous. A plasma ammonium level should be done as an emergency procedure in any child with these diseases who exhibits lethargy or vomiting of any degree. Parents should be taught that such symptoms are emergencies demanding immediate medical attention.

If the ammonium level approaches three times the upper limits of normal, the ammonium level should be repeated and venous plasma obtained for electrolytes, pH, PCO_2, and quantitative amino acids. Without waiting for the repeat ammonium value, the appropriate regimen described here should be followed as an emergency procedure.

Prescribed drugs may cause one or two vomiting episodes, usually toward the end of the 90-minute treatment period. Therapy with these drugs may also cause or exacerbate respiratory alkalosis.

All dietary or intravenous nitrogen intake should be discontinued. Because reduction of body protein breakdown is desirable, a high parenteral caloric intake should be provided with 10 to 15 percent glucose and Intralipid; for infants, the goal should be 80 to 100 cal/kg/d. Patients who can tolerate enteral feedings, should be given a formula consisting of 14 g of Mead Johnson Product 80056 in 100 ml of water, which supplies 20 cal/oz.

The hemodialysis team should be alerted. To avoid delays in establishing emergency vascular access, it may be most efficient to rely on cardiologists or intensivists to place the lines.

Plasma levels of ammonium, electrolytes, pH and PCO_2 should be measured 4 hours after completion of the priming infusion and every 8 hours thereafter, until plasma ammonium levels are normal or near normal. If intracranial pressure is elevated, begin conventional osmotherapy with mannitol. Corticosteroids may be contraindicated because they induce negative nitrogen balance. When the ammonium level is stable at normal or near-normal levels, oral medication may be gradually added as the intravenous medication is gradually reduced.

CARBAMYLPHOSPHATE SYNTHETASE OR ORNITHINE TRANSCARBAMYLASE DEFICIENCY

Priming infusion: to be given over 90 minutes in 25–35 ml/kg of 10% glucose or 400–600 ml/m^2 of 10% glucose (whichever is less). (see Notes below).

	g/kg/d	OR	g/m^2/d
Sodium benzoate[a]	0.250		5.5
Sodium phenylacetate[a]	0.250		5.5
10% Arginine HCl[b]	0.210 (2 ml/kg)		4.0

Sustaining infusion: to be given over 24 hours in maintenance fluids.

	g/kg/d	OR	g/m²/d
Sodium benzoate	0.250		5.5
Sodium phenylacetate	0.250		5.5
10% Arginine HCl[b]	0.210 (2 ml/kg)		4.0

Hemodialysis should be started as an emergency procedure, if plasma ammonium level does not decrease within 8 hours. If hemodialysis is necessary, it should be done with the largest catheters consistent with the patient's size (ammonium clearance is approximately equal to blood flow). Because both peritoneal dialysis and continuous arteriovenous hemofiltration produce ammonium clearances 10 percent that of hemodialysis, hemodialysis is the treatment of choice.

ARGININOSUCCINIC ACID SYNTHETASE DEFICIENCY

Priming infusion: to be given over 90 minutes in 25–35 ml/kg/d 10% glucose or 400–600 ml/m² of 10% glucose, whichever is less (see Notes below).

	g/kg/d	OR	g/m²/d
Sodium benzoate[a]	0.250		5.5
Sodium phenylacetate[a]	0.250		5.5
10% Arginine HCl[b]	0.660 (6 ml/kg)		12.0

Sustaining infusion: to be given over 24 hours in maintenance fluids.

	g/kg/d	OR	g/m²/d
Sodium benzoate[a]	0.250		5.5
Sodium phenylacetate[a]	0.250		5.5
10% Arginine HCl[b]	0.660 (6 ml/kg)		12.0

Hemodialysis should be started as an emergency procedure if plasma ammonium level does not decrease within 8 hours as described for CPSD and OTCD.

ARGININOSUCCINASE DEFICIENCY

Priming infusion: to be given over 90 minutes in 25–35 ml/kg of 10% glucose or 400–600 ml/m² of 10% glucose, whichever is less (see Notes below).

	g/kg/d	OR	g/m²/d
10% Arginine HCl	0.660 (6 ml/kg)		12.0

Sustaining infusion: to be given over 24 hours in maintenance fluids.

	g/kg/d	OR	g/m²/d
10% Arginine HCl	0.660 (6 ml/kg)		12.0

Hemodialysis should be started as an emergency procedure, if plasma ammonium level does not decrease within 8 hours as described above.

^aNote that 1 g of sodium benzoate contains 160 mg of sodium; 1 g of sodium phenylacetate contains 147 mg of sodium. Because urine potassium loss is enhanced by the excretion of the nonresorbable anions (hippurate and phenylglutamine) the plasma potassium levels should be monitored and treated when necessary.

^bNote because a hyperchloremic acidosis may ensue after high-dose arginine HCl, plasma levels of chloride and bicarbonate should be monitored and appropriate amounts of bicarbonate administered. 10% Arginine HCl is available as a sterile pyrogen-free solution from Kabivitrum (Clayton, NC).

REFERENCES

1. Brusilow SW, Tinker J, Batshaw ML: Amino acid acylation: A mechanism of nitrogen excretion in inborn errors of urea synthesis. Science 207:659, 1980
2. Brusilow SW, Horwich A: Urea cycle enzymes. p. 629. In Scriver S, Beaudet A, Sly W, Valle D (eds): The Metabolic Basis of Inherited Disease. 6th Ed. McGraw-Hill, New York, 1990
3. Brusilow SW: L-Arginine: An indispensable amino acid for patients with inborn errors of urea synthesis. J Clin Invest 74:2144, 1984
4. Fomon SJ: Infant Nutrition. p. 18. WB Saunders, Philadelphia, 1974
5. Pencharz PB, Steffee WP, Cochran W, et al: Protein metabolism in human neonates: Nitrogen balance studies, estimated obligatory losses of nitrogen and whole body turnover of nitrogen. Clin Sci Mol Med 52:485, 1977
6. Barness LA, Baker D, Guilbert P, et al: Nitrogen metabolism of infants fed human and cow's milk. J Pediatr 51:29, 1957
7. Callaway DH, Margan S: Variation in endogenous nitrogen excretion and dietary nitrogen utilization as determinants of human protein requirement. J Nutr 101:205, 1971
8. Cahill GF, Owen OE: The role of the kidney in the regulation of protein metabolism. p. 559. In Munro HN (ed): Mammalian Protein Metabolism. Vol. 4. Academic Press, San Diego, 1970
9. Waterlow JC: The partition of nitrogen in the urine of malnourished Jamaican infants. Am J Clin Nutr 12:235, 1963
10. Eisner T, Conner WE, Hicks K, et al: Stink of stinkpot turtle identified: Omega-phenylalkanoic acids. Science 196:1347, 1977
11. Brusilow SW: Phenylacetylglutamine may replace urea as a vehicle for waste nitrogen excretion. Pediatr Res 29:125, 1990
12. Iafolla AK, Gale DS, Roe CR: Citrate therapy in argininosuccinate lyase deficiency. J Pediatr 117:102, 1990
13. Lehninger AL: Biochemistry. Worth, New York, 1970

6

THERAPY FOR CYSTINOSIS

William A. Gahl

INTRODUCTION

Lysosomal storage disorders are notorious for their poor prognosis and limited avenues for therapy. Nephropathic cystinosis stands out as an exception to this rule, since therapy directed at the basic defect has had truly remarkable efficacy. To understand the effectiveness of therapy for cystinosis, one must be aware of the basic defect in cystinosis and its clinical manifestations. The following background information will serve to put the therapy of nephropathic cystinosis into perspective.

THE DISEASE

Nephropathic cystinosis is a rare, autosomal recessively inherited, lysosomal storage disorder.[1-4] The clinical manifestations result from the accumulation of cystine, the disulfide of cysteine, within cells. In many tissues and organs, including the kidney, liver, spleen, lymph nodes, intestines, and bone marrow and less often the thyroid, pancreas, and muscle, the cystine crystallizes because of its poor solubility. This results in clinical dysfunction. Cystine is stored specifically within lysosomes.[5] The function of these acidic vesicles is to degrade macromolecules into their component amino acids, sugars, or lipids. Normally, the small molecules produced will leave the lysosome by carrier-mediated transport across the lysosomal membrane. In the case of cystine, the disulfide moves to the cytosol, where it is rapidly reduced to cysteine. In cystinosis, cystine fails to be transported across the lysosomal membrane because its carrier system is defective.[6-8] As a result, cystine accumulates to 10- to 1000-fold normal levels, depending on the type of cell examined. The lysosomal membrane transport defect in cystinosis has been demonstrated in polymorphonuclear leukocytes,[6-8] lymphoblasts,[9] and cultured fibroblasts[10] in several different laboratories.

Clinical Manifestations

Patients with nephropathic cystinosis are normal at birth. They tend to have a pale complexion, blonde hair, and blue eyes (Fig. 6-1), but this is not true for blacks with cystinosis or for highly pigmented ethnic groups. By 6 to 12 months of age, failure to thrive accompanies the renal tubular Fanconi syndrome. Failure to resorb water, glucose, electrolytes, and minerals results in polyuria (2 to 6 L/d), polydipsia, dehydration, glucosuria, hypokalemia, acidosis, hypocalcemia, and hypophosphatemia. An occasional patient has hypomagnesemia. Episodes of acidosis and dehydration, often associated with a viral infection, generally bring patients to medical attention. The hypophosphatemia caused by phosphaturia can result in florid clinical rickets, with a rachitic rosary, frontal bossing, genu valgum, and failure to walk because of guarding. Patients with Fanconi syndrome also lose carnitine in their urine and have an associated muscle carnitine deficiency.[11] The generalized aminoaciduria and tubular proteinuria that are part of the Fanconi syndrome of cystinosis have no apparent clinical consequences.

Failure to grow can be a presenting finding in nephropathic cystinosis. By 12 months of age, the height of the average patient is at the third percentile,

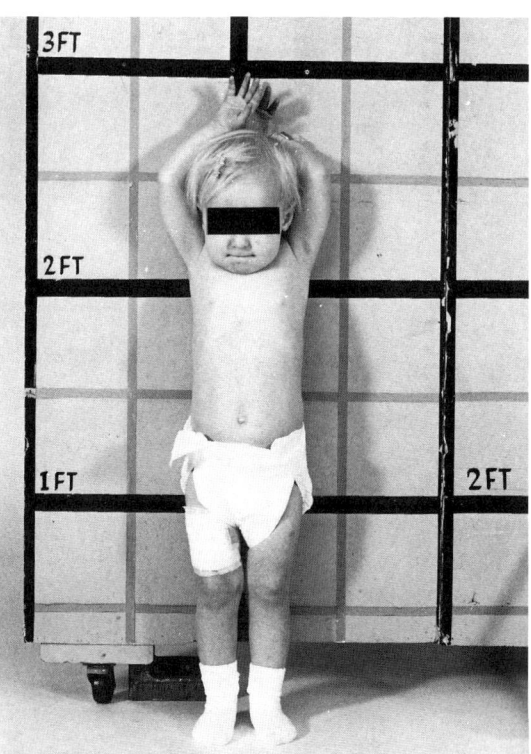

Fig. 6-1. Typical child with nephropathic cystinosis. This 3-year-old girl had fair hair and skin, as well as short stature.

and the child continues to grow at only about 50 to 60 percent of the normal rate.[12] The child has the height of a 4-year-old at age 8. Weight is proportional to height, but head circumference remains normal, giving the impression of macrocephaly.

While the renal tubular resorption problems occur early and remain stable, the renal glomerular damage progresses inexorably. Serum creatinine may not rise until approximately 5 years of age, but measurements of creatinine clearance demonstrate an early, continuous loss of filtration function. Studies of large numbers of cystinosis patients indicate that renal failure occurs between 9 and 10 years of age in untreated patients.[13] This can be hastened by episodes of renal hypoperfusion.

There is a good deal of ophthalmic pathology in cystinosis. A patchy retinal depigmentation occurs as early as the first year of life and may presage the retinal blindness that can occur during the second and third decades of life.[3,14,15] In patients with visual impairment, electroretinographic abnormalities confirm the presence of retinal damage. Corneal crystals accumulate late in the first year of life and progress to pack the cornea. This finding, apparent on slit-lamp examination (Fig. 6-2), is pathognomonic for cystinosis. The corneal crystals cause photophobia of variable onset and severity. During the second and third decades of life, corneal erosions cause severe pain. The corneas often become hazy; this opacification can interfere with vision. Occasionally, posterior synechiae have been present in older patients,[14] resulting in impaired pupillary constriction and enhanced photophobia.

Pretransplant patients have impaired exocrine gland function, including inability to sweat, tear, and salivate normally. The sweating problem has been examined most extensively.[16] While electrolyte concentrations are normal in the sweat of cystinosis patients, the amount of sweat is commonly reduced below the normal minimum (100 mg) produced on standard pilocarpine iontophoresis testing. This hypohidrosis causes heat avoidance, hyperthermia, and vomiting.

Primary hypothyroidism is another common complication of cystinosis,[17,18] especially after 10 years of age.[15,19] Cystine crystals occur in the thyroid gland, and fibrosis leads first to compensated and then to uncompensated hypothyroidism. Pancreatic endocrine insufficiency (diabetes mellitus) has been observed in a half-dozen patients.[20] At least two patients have exhibited pancreatic exocrine insufficiency,[21] treatable with pancreatic extracts. Approximately one-third of cystinosis patients have hepatomegaly or splenomegaly of unknown etiology.[1,19] Many patients have gastrointestinal symptoms, including vomiting, which is worse in the morning and on an empty stomach.

Puberty is delayed in both males and females, occurring at approximately 14 to 15 years of age in females and at 16 to 17 years of age in males.[3,15] Investigation into the etiology of the pubertal delay is complicated by the fact that most of these patients have had renal transplants and have received steroids during their adolescent years. Several males have primary hypogonadism, with fibrosis and crystal formation in their testes. In one male whose semen was analyzed, no sperm was present.

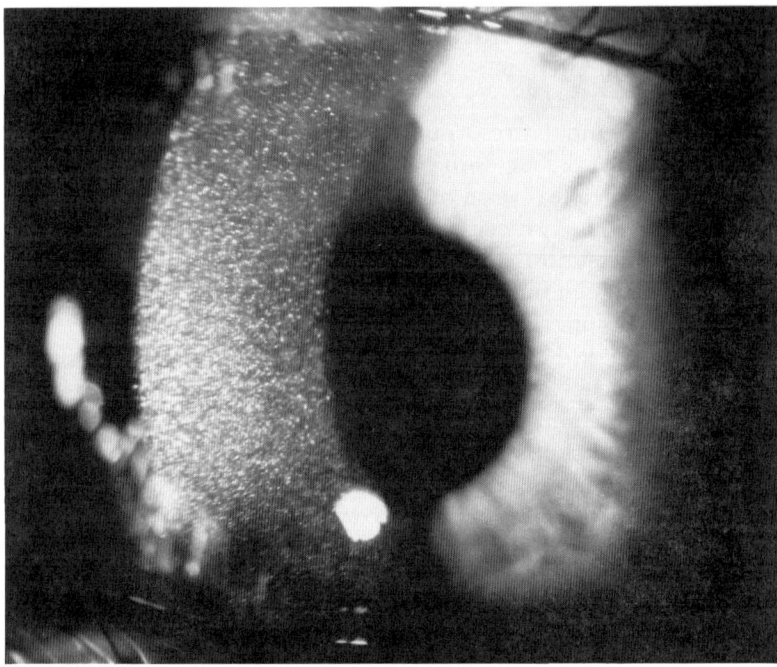

Fig. 6-2. Slit-lamp photograph of corneal cystine crystals. The cornea of a 5-year-old boy with cystinosis shown packed with crystals. (Courtesy of Dr. M.I. Kaiser-Kupfer, National Eye Institute.)

At least six patients examined at the National Institutes of Health (NIH) Clinical Center have had a myopathy associated with their cystinosis; one patient died of aspiration.[22] A primary manifestation of muscle damage appears to be difficulty in swallowing, which we have observed in several patients.[23] Several post-transplant patients have cerebral atrophy,[15,24-26] although most have a normal neurologic examination. There are individual reports of demyelination of the internal capsule,[27] calcification of the basal ganglia and paraventricular areas,[3,26] and nonabsorptive hydrocephalus.[28] Postmortem examinations have revealed cystine storage in various portions of the central nervous system.[29] All these late complications of cystinosis appear with increased frequency as the patients age.

Certain laboratory tests are characteristically abnormal in cystinosis. The erythrocyte sedimentation rate (ESR) is usually elevated, platelets are increased, and the total cholesterol concentration is high.[1,30] Liver enzymes are generally within the high-normal range.

Diagnosis

The diagnosis of cystinosis can be made by recognition of corneal crystals on slit-lamp examination by an experienced ophthalmologist.[1-4] These crystals are nearly always present after 1 year of age and may be seen as early as 3 months of age. Alternatively, elevated cystine levels in cultured fibroblasts or polymorphonuclear leucocytes are diagnostic.[31,32] Normally, the cystine levels are below 0.2 nmol of half-cystine per mg of cell protein. In cystinosis, the values are 5 to 10 nmol of half-cystine per mg of protein. Heterozygote values are less than 1.0 nmol of half-cystine per mg of protein. There is no need to take a biopsy of the kidney, bone marrow, or conjunctiva to diagnose cystinosis. Since fibroblasts store increased cystine, the diagnosis of cystinosis can be made prenatally on amniocytes.[33] Chorionic villus sampling has also been used successfully to diagnose cystinosis during the first trimester of pregnancy.[34]

Genetics

The incidence of cystinosis approximates 1 in 100,000 live births. Heterozygotes virtually never have any of the symptoms manifest by homozygotes. The inheritance is clearly autosomal recessive. There are variants of cystinosis, including adult or benign cystinosis,[1-4,35] in which the only clinical manifestation is the accumulation of cystine crystals in the cornea, conjunctiva, and bone marrow, and intermediate or late-onset cystinosis in which all the clinical manifestations of nephropathic cystinosis occur only later in life.[1-4,36] Onset of symptoms generally takes place during the second half of the first decade or later. The level of cystine storage in benign cystinosis is usually less than in late-onset cystinosis and much less than in nephropathic cystinosis.

THERAPY

Various therapeutic approaches to the major symptoms of nephropathic cystinosis are listed in Table 6-1.

Renal Fanconi Syndrome

Treatment of renal tubular Fanconi syndrome in cystinosis involves chronic replacement of the renal losses caused by the failure to resorb small molecules and nutrients. Since water itself is wasted by the cystinotic kidney, patients must be given free access to fluids to prevent dehydration. Patients also require alkalinization to compensate for renal bicarbonate losses. Polycitra (potassium/sodium citrate), which contains 2 mEq of citrate and 1 mEq of sodium and potassium per ml, serves as an excellent alkalinizing agent. Patients generally require 5 to 20 ml of Polycitra every 6 hours. Our goal has been to maintain the serum carbon dioxide above 20 mEq/L, although this often cannot be achieved in patients under 2 years of age.

Since children with cystinosis waste potassium, they are at risk of hypokalemic arrhythmias. Potassium should be supplemented as the gluconate, chloride, or phosphate salt. We attempt to maintain the serum potassium level above 3.0 mEq/L. This usually requires potassium supplementation equivalent to two to three times the daily maintenance requirements. The occasional patient who also has hyponatremia requires sodium chloride supplementation.

Table 6-1. Therapy for Nephropathic Cystinosis in the United States

Symptom	Goal	Therapy
Renal Fanconi syndrome	Replacement of tubular losses	Fluids and electrolytes (potassium, bicarbonate, sodium)
		Minerals (phosphate, calcium, magnesium)
		Carnitine (occasionally)
Renal failure	Preservation of function and improved growth	Oral cysteamine
	Replacement of renal function	Dialysis or transplant
Corneal erosions, photophobia	Pain relief, improved vision	Cysteamine eyedrops
	Removal of damaged ruined cornea	Penetrating keratoplasty
Nonrenal complications		
Hypohydrosis	Avoidance of hyperthermia	Avoid heat
Hypothyroidism	Maintenance of euthyroid state	L-Thyroxine
Hypogonadism (male)	Secondary sexual characteristics	Intramuscular testosterone
Diabetes mellitus	Normoglycemia	Insulin
Pancreatic exocrine insufficiency	Relief of steatorrhea	Pancreatic extracts
Swallowing difficulty	Prevention of aspiration	Dietary manipulation, oral exercises
Symptoms having potential therapies		
Growth retardation	Catch-up growth, normal height	Growth hormone
Anemia	Normal hemoglobin	Erythropoietin
Poor nutrition	Reduction of fluid intake	Indomethacin

Because of phosphaturia, cystinosis patients often have a vitamin D-resistant hypophosphatemic rickets. This condition will often resolve within approximately 3 months on treatment with sodium or potassium phosphate at doses of 2 to 4 g/d. A vitamin D supplement can assist intestinal absorption of phosphate but in general will not cure the rickets without phosphate supplementation.

An occasional patient has tetany caused by hypocalcemia and will benefit from calcium glubionate or another calcium salt. A typical dose may be 1 to 2 g bid to qid. Calcium and phosphate must be administered separately to avoid precipitation. Hypocalcemic tetany can occur after a dose of citrate because the alkalinization attendant to citrate therapy causes the ionized calcium in the serum to bind to protein. Rarely, magnesium needs to be replaced as well as calcium; both minerals must be present in adequate concentrations in order to avoid tetany.

L-Carnitine is a small molecule required for the transport of long-chain fatty acids into the mitochondria to undergo β-oxidation for energy production. The renal tubules handle carnitine like an amino acid; that is, they normally resorb 97 percent of the filtered load of carnitine.[11] Patients with renal Fanconi syndrome waste carnitine in their urine and, consequently, have low plasma carnitine concentrations. These translate into low muscle levels,[11] since the only supply of carnitine to the muscle is through the blood. Patients with cystinosis exhibit a cardinal sign of muscle carnitine deficiency on biopsy (i.e., lipid droplets on oil-red-0 stain). Repletion with oral L-carnitine at approximately 100 mg/kg/d in four divided doses can have beneficial effects in some patients.[30] However, the skeletal muscle represents a huge reservoir for carnitine, and this may take years to replete. Long-term therapeutic studies are currently under way to determine whether muscle carnitine deficiency can be prevented by early prophylactic supplementation.

While chronic replacement therapy is essential to the long-term health of children with cystinosis and Fanconi syndrome, acute therapy is crucial to survival in cases of dehydration and electrolyte imbalance. Great care must be taken not to underestimate the enormous replacement requirements of infants with ongoing renal losses. More than 4 L of water and 100 to 150 mEq of potassium per day may be needed to restore a dehydrated infant to normal fluid and electrolyte status. A previous review[1] gives an example of the type of replacement typically required for an infant with cystinosis and dehydration.

A major concomitant of renal tubular Fanconi syndrome appears to be growth retardation. We have determined that, for patients with cystinosis to grow at a normal rate, three forms of therapy are required. One is adequate nutrition. A second is sufficient phosphate supplementation to prevent rickets. The final required therapy is oral cysteamine,[12] which is discussed below.

We do not know all the nutrients that must be resorbed by the renal tubule to maintain adequate concentrations in the human body. As these are discovered, they may prove deficient in cystinosis patients with tubular Fanconi syndrome. Once identified, the wasted nutrients can be replaced in children with Fanconi syndrome, who might then thrive even more than they do today.

Renal Glomerular Dysfunction

The glomerular damage caused by cystinosis makes it a fatal disease. Patients born before 1960 were almost certain to die of renal failure. Patients born after 1960 could survive long enough to receive hemodialysis[37] or a renal transplant,[38–41] first performed in a child with cystinosis in 1968. Patients born after 1975 could benefit from specific cystine-depleting therapy with cysteamine,[12] which came into fairly wide usage in 1978 and whose efficacy was well accepted by the late 1980s.

The natural course of cystinosis includes renal failure at a mean age of 9.2 years,[13] with a range of a few years on either side. Once uremic symptoms occur, hemodialysis or peritoneal dialysis can provide temporary relief while the patient awaits a renal transplant. Renal allografts fare well in cystinosis patients, often lasting 15 to 20 years. Living related donor grafts have proved superior to cadaver transplants.[15] This is true despite the fact that the donor is often a heterozygote for cystinosis, with 50 percent of the normal lysosomal cystine carrying capacity. We know of patients who have had their kidneys for more than 17 years, but we also have patients who have received several renal allografts. Improvements in antihypertensive and immunosuppressive agents such as cyclosporine continue to reduce rejection rates for transplanted kidneys. The recent availability of erythropoietin may make renal allograft procedures feasible for a large group of patients who could not previously receive one because of poor matching. In 1985 there were more than 70 cystinosis patients over the age of 10 in the United States and Canada who were alive by virtue of renal transplantation.[19] Many European patients have also received a renal allograft,[42] and the number of transplanted patients continues to increase.

Although a transplanted kidney may contain increased cystine because of infiltration by host mononuclear cells,[43] the disease cystinosis does not recur in the donor kidney. Post-transplant patients do not have renal Fanconi syndrome or progressive renal failure, but neither does the allograft prevent cystine accumulation from continuing in nonrenal tissues. Growth after kidney transplantation varies considerably among cystinosis patients.[15,42]

The intent of all directed therapy in cystinosis has been to prevent renal glomerular damage. In 1976, it was proposed that cysteamine (β-mercaptoethylamine) could deplete cells of cystine sufficiently to prevent further damage to parenchymal tissue.[44] In fact, a 90 to 95 percent reduction in cellular cystine levels was achieved both in vitro and in vivo with the use of cysteamine. The mechanism of action of cysteamine recently has been elucidated. The aminothiol traverses the plasma membrane and the lysosomal membrane by virtue of its amine group and is concentrated within the acidic lysosome because the amine becomes positively charged. The free thiol group then participates in a disulfide interchange reaction with lysosomal cystine, to produce cysteine and the mixed disulfide cysteine-cysteamine.[44] Cysteine can leave the cystinotic lysosome quite freely,[6] and the mixed disulfide cysteine-cysteamine can also leave the cystinotic lysosome by a process that does not require the mutant cystine carrier.[45] In fact, that process involves a lysine carrier in the lysosomal membrane, which is functional in cystinotic cells and which recognizes the mixed disulfide,[10] a structural analogue of lysine itself. In this fashion, the cystine content of cystinotic cells is reduced to that of heterozygotes, who are never affected with any of the pathology of cystinosis.

After a 1976 to 1978 study demonstrated that ascorbic acid lacked efficacy in preserving renal function in cystinosis,[46] a national trial of oral cysteamine therapy was begun in 1978.[12] An historical control group consisted of 55 patients treated with either ascorbic acid or placebo between 1976 and 1978.[46] The treatment group consisted of 93 children with cystinosis receiving oral cysteamine between 1978 and 1985. Cysteamine hydrochloride was given orally as a solution of 50 mg of free base per ml. The dose was approximately 50 mg of free base/kg/d, every 6 hours, with a maximum dose of 90 mg/kg/d. The doses were titrated in an attempt to maintain the leukocyte cystine level below 1 nmol of half-cystine per mg of protein. On average, an 82 percent depletion of leukocyte cystine was achieved.[12] In general, the greatest cystine depletion occurred 1 hour after a dose and by 5 to 6 hours after a dose the degree of depletion was waning.

The cysteamine-treated patients were studied for an average of 2.8 years. Their mean creatinine clearance remained unchanged (38.4 ± 2.0 ml/min/1.73 m^2 at the start of the study; 38.5 ± 2.5 ml/min/1.73 m^2 at the end of the study). Among the controls, the mean creatinine clearance decreased 15 percent over only 1.3 years (34.9 ± 1.8 to 29.7 ± 2.0 ml/min/1.73 m^2). At the end of the study, the cysteamine-treated patients were older, yet their mean creatinine clearance was greater than that of the historical control group.[12] This finding contradicts the natural history of the disorder, in which glomerular function decreases with age. Furthermore, of all patients analyzed at age 6 years, 17 of 27 children treated with cysteamine for at least 1 year had a serum creatinine of less than 1 mg/dl, compared with only 2 of 17 in the historical control group ($P = 0.002$). At the National Institute of Child Health and Human Development, we now follow a handful of classic cystinosis patients with normal serum creatinine concentrations at 10 to 12 years of age.

Although anecdotal reports of cysteamine use during the early 1980s were inconclusive,[47] current evidence for the efficacy of cysteamine vis-à-vis glomerular function is incontrovertible. Not only have serum creatinine and creatinine clearance parameters demonstrated the benefits of cysteamine, but so also has analysis using the predicted reciprocal serum creatinine at age 10, a powerful gauge of renal function.[48] By all these parameters, damage already done to kidneys appears to be irreversible with oral cysteamine therapy, but progression of renal deterioration can be retarded, perhaps indefinitely. A new study of standard-dose (1.3 g/m^2/d) or high-dose (1.95 g/m^2/d) cysteamine or phosphocysteamine (see below) was initiated in 1987. Preliminary results have bolstered optimism that chronic cystine depletion therapy will benefit renal function in cystinosis.

Growth has also been markedly improved by cysteamine therapy.[12] For example, at 2 to 3 years of age, patients treated with cysteamine grew at 93 percent the normal rate, whereas children with cystinosis who did not receive cysteamine grew at only 54 percent the normal rate. The benefit to growth was appreciated only in the group of children under 6 years of age, although several children begun on cysteamine during the first 2 to 3 years of life have continued to grow at a normal rate through at least 10 years of age.

One patient, diagnosed because of an affected sibling, was treated from birth with oral cysteamine.[49] After several years, renal glomerular as well as tubular function appeared to be preserved. We have treated a boy with cysteamine from 2 weeks of age and noted a markedly attenuated Fanconi syndrome compared with his affected older brother, but hypophosphatemic rickets and aminoaciduria nevertheless occurred at approximately 1 year of age.[50] In our experience, early intervention with cysteamine can modify, but not prevent, renal tubular Fanconi syndrome in cystinosis.

Cysteamine has clinically significant side effects, largely related to the gastrointestinal system. Approximately 14 percent of patients do not tolerate the drug because of increased vomiting or nausea.[12] There is an isolated report of reversible leukopenia, rash, and lethargy in three different patients receiving very high doses of cysteamine, without incremental dosing.[51] Recently, modification of the cysteamine molecule with placement of a phosphate ester over the thiol group has improved patient compliance. The resulting drug, called phosphocysteamine, lacks the odor and bad taste of cysteamine but is converted to cysteamine in the body and has the same cystine-depleting efficacy as cysteamine.[3,52,53] Patients are now given the option of receiving oral cysteamine or phosphocysteamine, and work is under way to prepare a cysteamine preparation in capsular form. The appropriate formulation for oral cysteamine will depend on future kinetic studies. For example, a certain peak concentration of plasma cysteamine may be required to achieve cellular cystine depletion.

Cysteamine is currently an Investigational New Drug, meaning that it cannot be prescribed by all physicians but can be used only with Food and Drug Administration (FDA) approval according to a protocol sanctioned by an Institutional Review Board. Efforts are under way to secure New Drug Approval for the drug, which would effectively bring cysteamine to market. Because the market is so limited, and the drug so essential, cysteamine has been granted Orphan Drug status in the eyes of the federal government.

Ophthalmic Therapy

Cystine accumulation in cystinosis causes damage to the retina (decreased visual acuity), lens (crystal-containing membrane, posterior synechiae), and cornea (erosions).[14] The only hope for the retinal and lens pathology is for oral cysteamine to be delivered to these tissues to prevent further damage. By contrast, the corneal symptomatology in cystinosis can be treated by a number of different modalities.

First, the photophobia can be relieved by the use of dark glasses and the avoidance of sun exposure. Second, many older patients have decreased tearing, and their dry eyes can be soothed by frequent administration of standard, over-the-counter eyedrops. An occasional patient will benefit from the use of soft contact lenses, which prevent movement of the lid over erosions in the cornea. One patient had such severe recurrent corneal erosions and intractable pain that he required a penetrating keratoplasty in one eye.[54] This corneal transplant offered him enormous benefit in the form of pain relief and allowed him to resume a functional life.

The mainstay of therapy for cornea crystals is proving to be cysteamine eyedrops (i.e., a topical cysteamine solution). While long-term oral cysteamine had no effect on corneal crystal formation in any patients,[12] initial studies of topical cysteamine use in young children indicated that the crystals could be entirely removed by 0.1 percent cysteamine in normal saline administered every hour while awake for 4 to 6 months.[55] This result was demonstrated by a placebo-controlled, double-masked protocol; after correctly breaking the code masking the cysteamine-treated eye, both eyes were treated with cysteamine eyedrops. Subsequent toxicity studies in rabbits verified that up to 0.5 percent cysteamine could safely be given topically according to the above regimen.[56] We have recently confirmed the safety and efficacy of 0.5 percent cysteamine eyedrop therapy in all ages of patients.[57] At least two 6-year-old patients now have corneas that are clear of crystals by virtue of topical cysteamine therapy. One 22-year-old woman had the haziness removed from her corneas after 4 months of topical cysteamine therapy, and her photophobia was markedly relieved. A 16-year-old girl experienced similar relief of photophobia and corneal opacity. She also had improved vision. A total of 16 patients have exhibited such an obvious difference, as seen by slit-lamp examination, between cysteamine-treated and placebo-treated eyes that the code could be broken, and in every case the treated eye was correctly identified.

The reason for the enhanced efficacy of topical compared with oral cysteamine may be that oral cysteamine cannot reach the poorly vascularized cornea. By contrast, a high concentration of cysteamine can be delivered by topical eyedrop therapy. Just how frequently this concentrated cysteamine must be administered to the cornea still needs to be determined. Therapy is now being individualized, with the consideration that crystal removal may require more frequent administration than prophylaxis against crystal formation.

Other Organs

Patients with inadequate sweating,[16] or hypohidrosis, should be counseled to avoid heat. Specifically, they should not remain in the hot sun for more than 20 to 30 minutes. When the thyroid-stimulating hormone (TSH) level rises significantly in cystinosis patients, replacement therapy with L-thyroxine should be initiated. Male patients with primary hypogonadism can be treated with intramuscular testosterone injections, to enhance secondary sexual characteristics. Patients with diabetes mellitus should be treated with insulin; pancreatic exocrine insufficiency can be treated with pancreatic extracts. For patients with swallowing difficulties, dietary manipulations and oral exercises can help guard against aspiration.[23]

Potentially beneficial therapies include the use of growth hormone, indomethacin, and erythropoietin (Table 6-1). Efforts are currently under way to evaluate systematically the use of growth hormone supplementation in cystinosis. Anecdotal reports already indicate short-term benefits attributable to regular growth hormone injections,[58] but long-term efficacy will require many years to demonstrate.

Indomethacin is used in some European countries to reduce urine output, permitting increased caloric intake in the form of solid foods rather than fluids. Serum creatinine can rise with indomethacin administration, but this effect is considered reversible. Indomethacin is not widely used in the United States because serum creatinine provides an important outcome parameter for ongoing controlled cysteamine trials. A study of the safety and efficacy of indomethacin, especially with respect to growth enhancement, may be indicated.

Erythropoietin, a renal hormone that stimulates erythropoiesis, can counter the chronic anemia that accompanies the renal failure of cystinosis. Therapy with erythropoietin can improve the quality of life for cystinosis patients on dialysis and perhaps make them better candidates for renal allograft procedures.

For long-term involvement of crucial organs, the only hope of therapy outside symptomatic treatment lies in intracellular cystine depletion. Since cysteamine apparently depletes kidney cells of cystine,[12] it will likely do the same for muscle, bone marrow, liver, and perhaps even the central nervous system, if it crosses the blood-brain barrier. This comprises the hope for the future. We expect that examination of patients treated with cysteamine from infancy will, 10 to 15 years from now, reveal preservation of organ systems rather than their deterioration. Specific cystine-depleting therapy is most likely the way that the natural history of the disease will be altered, rather than by gene replacement. Gene therapy appears remote, since the gene itself has not yet been identified, nor is there a method to target the gene to all the various cell types involved. In addition, prenatal therapy is not a reasonable alternative now because of the possible teratogenicity of cysteamine.

Despite these reservations, it is truly remarkable to have an effective therapy for a lysosomal storage disease. Cysteamine has changed the lives of cystinosis families from ones of despair and certain death to lives filled with expectations for achieving normal human aspirations. Partners in this transformation have been the patients who valiantly participate in treatment protocols, the local physicians who care for chronically ill children, and the Cystinosis Foundation, a family support group that provides a rallying point and common ground for all involved. (The Cystinosis Foundation can be contacted at 17 Lake Ave., Piedmont, CA 94611.)

REFERENCES

1. Gahl WA: Cystinosis coming of age. Adv Pediatr 33:95, 1986
2. Gahl WA, Renlund M, Thoene JG: Lysosomal transport disorders: Cystinosis and sialic acid storage disorders. p. 2619. In Scriver CR, Beaudet AL, Sly WS, Valle DL (eds): Metabolic Basis of Inherited Disease. 6th Ed. McGraw-Hill, New York, 1989
3. Gahl WA, Thoene JG, Schneider JA, et al: Cystinosis: Progress in a prototypic disease. Ann Intern Med 109:557, 1988
4. Adamson MD, Andersson HC, Gahl WA: Cystinosis. Semin Nephrol 9:147, 1989
5. Schulman JD, Bradley KH, Seegmiller JE: Cystine: Compartmentalization within lysosomes in cystinotic leukocytes. Science 166:1152, 1969
6. Gahl WA, Tietze F, Bashan N, et al: Defective cystine exodus from isolated lysosome-rich fractions of cystinotic leukocytes. J Biol Chem 257:9570, 1982
7. Gahl WA, Bashan N, Tietze F, et al: Cystine transport is defective in isolated leukocyte lysosomes from patients with cystinosis. Science 217:1263, 1982
8. Gahl WA, Tietze F, Bashan N, et al: Characteristics of cystine countertransport in normal and cystinotic lysosome-rich leucocyte granular fractions. Biochem J 216:393, 1983
9. Jonas AJ, Smith ML, Schneider JA: ATP-dependent lysosomal cystine efflux is defective in cystinosis. J Biol Chem 257:13185, 1982
10. Pisoni RL, Thoene JG, Christensen HN: Detection and characterization of carrier-mediated cationic amino acid transport in lysosomes of normal and cystinotic human fibroblasts. Role in therapeutic cystine removal. J Biol Chem 260:4791, 1985

11. Bernardini L, Rizzo WB, Dalakas M, et al: Plasma and muscle free carnitine deficiency due to Fanconi syndrome. J Clin Invest 75:1124, 1985
12. Gahl WA, Reed GF, Thoene JG, et al: Cysteamine therapy for children with nephropathic cystinosis. N Engl J Med 316:971, 1987
13. Gretz N, Manz F, Augustin R, et al: Survival time in cystinosis: A collaborative study. Proc Eur Dial Transplant Assoc 19:582, 1982
14. Kaiser-Kupfer MI, Caruso RC, Minckler DS, et al: Long-term ocular manifestations in nephropathic cystinosis. Arch Ophthalmol 104:706, 1986
15. Gahl WA, Kaiser-Kupfer MI: Complications of nephropathic cystinosis after renal failure. Pediatr Nephrol 1:260, 1987
16. Gahl WA, Hubbard VS, Orloff S: Decreased sweat production in cystinosis. J Pediatr 6:904, 1984
17. Chan AM, Lynch MJG, Bailey JD, et al: Hypothyroidism in cystinosis. Am J Med 48:678, 1970
18. Lucky AW, Howley PM, Megylesi K, et al: Endocrine studies in cystinosis: Compensated primary hypothyroidism. J Pediatr 91:204, 1977
19. Gahl WA, Schneider JA, Thoene JG, et al: Course of nephropathic cystinosis after age 10 years. J Pediatr 109:605, 1986
20. Fivush B, Green OC, Porter CC, et al: Pancreatic endocrine insufficiency in post-transplant cystinosis. Am J Dis Child 141:1087, 1987
21. Fivush B, Flick JA, Gahl WA: Pancreatic exocrine insufficiency in a patient with nephropathic cystinosis. J Pediatr 112:49, 1988
22. Gahl WA, Dalakas M, Charnas L, et al: Myopathy and cystine storage in muscles in a patient with nephropathic cystinosis. N Engl J Med 319:1461, 1988
23. Sonies B, Ekman EF, Andersson H, et al: Oropharyngeal dysfunction in nephropathic cystinosis. N Engl J Med 323:565, 1990
24. Cochat K, Drachman R, Gagnadoux MF, et al: Cerebral atrophy and nephropathic cystinosis. Arch Dis Child 61:401, 1986
25. Schnaper WH, Cole BR, Hodges FJ, et al: Cerebral cortical atrophy in pediatric patients with end-stage renal disease. Am J Kidney Dis 2:645, 1983
26. Fink JK, Brouwers P, Barton N, et al: Neurologic complications in longstanding nephropathic cystinosis. Arch Neurol 46:543, 1989
27. Levine S, Paparo G: Brain lesions in a case of cystinosis. Acta Neuropathol (Berl) 57:217, 1982
28. Ross DL, Strife CF, Towbin R, et al: Nonabsorptive hydrocephalus associated with nephropathic cystinosis. Neurology 32:1330, 1982
29. Jonas AJ, Conley SB, Marshall R, et al: Nephropathic cystinosis with central nervous system involvement. Am J Med 83:966, 1987
30. Gahl WA, Bernardini I, Dalakas M, et al: Oral carnitine therapy in children with cystinosis and renal Fanconi syndrome. J Clin Invest 81:549, 1988
31. Schneider JA, Bradley KH, Seegmiller JE: Increased cystine in leukocytes from individuals homozygous and heterozygous for cystinosis. Science 157:1321, 1967
32. Schneider JA, Rosenbloom FM, Bradley KH, et al: Increased free-cystine content of fibroblasts cultured from patients with cystinosis. Biochem Biophys Res Commun 29:527, 1967
33. Boman H, Schneider JA: Prenatal diagnosis of nephropathic cystinosis. Acta Paediatr Scand 70:389, 1981
34. Smith ML, Pellet OL, Cass MMJ, et al: Prenatal diagnosis of cystinosis utilizing chorionic villous sampling. Prenatal Diagn 7:23, 1987

35. Brubaker RF, Wong VG, Schulman JD, et al: Benign cystinosis: The clinical, biochemical and morphologic findings in a family with two affected siblings. Am J Med 4:546, 1970
36. Goldman H, Scriver CR, Aaron K, et al: Adolescent cystinosis: Comparisons with infantile and adult forms. Pediatrics 47:979, 1971
37. Mahoney CP, Manning GB, Hickman RO, et al: Hemodialysis in a patient with cystinosis. Am J Dis Child 112:65, 1966
38. Mahoney CP, Striker GE, Hickman RO, et al: Renal transplantation for childhood cystinosis. N Engl J Med 283:397, 1970
39. Malekzadeh MH, Neustein HB, Schneider JA, et al: Cadaver renal transplantation in children with cystinosis. Am J Med 63:525, 1977
40. West JC, Goodman SL, Schroter GP, et al: Pediatric kidney transplantation for cystinosis. J Pediatr Surg 12:651, 1977
41. Langlois RP, O'Regan S, Pelletier M, et al: Kidney transplantation in uremic children with cystinosis. Nephron 28:273, 1981
42. Broyer M, Guillot M, Gubler M-C, et al: Infantile cystinosis: A reappraisal of early and late symptoms. p. 137. In Hamburger J, Crosnier J, Grünfeld J-P, et al. (eds): Advances in Nephrology. Year Book Medical Publishers, Chicago, 1981
43. Goodman SL, Hambidge KM, Mahoney CP, et al: Renal homotransplantation in the treatment of cystinosis. p. 225. In Schulman JD (ed): Cystinosis. DHEW Publication No (NIH) 72-249. Government Printing Office, Washington, DC, 1973
44. Thoene JG, Oshima RG, Crawhall JG, et al: Intracellular cystine depletion by aminothiols in vitro and in vivo. J Clin Invest 58:180, 1976
45. Gahl WA, Tietze F, Butler JD, et al: Cysteamine depletes cystinotic leukocyte granular fractions of cystine by the mechanism of disulfide interchange. Biochem J 228:545, 1985
46. Schneider JA, Schlesselman JJ, Mendoza SA, et al: Ineffectiveness of ascorbic acid therapy in nephropathic cystinosis. N Engl J Med 300:756, 1979
47. Yudkoff M, Foreman JW, Segal S: Effects of cysteamine therapy in nephropathic cystinosis. N Engl J Med 304:141, 1981
48. Gahl WA, Schneider JA, Schulman JD, et al: Predicted reciprocal serum creatinine at age ten as a measure of renal function in children with nephropathic cystinosis treated with oral cysteamine. Pediatr Nephrol 4:129, 1990
49. Da Silva VA, Zurbrugg RP, Lavanchy P, et al: Long-term treatment of infantile nephropathic cystinosis. N Engl J Med 313:1460, 1985
50. Adamson MD, Schneider JA, Bernardini IM, et al: Attenuation but not prevention of renal Fanconi syndrome by cysteamine in nephropathic cystinosis. Pediatr Res 25:334A, 1989
51. Corden BJ, Schulman JD, Thoene JG, et al: Adverse reactions to oral cysteamine use in nephropathic cystinosis. Dev Pharmacol Ther 3:25, 1981
52. Thoene JG, Lemons R: Cystine depletion of cystinotic tissues by phosphocysteamine (WR 638). J Pediatr 96:1043, 1980
53. Smolin LA, Clark KF, Thoene JG, et al: A comparison of the effectiveness of cysteamine and phosphocysteamine in elevating plasma cysteamine concentration and decreasing leukocyte free cystine in nephropathic cystinosis. Pediatr Res 23:616, 1988
54. Kaiser-Kupfer MI, Datiles MB, Gahl WA: Corneal transplant in a twelve-year old boy with nephropathic cystinosis. Lancet 1:331, 1987
55. Kaiser-Kupfer MI, Fujikawa L, Kuwabara T, et al: Removal of corneal crystals by topical cysteamine in nephropathic cystinosis. N Engl J Med 316:775, 1987

56. Jain S, Kuwabara T, Gahl WA, et al: Range of toxicity of topical cysteamine in rabbit eyes. J Ocul Pharmacol 4:127, 1988
57. Kaiser-Kupfer MI, Gazzo MA, Datiles MB, et al: A randomized placebo-controlled trial of cysteamine eyedrops in nephropathic cystinosis. Arch Ophthalmol 108:689, 1990
58. Wilson DP, Jelley D, Stratton R, Coldwell JG: Nephropathic cystinosis: Improved linear growth after treatment with recombinant human growth hormone. J Pediatr 115:758, 1989

7

THERAPY FOR X-LINKED ADRENOLEUKODYSTROPHY

Hugo W. Moser · Patrick Aubourg ·
David Cornblath · Janet Borel ·
Yan-Wan Wu · Ann Bergin ·
Sakkubai Naidu · Ann B. Moser

INTRODUCTION

Adrenoleukodystrophy (ALD) is an X-linked disorder associated with the accumulation of very long chain fatty acids (VLCFA), particularly hexacosanoic acid (C26:0), caused by an impaired capacity to degrade these substances. The disorder manifests with adrenocortical insufficiency, treated successfully by steroid replacement, and with a progressive neurologic disability that until now resisted all forms of therapy. Two recently introduced therapeutic approaches offer hope. The first is a dietary regimen that involves oral administration of mono-unsaturated oils (glycerol trioleate and trierucate) combined with dietary restriction of VLCFA intake. This regimen normalizes plasma levels of C26:0 within 4 weeks. We present evidence that it leads to a statistically significant improvement of peripheral nerve function in men with adrenomyeloneuropathy (AMN), the adult form of the disease. The regimen also normalizes the microviscosity of the red blood cell membranes. Microviscosity is increased in untreated patients and may be indicative of a generalized membrane function disturbance that may have pathogenetic significance. The second therapeutic approach is bone marrow transplantation. Recent experience indicates that bone marrow transplantation normalizes plasma VLCFA levels, suggesting that it may, for the first time, reverse neurologic disability in patients with early neurologic involvement. The type of therapy to be recommended varies with the patient's age and neurologic phenotype. ALD may eventually be a prime candidate for gene therapy.

Until only a few years ago ALD, along with the other leukodystrophies and related disorders, were placed in the "no specific treatment" category. Management had to rely on general supportive, albeit important, measures. While specific therapeutic approaches to ALD are still in their early stages and experimental, they have a rational basis, and there are early indications of favorable clinical effects in some settings. At least today, there are therapeutic choices to be debated and tested, whereas only 3 years ago there were none.

NATURE AND CLINICAL SPECTRUM OF ADRENOLEUKODYSTROPHY

ALD has been mapped to Xq28, the terminal segment of the long arm of the X-chromosome, the region that also contains the genes for glucose 6-phosphate dehydrogenase (G6PD), hemophilia, and the red and green color vision pigments.[1-5] The main biochemical abnormality is the accumulation of saturated unbranched VLCFA, particularly hexacosanoic (C26:0) and tetracosanoic (C24:0), because of the impaired capacity to oxidize these substances,[6] a function that normally takes place in the peroxisome.[7] There is evidence that the basic defect is the deficient activity of lignoceroyl-CoA ligase, the enzyme that catalyzes the formation of the CoA derivative of VLCFA.[8,9] Accumulation of VLCFA is most striking in the adrenal cortex and cerebral white matter,[10] but some degree of excess is demonstrable in all tissues and body fluids, including plasma,[11] red blood cells,[12] and cultured amniocytes.[13] Studies of these accessible tissues and body fluids have facilitated the diagnosis. X-linked ALD must be distinguished sharply from neonatal ALD.[14] This disorder has an autosomal-recessive mode of inheritance, resembles the Zellweger cerebrohepatorenal syndrome, and is characterized by defective formation of peroxisomes. The therapeutic options discussed here do not apply to this disorder.

Varying Phenotypes of X-Linked Adrenoleukodystrophy

A wide range of phenotypic expression exists for X-linked ALD. Since this variable expression is observed frequently among members of the same kindred,[15,16] it is thought to be related to the influence of modifier genes or unknown environmental factors. The first case report[17] described what is now referred to as childhood ALD, the most common and serious form of the illness.[16,18,19] Boys develop normally until 4 to 8 years of age, then develop progressive behavioral and cognitive deficits, impaired vision and hearing, and motor function deficits that may lead to an apparently vegetative state within 2 years.

Adrenomyeloneuropathy[20-22] is the second most common phenotype. It involves mainly the spinal cord and peripheral nerves and causes progressive spastic paraparesis, as well as sensory and sphincter disturbances, with an average age of onset of 28 years and a progressive course that extends over decades. The designation *Addison-only* refers to patients who have adrenal insufficiency but who are neurologically intact or nearly so. This form of ALD is more common than had been previously realized.[23] Possibly as many as 40 percent of male addisonian patients have the biochemical defect of ALD, demonstrable by the plasma VLCFA assay.[11] In most of these patients AMN eventually develops, but there may be long intervals (in one instance, 32 years) between involvement of the adrenal and nervous systems. Approximately 8 percent of persons with the biochemical defect of ALD are apparently free of endocrine or neurologic symptoms. These include not only young children who are at risk of the development of the childhood form of ALD, but also adolescents and a few adults. Table 7-1 lists the frequency of the various phenotypes among ALD hemizygotes tested at the Kennedy Institute. Approximately 15 percent of women heterozygous for ALD show the development of neurologic disability that resembles AMN, but it is usually milder and of later onset.[19] Boys with the rapidly progressive childhood ALD show an acute cerebral demyelinative process. AMN patients show involvement of dorsal columns and corticospinal tracts in the spinal cord, as well as axonal neuropathy and varying degrees of cerebral demyelination. The adrenal glands show an accumulation of VLCFA containing cholesterol ester in the zona fasciculata and, later, atrophy of these cells.[24]

Pathogenesis of Adrenal and Brain Lesions

The pathogenesis of ALD is presumably related in some way to the principal biochemical abnormality, namely the accumulation of VLCFA, but a causal relationship has not been established. One plausible mechanism is suggested by the demonstration that the microviscosity of red blood cell membranes of

Table 7-1. Phenotype Distribution Among 1217 Male Adrenoleukodystrophy Patients Belonging to 571 Kindreds Tested at the Kennedy Institute

Phenotype	%
Childhood ALD	45.0
Neurologic symptoms before age 10 years	
Adolescent ALD	4.9
Cerebral symptoms beginning between ages 10 and 21 years	
Adrenomyeloneuropathy (AMN)	23.0
Spinal cord and peripheral nerve disability beginning in late adolescent or adult	
Adult cerebral	2.7
Cerebral symptoms beginning after age 21 years	
Addison-only	9.0
Asymptomatic	7.6
Prenatal	2.4
Phenotype uncertain	5.4

Abbreviation: ALD, adrenoleukodystrophy.

ALD patients is significantly higher than in aged-matched control subjects,[25] attributed to the excess of the rigid saturated VLCFA. That this degree of increased microviscosity may have deleterious physiologic effects is suggested by studies with cultured human adrenocortical cells.[26] The addition of C26:0 fatty acid to the culture medium in a concentration equivalent to that in the plasma of ALD patients caused an impairment in the capacity of these cells to produce cortisol in response to adrenocorticotropin (ACTH). It was suggested that the VLCFA excess interfered with the function of the ACTH receptor, and it is postulated that such an alteration in membrane structure may have other deleterious effects. In a morphologic study of postmortem ALD adrenal glands, it was found that the degree of VLCFA excess (as judged by the number of lamellar cytoplasmic inclusions) correlated with the extent of cell dysfunction,[27] again suggesting that VLCFA excess had a direct toxic effect.

While a direct toxic action of VLCFA may account for the adrenocortical dysfunction, the pathogenesis of the central nervous system (CNS) lesions is more complex. There is no correlation between the plasma VLCFA levels and the severity of neurologic manifestations,[19] and some persons who have the biochemical defect of ALD escape neurologic disability altogether. We postulate that the VLCFA excess is necessary but not sufficient for the causation of the ALD CNS lesions and that additional factors are required. One possibility is that autoimmune mechanisms are involved. Striking perivascular lymphocytic infiltration is a characteristic feature of ALD brain lesions[24] and is not observed in metachromatic leukodystrophy[28] or in ALD adrenal glands.[27] The perivascular lymphocytes in ALD brain tissue contain T cells (34 percent), T-8 cells (16 percent), B cells (24 percent), and monocyte/macrophages (11 percent).[29] This pattern is similar to that observed in the CNS during a cellular immune response. We hypothesize that the altered brain lipid composition presumed to be present in all ALD patients leads to an autoimmune response in some, but not all, ALD patients.[16,19]

Source of VLCFA in Adrenoleukodystrophic Tissues

There is little doubt that the VLCFA excess in ALD is caused by the impaired capacity to degrade these substances. The VLCFA are derived both from the diet and from endogenous synthesis. The average American diet was estimated to contain 12 to 40 mg of C26:0.[30] In a study in which 10 mg/d of deuterium-labeled C26:0 was administered through a nasogastric tube to a terminally ill ALD patient during the last 100 days of his life, it was shown that, in certain portions of postmortem brain tissue, 90 percent of the C26:0 was deuterium labeled, indicating that it had been derived from what the patient had eaten during that final period.[31] In addition to the dietary source, VLCFA are also synthesized by a microsomal system that elongates fatty acids with a 16-carbon chain length or longer.[32] The same enzyme system appears to elongate saturated and mono-unsaturated fatty acids,[33] a finding that is crucial for dietary therapy. Cultured skin fibroblasts from normal persons or ALD patients synthesize VLCFA.[34,35] Dietary studies in ALD patients suggest that endogenous synthesis is quantitatively a more important source of VLCFA than is dietary intake. Dietary restriction alone failed to reduce plasma VLCFA levels in ALD patients, but administration of certain mono-unsaturated fatty acids that reduce the synthesis of VLCFA normalized these levels within 4 weeks.

THERAPY

Approaches Before 1987

Replacement therapy with corticosteroids corrects the adrenocortical insufficiency. Deficiencies mainly involve glucocorticoids. The importance of such replacement therapy cannot be overemphasized. Almost all affected boys and 60 percent of men with AMN (but almost none of the heterozygotes) have impaired adrenal reserve. Left untreated, these patients may succumb to adrenal crisis. Adrenal replacement therapy does not appear to alter the course of the neurologic deterioration. It is the neurologic disability that has resisted therapeutic efforts until now.[16,19,36,37]

Effect of Dietary Therapy on Plasma VLCFA Levels

Failure of Dietary VLCFA Restriction and Administration of Carnitine or Clofibrate to Alter Plasma VLCFA Level

Previously cited data that VLCFA accumulating in ALD brain tissue are at least in part of dietary origin[31] caused us to devise a diet that would reduce the daily C26:0 intake to less than 3 mg, compared with the usual 12 to 40 mg contained in the typical American diet.[30] A 274-page recipe book is published by the United Leukodystrophy Foundation.[38] This approach proved unsuccessful. It failed to alter either the plasma C26:0 level or the clinical course.[39] Administration of carnitine and clofibrate also appeared to be ineffective.

Attainment of 50 Percent Reduction of Plasma VLCFA Levels With Addition of Glycerol Trioleate Oil to VLCFA-Restricted Diet

In 1986 Rizzo et al.[34] made the important observation that the levels of C26:0 in cultured ALD fibroblasts were reduced by 50 percent when oleic acid was altered to the culture medium. This reduction is attributed to substrate competition, since saturated and mono-unsaturated fatty acids appear to be chain lengthened by the same microsomal elongating system.[33] This observation led to therapeutic trials in which glycerol trioleate (GTO) in a daily dosage of 1 to 2.5 g/kg body weight (approximately 45 to 90 ml) was added to the VLCFA-restricted diet. GTO provided approximately 25 percent of total calories. Close monitoring and cooperation ensured a nutritionally adequate diet. Patient cooperation was excellent, and the diet was found to be manageable. Rizzo et al.[40] and our group found that this regimen reduced plasma C26:0 levels by about 50 percent after 4 months of dietary use (Fig. 7-1). While some degree of dietary VLCFA restriction is necessary to achieve this result,[41] the permissible levels have not been defined.

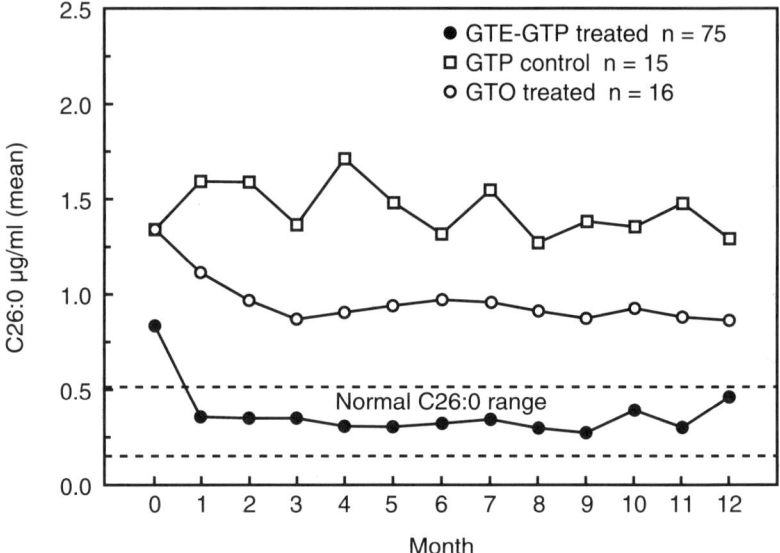

Fig. 7-1. Adrenoleukodystrophy. Comparison of GTE–GTO and GTO oil on plasma C26:0 levels.

Normalization of Plasma C26:0 Levels With Addition of Glycerol Trierucate Oil to Previous Regimen

Erucic acid (*cis*-13-docosenoic acid) (C22:1) is a component of rapeseed oil that has been the topic of extensive nutritional, epidemiologic, and biochemical studies.[42] Initially, there was concern about the administration of this oil, since in rodents erucic acid produces a cardiac lipidosis. Such a phenomenon had never been observed in humans or primates, and there has been no evidence of cardiac involvement in any of the erucic acid-treated ALD patients. One of the ALD patients died of pneumonia secondary to aspiration of gastric contents after 10 months of glycerol trierucate (GTE) therapy. Postmortem examination showed no evidence of cardiac involvement or lipidosis.[43] The GTE oil is administered together with the GTO oil, in a ratio of 1 part GTE oil to 4 parts GTO oil. The daily GTE intake was approximately 11 ml. The mixture is referred to as Lorenzo's oil, in recognition of the role of Augusto and Michaela Odone in developing this mixture for the benefit of their son. In our clinic, the GTE–GTO oil is combined with a diet that limits C26:0 intake to approximately 3 mg/d.[30,38,41]

The use of the GTE–GTO normalizes plasma C26:0 levels, usually within 4 weeks[43] (Fig. 7-1). The reduction of saturated VLCFA is accompanied by an increase in mono-unsaturated VLCFA, particularly erucic acid itself. Except for moderate and transient reduction in platelet count in 12 of 126 patients, there have been no side effects. No abnormal bleeding occurred, and the platelet counts increased again without interruption of the administration of the oil. We have noted that essential fatty acid levels may fall in GTE–GTO treated patients and recommend that essential fatty acid supplements be provided and their levels monitored. Of the 126 patients enrolled originally, 5 have dropped out. The others have found it manageable and have complied.

The rapid normalization of plasma C26:0 levels has prompted us to initiate a large-scale clinical trial that will eventually involve more than 200 patients. The most urgent question is whether normalization of plasma C26:0 levels can *prevent* the onset of neurologic disability in asymptomatic boys who are treated early in life. The answer to this question will not be available for several years. We report here early and somewhat encouraging findings of a randomized prospective study of GTO oil in AMN patients. Clinical evaluation of the GTE–GTO regimen is not yet possible, since it was initiated only recently.

Effect of GTO Therapy on Peripheral Nerve Function in AMN Patients

We have completed a randomized prospective study involving a total of 34 men with AMN and 16 women who are heterozygous for ALD and who have neurologic disability attributable to this condition. We assembled 25 pairs, matched according to sex, age (within 5 years), and severity of spinal cord involvement. One member of each pair was assigned to a treatment versus a control group. The treatment group received the GTO oil and VLCFA-restricted diet as described previously.[41] The control group continued their customary diet. Patients were admitted to the Clinical Research Unit at the Johns Hopkins Hospital for 3 days at baseline, at 6 months, and at 12 months. Observers were blinded to the patients' dietary status. Clinical evaluations included neurologic examination, cognitive function; magnetic resonance imaging (MRI) of brain and spinal cord; endocrine, nutritional, and biochemical studies; and somatosensory, visual, and brainstem auditory-evoked responses, as well as studies of peripheral nerve function.

This report focuses on the changes in peripheral nerve function. We have emphasized this parameter because of the results obtained in the therapy of Refsum disease. Refsum disease is a lipidosis in which phytanic acid accumulates in the tissues and body fluids.[44] Dietary restriction of phytanic acid has been shown to normalize plasma phytanic acid levels and to improve conduction velocity and other aspects of peripheral nerve function.[44] Although vision, hearing, and CNS functions did not improve, further deterioration was prevented. Since untreated AMN patients show slow neurologic worsening over a span of one or more decades, we considered it unlikely that a 1-year period of observation would permit detection of a change in the rate of this slow progression. However, if treatment were to improve peripheral nerve function, the experience with Refsum disease suggested that this would be demonstrable during a 1-year therapeutic trial.

For each participant in the study, nerve conduction studies of the right sural and median sensory nerves; right peroneal and tibial motor nerves, including F-wave latencies; and right tibial H reflex were recorded using techniques standard in our laboratory. Most of the conduction studies were performed in the afternoon, and all were performed with an infrared temperature control system to maintain surface limb temperature (lateral border of the right foot and tip of digit 3) at 34°C. For sensory nerve studies, action potential amplitudes were measured peak-to-peak and velocity calculated from the onset latency. For motor conduction studies, compound muscle action potential (CMAP) amplitudes were recorded as baseline-to-peak amplitudes. For recording of the F-wave latency, 10 supramaximal stimulations were given distally and the shortest F-wave latency recorded. For each patient at each nerve conduction study, a total of 12 attributes of nerve conduction were recorded. To determine whether changes in nerve conduction measurements were a result of therapy or variability in the measurement, we determined the intraobserver variability for the 12 nerve conduction attributes (Table 7-2).

Evaluation of Peripheral Nerve Function Changes

Since the rationale and aim of the dietary therapy are to reduce the levels of saturated VLCFA, we chose as the independent variable the extent to which this was achieved.

Table 7-3 shows that in the 21 patients whose plasma C26:0 levels fell to 70 percent or less of control, 7 of the 12 attributes of peripheral nerve function improved; in 4, changes were equivocal or absent; and in 1 (median nerve sensory amplitude), there was a slight deterioration in the treated group. For two of the attributes, the changes reached statistical significance (Table 7-4). The P value was 0.04 for sural nerve amplitude, and 0.009 for peroneal nerve amplitude. Figure 7-2 illustrates the results for peroneal nerve amplitude. Note that 12 months after beginning of GTO therapy, the mean amplitude in the respon-

Table 7-2. Intraobserver Variability of Peripheral Nerve Conduction Studies[a]

NCS Value	Mean	SD	Correlation Coefficient
Sural Amp	20 µV	13	0.96
Sural CV	44 m/s	9	0.93
Peroneal DL	4.9 ms	1	0.90
Peroneal Amp	3.9 mV	2	0.95
Peroneal CV	38 m/s	8	0.97
Peroneal F	57 ms	12	0.98
Med Sens Amp	23 µV	15	0.99
Median SCV	54 m/s	11	0.94
Median DL	3.9 ms	0.9	0.92
Median Amp	10 mV	3	0.94
Median CV	52 m/s	8	0.97
Median F	31 ms	7	0.98

[a] A single observer (DRC) performed two sets of measurements of peripheral nerve function in 30 patients at 24-hour intervals. At the time of the second study, the results of the first study were not available. Note that the correlation coefficients were 0.90 or better for each measurement.

Abbreviations: Amp, amplitude; CV, conduction velocity; DL, distal latency; F, F-wave latency; Med, median; Sens, sensory; SCV, sensory conduction velocity.

Table 7-3 Glyceryl Trioleate Therapy for Adrenoleukodystrophy: Correlation Between Peripheral Nerve Function and Reduction of Plasma C26:0 Level to Less Than 70% of Baseline[a]

Amplitude	
Sural	+ +
Peroneal	+ + +
Median motor	+
Median sensory	−
Latency	
Peroneal	+ +
Median	+ +
Peroneal F wave	+
Median F wave	+/−
Velocity	
Sural	0
Peroneal	+/−
Median sensory	+
Median motor	0

[a] Qualitative trends of alterations in peripheral nerve attributes as a function of changes in plasma C26:0 levels during the randomized prospective study of the GTO diet described in the text. Figure 7-1 shows that the mean C26:0 level in the GTO group diminished during the first 4 months and then remained relatively constant at approximately 50% of that in the untreated patients for the remaining 8 months of the trial. To determine the correlations listed here, we compared the baseline and 12 months nerve function values in the 21 GTO-treated patients whose mean plasma C26:0 levels during the last 8 months of therapy were 70% of their plasma levels at baseline. These patients were considered to be "biochemical" diet responders. Seven of twelve peripheral nerve attributes improved in these responders, four remained essentially unchanged, and one became worse. Two of the improvements were statistically significant (see Table 7-4).

Table 7-4. Glyceryl Trioleate Therapy for Adrenomyeloneuropathy T-Tests on Peripheral Nerve Data[a]

Amplitude	No.	Mean	SD	P
Sural				
Baseline				
LE 70% of baseline	11	23	9	0.24
GT 70% of baseline	40	19	11	
1 yr				
LE 70% of baseline	9	27	10	0.04
GT 70% of baseline	20	18	9	
Peroneal				
Baseline				
LE 70% of baseline	11	4091	1667	0.58
GT 70% of baseline	39	3748	2109	
1 yr				
LE 70% of baseline	9	5778	2279	0.009
GT 70% of baseline	21	3160	1763	

[a] T-tests of sural and peroneal amplitude data at baseline and 12 months. Patients were split into two groups based on the degree of diminution of plasma C26:0 levels in association with GTO therapy. Those patients whose plasma C26:0 levels fell to <70% of baseline (LE 70%) are considered to be "biochemical responders" and those whose 12-months levels were ≥70% (GE 70%) to be nonresponders. Note that at baseline there was no significant difference between the two groups, whereas at 12 months the biochemical responders had significantly higher amplitude.

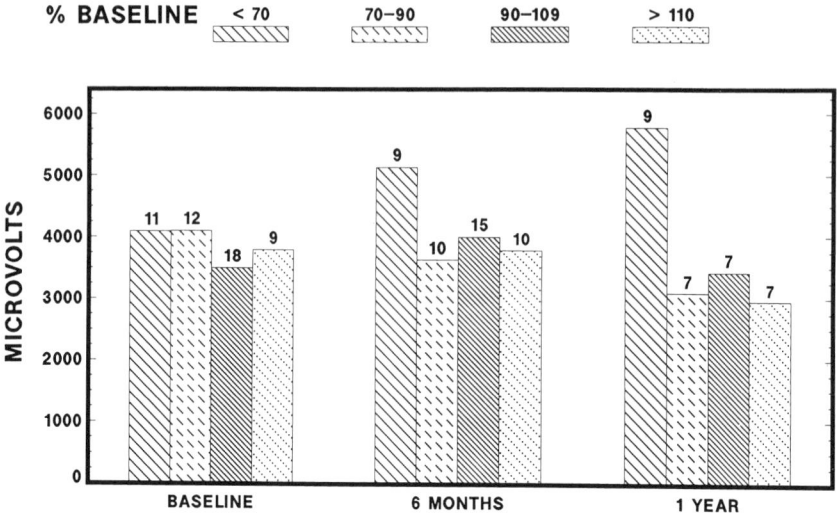

Fig. 7-2. Changes in peroneal nerve amplitude as a function of reduction in plasma C26:0 levels.

ders (plasma C26:0 level less than 70 percent of baseline) increased to 5900 µV from 4200 µV, a change far beyond that attributable to test variability (Fig. 7-2). The patients whose plasma C26:0 levels increased during the period of observation showed a diminished amplitude at 12 months, whereas the amplitude remained essentially unchanged in those patients whose plasma C26:0 levels remained relatively constant. We consider this suggestive evidence that lowering of C26:0 level achieved with the GTO regimen did have a favorable effect on peripheral nerve function in patients with AMN and in heterozygotes whose neurologic deficit resembled that of male AMN patients. This represents the first objective evidence of neurologic function. Note that this effect was obtained with the GTO oil, a regimen that has a less powerful biochemical effect than that of the more recently introduced GTE–GTO regimen (Fig. 7-1). Data on the effects of the latter regimen on peripheral nerve function are not yet available.

Clinical Effects of GTE–GTO

An international therapeutic trial to assess the clinical efficacy of GTE–GTO oil is in progress. The trial involves three groups of ALD patients.

1. *Boys who have the rapidly progressive form of childhood ALD.* Uziel et al.[45] reported disappointing results in this group, and our own experience with 10 patients is similar. Even though data analysis is not yet complete, the initial data suggest that the rapid rate of worsening is not altered significantly, even though plasma C26:0 levels are normalized. This negative result may indicate that the rationale behind the approach is faulty. An alternate, and more optimistic, interpretation is that the rapid pro-

gression of this form of the disease devastates the CNS before the effect of intervention is able to assert itself. It is our current hypothesis that immunopathogenetic mechanisms (see under Pathogenesis) may be operative and continue to do damage, even when the primary event (VLCFA excess) is mitigated. Together with Dr. Rizzo, we have initiated a study in which GTE–GTO is combined with intravenous administration of γ-globulin. IV γ-globulin has been reported to be of benefit in thrombocytopenic purpura and other autoimmune disorders[46] and appeared to benefit one ALD patient.[47] We are encouraged by a partial reversal of neurologic disability in one 5-year-old boy who has been treated with this regimen for 6 months. A patient with the adult cerebral form has continued to progress. Four other patients have just begun therapy.

2. *Men with AMN and symptomatic ALD heterozygotes.* The suggestive evidence that GTO therapy improves peripheral nerve function in AMN patients, and the observation that GTE–GTO has a more powerful biochemical effect has prompted us to undertake a large-scale therapeutic trial of GTE–GTO therapy in this group. Eighty-five patients are enrolled in the study. Clinical data are not yet available.

3. *Asymptomatic or neurologically intact persons.* This group includes persons with the biochemical defect of ALD who are free of disability, as well as patients with Addison disease who have the ALD biochemical defect but who are neurologically intact. This group is of greatest interest. If VLCFA accumulation is the main cause or trigger of neurologic disability, normalization of VLCFA levels may succeed in preventing the onset of neurologic dysfunction. The fact that these boys are entirely normal until 4 to 8 years old provides a valuable window of opportunity. Since only up to 7 percent of ALD patients are new mutations,[48] most asymptomatic ALD patients can be identified by screening at-risk families with the plasma VLCFA assay.[11] We also recommend plasma VLCFA assays for all male addisonian patients. Because of the serious outlook for ALD and the apparent safety of the diet, we favor an active search process and recommend that persons with positive results participate in the trial. In most of the neurologically intact biochemically affected persons identified so far, plasma C26:0 levels have normalized with this regimen. As is true for other X-linked disorders, persons at risk may include distant relatives who may not be aware that they are related to the proband and may even live in different continents. Communications require a great deal of tact and sensitivity to ethical issues. It is essential to emphasize that it is not established that the preventive effort will be successful. As shown in Table 7-1 up to 50 percent of asymptomatic persons with the ALD biochemical defect will escape the severe childhood form of the disease, even without therapy. Proof of success requires follow-up of a significantly large number of persons (50 or more) for at least 5 years and the demonstration that substantially more than 50 percent of the treated group remain free of neurologic disability.

Normalization of Red Blood Cell Microviscosity in ALD Patients by GTE–GTO Oil Administration

Table 7-5 shows the effect of GTE–GTO therapy on red blood cell membrane microviscosity as measured by fluorescence polarization using the lipid probe 1,6-diphenylhexatriene as described by Knazek et al.[25] Consonant with the previous report,[25] microviscosity was increased in untreated ALD patients. GTE–GTO therapy normalized the microviscosity values; these changes were statistically significant. This result is encouraging, since there is evidence that the increased membrane microviscosity is involved in the pathogenesis of the adrenal dysfunction.[26] It is of interest that the membrane microviscosity was normalized in spite of the GTE–GTO-treated ALD patients having increased levels of mono-unsaturated VLCFA. This finding suggests that the increased viscosity in untreated patients is mainly a result of the accumulation of saturated VLCFA. As shown in Figure 7-2 these levels are normalized by the dietary therapy.

Bone Marrow Transplantation

A recent favorable experience with bone marrow transplantation[49] has important implications for the therapy of ALD. An 8-year-old boy who had early neurologic involvement was transplanted 2 years ago. His nonidentical twin was the donor. The donor was unaffected by ALD but was HLA-identical. Two years after the transplant, the neurologic deficit has disappeared, the MRI scan has become normal (Fig. 7-3), and cognitive function, which had been slightly impaired before transplant, is now again equal to that of the twin. Plasma VLCFA levels are normal even on a regular diet (Fig. 7-4).

It is important to assess the significance of this finding. Walsh[50] has reported spontaneous fluctuations in the neurologic manifestations of two ALD patients. In the first patient, striking but relatively brief fluctuations appeared to be related to steroid therapy and were followed by neurologic deterioration, which caused death 2 years later. The second patient, who was severely disabled

Table 7-5. Effect of Glyceryl Trierucate Therapy on Red Blood Cell Membrane Microviscosity in Adrenoleukeukodystrophy (ALD) Patients[a]

	No.	Mean	SD	P
All ALD patients	24	0.273	0.018	0.03 (vs. normal control subjects)
Heterozygotes	6	0.279	0.008	0.0044 (vs. normal control subjects)
Treated males	19	0.265	0.017	0.0122 (vs. untreated males)
Normal control subjects	24	0.265	0.012	0.462 (vs. treated males)

[a] Microviscosity was measured by fluorescence polarization using the same technique and instrument as described previously.[26] Microviscosity of untreated male AMN patients and female heterozygotes was significantly higher than that of control subjects. Post-therapy measurements were obtained 2 to 8 months after initiation of the diet at times when the plasma C26:0 levels had been normalized. Post-therapy microviscosity values are similar to those of control subjects and are significantly lower than in the untreated patients.

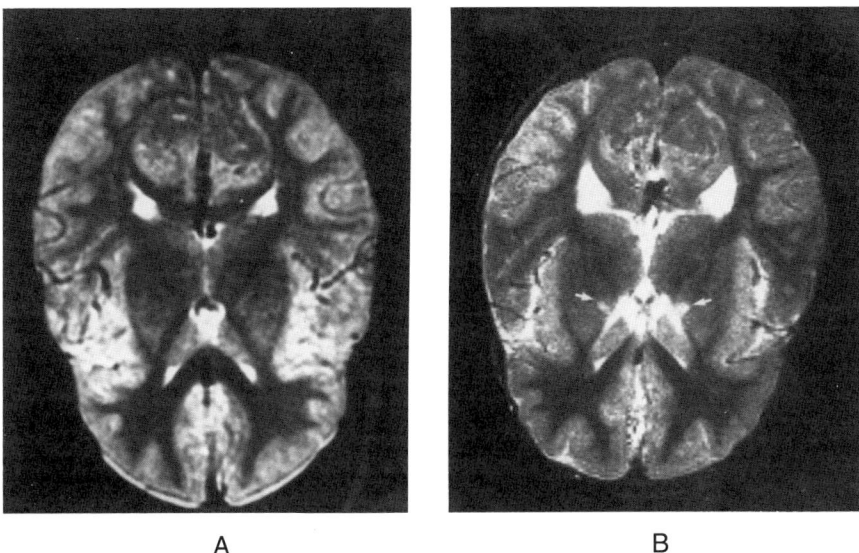

Fig. 7-3. T_2-weighted MRI scans, showing signals of increased intensity (arrows) within both capsules, the pallidum, and the caudate nuclei before bone marrow transplantation (**A**) and their complete disappearance 18 months thereafter (**B**).

at 6½ years of age, made what appears to have been a spontaneous recovery. Sixteen months later, his motor function and school performance were normal, but computed tomography (CT) scans continued to show parieto-occipital white matter lesions; bilateral optic nerve atrophy was noted, and deep tendon reflexes were abnormally brisk. Even though these remissions were striking and significant, they were not as long-lasting or complete as that observed in the transplanted patient. We have reported one adolescent with the biochemical defect of ALD who had a scotoma and impaired visual acuity that recovered spontaneously.[51] Except for the visual disturbance, the patient had been neurologically intact. Apart from these cases, we are not aware of other instances of documented neurologic remissions in the more than 1200 ALD patients who are known to us. We consider it likely that the striking and apparently complete clinical recovery is indeed attributable to the transplant.

The finding that plasma levels of VLCFA were normalized even when the patient consumed a regular diet indicates that the provision of enzymatically competent bone marrow-derived cells was sufficient to affect the "overall economy" of VLCFA metabolism. We had reached the same conclusion in an earlier study of a bone marrow transplant in an ALD patient.[52] This finding suggests that the enzyme that is deficient in ALD is normally present in excess and that partial replacement will be sufficient to influence the course of the disease. Of greatest interest is the reversal of the early neurologic lesions. The mechanism of this reversal is not clear. It suggests that bone marrow-derived cells do reach the CNS, enhanced perhaps by the perivascular lymphocytic infiltration that is such a characteristic feature, and that in some way the enzymatic capacity of

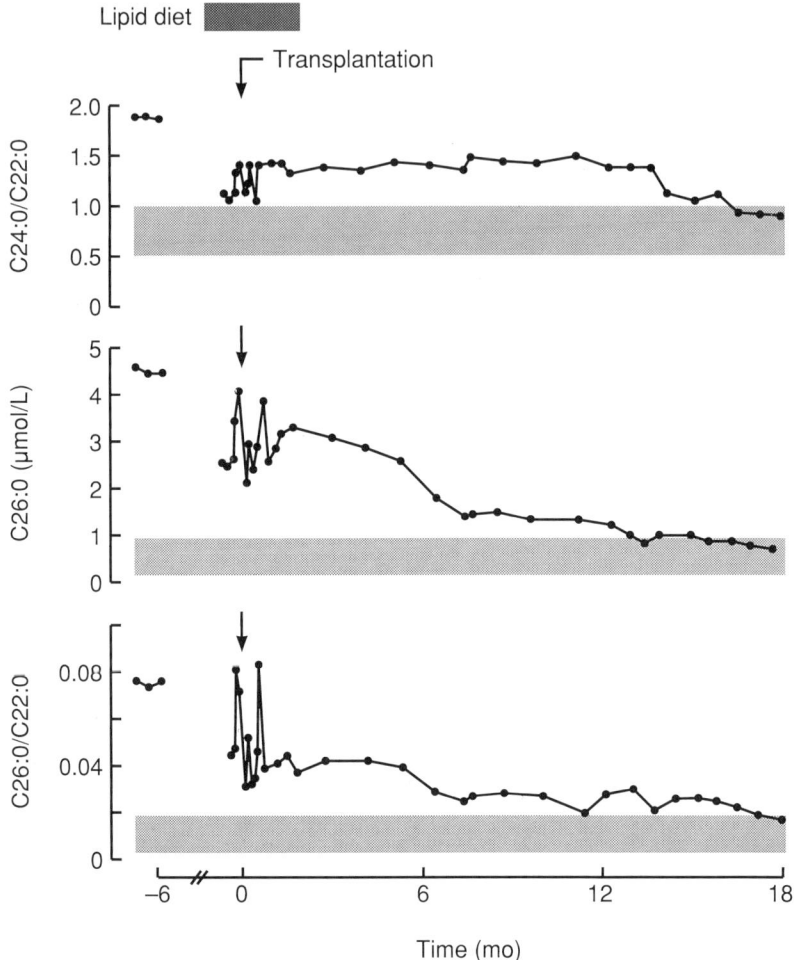

Fig. 7-4. Effect of bone marrow transplantation on the plasma levels of saturated very long chain fatty acids. Shaded bands in the three panels indicate the normal mean ±SD. The period during which a lipid diet was consumed is shown by the shaded area at the top.

these cells has a favorable effect. Other mechanisms must also be considered, such as the effect of the intense immunosuppression and the normalization of the VLCFA levels in plasma. In any case, the favorable CNS effects provide the hope that equally favorable results could be obtained by introducing the normal gene into the patient's own bone marrow cells and that ALD patients may become candidates for gene therapy.

While the effect of bone marrow transplant in this patient was dramatic and the procedure was tolerated well, it must be kept in mind that this procedure is associated with significant morbidity and mortality. Furthermore, it has been our experience[52] and that of others[53] that in patients with more advanced disease the transplant procedure increases the neurologic deficit, and we consider the procedure contraindicated under such circumstances.

Current Therapeutic Recommendations for ALD Patients

Current therapeutic recommendations for ALD patients depend on the patient's phenotype.

1. *Childhood ALD in the rapidly progressive stage.* This is the most severe form of the illness. We recommend the GTE–GTO diet combined with monthly IV γ-globulin administration according to the regimen currently being used for thrombocytopenic purpura.[54] Bone marrow transplantation is not recommended because at this stage of the illness it appears to aggravate the neurologic deficit. GTE–GTO dietary therapy alone does not appear to alter the progression of the disease; for this reason, we recommend the combination with an approach that may modify the postulated immunopathogenetic mechanisms.
2. *Men with AMN and heterozygotes with an AMN-like syndrome.* Because of the suggestive evidence that reduction of plasma levels of VLCFA improves peripheral nerve function, we recommend that these patients be placed on the GTE–GTO regimen. As this is still an investigative approach, careful monitoring for side effects and clinical effectiveness is required by the U.S. Food and Drug Administration (FDA). Because of the relatively lesser severity of this syndrome and the suggestive evidence that dietary therapy is of benefit, it is our current view that the risks associated with bone marrow transplantation are not warranted for this group.
3. *Neurologically intact asymptomatic boys or men or Addison-only patients who have the biochemical abnormality of ALD.* It is recommended strongly that such persons be placed on the GTO–GTE regimen. Careful nutritional and medical supervision is essential and required by the FDA. Close communication with the family, physician, and school is essential. Ethical concerns must be handled with care, and psychological counseling may be required. We place in this group only persons who are neurologically intact as judged by neurologic examination, MRI studies, cognitive tests, behavior, and school performance. There is a 25 to 50 percent chance that such persons may escape CNS damage in childhood and adolescence, even without therapy. It is our hope, but not yet supported by evidence, that dietary therapy will improve the prognosis. Because of the possibility that a significant proportion of persons in this group are not "destined" to develop serious neurologic disability, we believe that the risks of bone marrow transplantation are not warranted here.
4. *Boys or men with early or minimal signs of neurologic involvement.* This group presents a challenge in respect to identification, prognosis, and recommendation for therapy. These persons are usually identified by screening relatives of symptomatic ALD or AMN patients or by performing neurologic evaluations in addisonian patients who have the biochemical defect of ALD. Our quandary is that we do not know the prognostic significance of a subtle neurologic deficit or MRI abnormality in persons with the biochemical defect of ALD. The data that indicate rapid neuro-

logic progression in childhood ALD[19] are based on the follow-up evaluation of patients who were diagnosed because of unequivocal neurologic symptoms. These prognostications may not be relevant for persons who have minimal or clinically inapparent neurologic involvement that was detected only by MRI studies or detailed examinations. The quandary is compounded by the fact that bone marrow transplant may be effective only at that very early stage of the illness. It is our current recommendation that boys at this stage of the illness be considered for bone marrow transplantation. The final decision will take into account the availability of matched donors and the wishes of the patient and the family. It is likely that the present uncertainties will be resolved, at least in part, by the observations and studies being carried out in many parts of the world. Because the therapeutic considerations for the neurologically intact group differ from those for the neurologically minimally involved group, it is important to distinguish between them. We recommend that the neurologically intact children, most of whom will be receiving the GTE-GTO dietary regimen, be monitored every 6 months by neurologic examination and MRI studies and that those who show early neurologic involvement be considered for bone marrow transplantation.

5. *Asymptomatic heterozygotes.* Because of the relatively benign prognosis for these persons, we do not recommend any therapeutic intervention for this group.

CONCLUSION

It is encouraging that several therapeutic options are now being developed and evaluated for patients with ALD. Knowledge is advancing rapidly and there is reason to hope that effective approaches will become available during the next 5 to 10 years.

ACKNOWLEDGMENTS

This work was supported in part by grant 000227 from the Food and Drug Administration, and by grants HD10981, HD24061, HD26371, RR00035, RR00052, RR00722, and DK33914 from the U.S. Public Health Service. The United Leukodystrophy Foundation provided invaluable moral, communication, and financial support. We thank Ms. Jan Shankroff for biostatistical analyses and Ms. Ann Snitcher for expert secretarial assistance.

REFERENCES

1. Migeon BR, Moser HW, Moser AB, et al: Adrenoleukodystrophy: Evidence for X-linkage, inactivation and selection favoring the mutant allele in heterozygous cells. Proc Natl Acad Sci USA 78:5066, 1981

2. Aubourg P, Sack GH, Meyers DA, et al: Linkage of adrenoleukodystrophy to a polymorphic DNA probe. Ann Neurol 21:240, 1987
3. Aubourg P, Sack GH, Moser HW: Frequent alteration of visual pigment genes in adrenoleukodystrophy. Am J Hum Genet 42:408, 1988
4. Sack GH, Raven MB, Moser HW: Color vision defects in adrenoleukodystrophy. Am J Hum Genet 44:794, 1989
5. Aubourg PB, Feil R, Guidoux S, et al: The red-green visual pigment gene region in adrenoleukodystrophy. Am J Hum Genet 46:459, 1990
6. Singh I, Moser AB, Moser HW, Kishimoto Y: Adrenoleukodystrophy: Impaired oxidation of very long chain fatty acids in white blood cells, cultured skin fibroblasts and amniocytes. Pediatr Res 18:286, 1984
7. Singh I, Moser AB, Goldfischer S, Moser HW: Lignoceric acid is oxidized in the peroxisomes: Implications for the Zellweger cerebro-hepato-renal syndrome and adrenoleukodystrophy. Proc Natl Acad Sci USA 81:4203, 1984
8. Lazo O, Contreras M, Hashmi M, et al: Peroxisomal lignoceroyl-CoA ligase deficiency in childhood adrenoleukodystrophy and adrenomyeloneuropathy. Proc Natl Acad Sci USA 85:7647, 1988
9. Wanders RJA, van Roermund CWT, van Wijland MJA, et al: Direct evidence that the deficient oxidation of very long chain fatty acids in X-linked adrenoleukodystrophy is due to an impaired ability of peroxisomes to activate very long chain fatty acids. Biochem Biophys Res Commun 153:618, 1988
10. Igarashi M, Schaumburg HH, Powers J, et al: Fatty acid abnormality in adrenoleukodystrophy. J Neurochem 26:851, 1976
11. Moser HW, Moser AB, Frayer KK, et al: Adrenoleukodystrophy: Increased plasma content of saturated very long chain fatty acids. Neurology 31:1241, 1981
12. Tsuji S, Suzuki M, Ariga T, et al: Abnormality of long-chain fatty acids in erythrocyte membrane sphingomyelin from patients with adrenoleukodystrophy. J Neurochem 36:1046, 1981
13. Moser HW, Moser AB, Powers JM, et al: The prenatal diagnosis of adrenoleukodystrophy. Demonstration of increased hexacosanoic acid in cultured amniocytes and fetal adrenal gland. Pediatr Res 16:172, 1982
14. Kelley RI, Datta NS, Dobyns WB, et al: Neonatal adrenoleukodystrophy: New cases, biochemical studies, and differentiation from Zellweger and related peroxisomal polydystrophy syndromes. Am J Med Genet 23:869, 1986
15. Moser HW, Moser AB, Singh I, O'Neill BR: Adrenoleukodystrophy: Survey of 303 cases: Biochemistry, diagnosis and therapy. Ann Neurol 16:628, 1984
16. Moser HW: X-linked adrenoleukodystrophy. p. 1511. In Scriver CR, Beaudet AL, Sly WS, Valle D (eds): The Metabolic Basis of Inherited Disease. 6th Ed. McGraw-Hill, New York, 1989
17. Siemerling E, Creutzfeldt HG: Bronzekrankheit und sklerosierende encephalomyelitis. Arch Psychiatr Nervenkr 68:217, 1923
18. Schaumburg HH, Powers JM, Suzuki K, Raine CS: Adrenoleukodystrophy (sex-linked Schilder disease): Ultrastructural demonstration of specific cytoplasmic inclusions in the central nervous system. Arch Neurol 31:210, 1974
19. Moser HW, Naidu S, Kumar AJ, Rosenbaum AE: The adrenoleukodystrophies. CRC Crit Rev Neurobiol 3:29, 1987
20. Budka H, Sluga E, Heiss WD: Spastic paraplegia associated with Addison's disease: Adult variant of adrenoleukodystrophy. J Neurol 213:237, 1976
21. Griffin JW, Goren E, Schaumburg H, et al: Adrenomyeloneuropathy: A probable variant of adrenoleukodystrophy. Neurology 27:1107, 1977

22. Schaumburg H, Powers JM, Raine CS, et al: Adrenomyeloneuropathy: A probable variant of adrenoleukodystrophy. Neurology 27:1114, 1977
23. Sadeghi-Nejad A, Senior B: Adrenomyeloneuropathy presenting as Addison's disease in childhood. N Engl J Med 322:13, 1990
24. Powers JM: Adrenoleukodystrophy (adreno-testiculo-leuko-myelo-neuropathic-complex). Clin Neuropathol 4:181, 1985
25. Knazek RA, Rizzo WB, Schulman JD, Dave JR: Membrane microviscosity is increased in the erythrocytes of patients with adrenoleukodystrophy and adrenomyeloneuropathy. J Clin Invest 72:245, 1983
26. Whitcomb RW, Linehan WR, Knazek RA: Effects of long-chain, saturated fatty acids on membrane microviscosity and adrenocorticotropin responsiveness of human adrenocortical cells in vitro. J Clin Invest 81:185, 1988
27. Powers JM, Schaumburg HH, Johnson AB, Raine CS: A correlative study of the adrenal cortex in adreno-leukodystrophy—Evidence for a fatal intoxication with very long chain saturated fatty acids. Invest Cell Pathol 3:353, 1980
28. Adams RD, Kubik CS: The morbid anatomy of the demyelinative diseases. Am J Med 12:510, 1952
29. Griffin DE, Moser HW, Mendoza Q, et al: Identification of the inflammatory cells in the nervous system of patients with adrenoleukodystrophy. Ann Neurol 18:660, 1985
30. Van Duyn MA, Moser AB, Brown FR III, et al: The design of a diet restricted in saturated very long chain fatty acids: Therapeutic application in adrenoleukodystrophy. Am J Clin Nutr 40:277, 1984
31. Kishimoto Y, Moser HW, Kawamura N, et al: Evidence that abnormal very long chain fatty acids of brain cholesterol esters are of exogenous origin. Biochem Biophys Res Commun 96:69, 1980
32. Murad S, Kishimoto Y: Chain elongation of fatty acid in brain: A comparison of mitochondrial and microsomal enzyme activities. Arch Biochem Biophys 185:300, 1978
33. Bourre J-M, Daudu O, Baumann N: Nevonic acid biosynthesis by erucyl-CoA elongation in normal and quaking mouse brain microsomes. Elongation of other unsaturated fatty acyl-CoAs (mono and polyunsaturated). Biochim Biophys Acta 424:1, 1976
34. Rizzo WB, Watkins PA, Phillips MW, et al: Adrenoleukodystrophy: Oleic acid lowers fibroblast saturated C22-C26 fatty acids. Neurology 36:357, 1986
35. Tsuji S, Ohno T, Miyatake T, et al: Fatty acid elongation activity in fibroblasts from patients with adrenoleukodystrophy (ALD). J Biochem (Tokyo) 96:1241, 1984
36. Stumpf DA, Hayward A, Haas R, Schaumburg HH: Adrenoleukodystrophy. Failure of immunosuppression to prevent neurological progression. Arch Neurol 38:48, 1981
37. Naidu S, Bresnan MJ, Griffin D, et al: Childhood adrenoleukodystrophy: Failure of intensive immunosuppression to arrest neurologic progression in childhood adrenoleukodystrophy. Arch Neurol 45:846, 1988
38. Borel J, Cohen J (eds): ALD/AMN Diet Cookbook. 2nd Ed. United Leukodystrophy Foundation/The Kennedy Institute, DeKalb, IL/Baltimore, 1990
39. Brown FR III, Van Duyn MA, Moser AB, et al: Adrenoleukodystrophy: Effects of dietary restriction of very long chain fatty acids and of administration of carnitine and clofibrate on clinical status and plasma fatty acids. Johns Hopkins Med J 151:164, 1982

40. Rizzo WB, Phillips MW, Dammann AL, et al: Adrenoleukodystrophy: Dietary oleic acid lowers hexacosanoate levels. Ann Neurol 21:232, 1987
41. Moser AB, Borel J, Odone A, et al: A new dietary therapy for adrenoleukodystrophy: Biochemical and preliminary clinical results in 36 patients. Ann Neurol 21:240, 1987
42. Kramer J, Sauer FD, Pidgen WJ: High and Low Erucic Acid Rapeseed Oils: Production, Usage, Chemistry and Toxicological Evaluations. Academic Press, San Diego, 1983
43. Rizzo WB, Leshner RT, Odone A, et al: Dietary erucic acid therapy for X-linked adrenoleukodystrophy. Neurology 30:1415, 1989
44. Steinberg D: Refsum Disease. p. 1533. In Scriver CR, Beaudet AL, Sly WS, Valle D (eds): The Metabolic Basis of Inherited Disease. 6th Ed. McGraw-Hill, New York, 1989
45. Uziel G, Bertini E, Rimoldi M, Gambetti M: Italian multicentric dietary therapeutical trial in adrenoleukodystrophy. p. 163. In Uziel G, Wanders RJA, Cappo M (eds): Adrenoleukodystrophy and Other Peroxisomal Disorders: Clinical, Biochemical, Genetic and Therapeutic Aspects. Excerpta Medica, 1990
46. Gelfand EW: Intervention in autoimmune disorders: Creation of a niche for intravenous γ-globulin therapy. Clin Immunol Immunopathol 53:S1, 1989
47. Miike T, Taku K, Tamura T, et al: Clinical improvement of adrenoleukodystrophy following intravenous gammaglobulin therapy. Brain Dev 11:134, 1989
48. Moser HW, Moser AB, Naidu S, et al: X-linked adrenoleukodystrophy: Epidemiology, pathogenesis and therapy. p. 127. Excerpta Medica International Congress, Amsterdam, 1989
49. Aubourg P, Blanche S, Jambaque I, et al: Reversal of early neurologic and neuroradiologic manifestations of X-linked adrenoleukodystrophy by bone marrow transplantation. N Engl J Med 322:1860, 1990
50. Walsh PJ: Adrenoleukodystrophy: Report of two cases with relapsing and remitting courses. Arch Neurol 37:448, 1980
51. Carpenter GG, Graziani LJ, Moser HW, Schaumburg HH: Adrenoleukodystrophy (ALD) with transient amaurosis without clinical degeneration: A disease variant or third allele. Pediatr Res 18:374a, 1984
52. Moser HW, Tutschka PJ, Brown FR III, et al: Bone marrow transplant in adrenoleukodystrophy. Neurology 34:1410, 1984
53. Weinberg K, Moser A, Watkins P, et al: Bone marrow transplantation (BMT) for adrenoleukodystrophy (ALD). Pediatr Res 23:334A, 1988
54. Kobayashi RH, Kobayashi AD, Lee N, et al: Home self-administration of intravenous immunoglobulin therapy in children. Pediatrics 85:705, 1990

8

ENZYME REPLACEMENT: OVERVIEW AND PROSPECTS

Mario C. Rattazzi · Kostantin Dobrenis

INTRODUCTION

Enzyme replacement therapy (ERT) for lysosomal storage diseases received considerable attention during the 1970s.[1,2] Interest in this logical and seemingly straightforward therapeutic approach waned considerably until the recent clinical successes of adenosine deaminase replacement in severe combined immunodeficiency and of β-glucosidase replacement in the non-neuronopathic type I Gaucher disease (see Chs. 9 and 10). These successes have stimulated renewed interest in this therapeutic strategy. Therefore, it seems important to highlight the principles and problems underlying ERT, with particular emphasis on efforts to target enzyme to neural cells for the treatment of disorders with primary neurologic involvement. It is our contention that we are ready for a new look at this old problem, as we now have considerably better chances of succeeding where we failed 15 years ago. Our work[3-7] has used human β-D-N-acetylhexosaminidase A (Hex A) as the exogenous enzyme for replacement in the feline homologue of human GM2 gangliosidosis.[8] We have explored several aspects of ERT, and data from our work may serve to illustrate several points. Focusing on ERT as a potential therapeutic intervention, however, is not meant to exclude the development of other approaches, such as bone marrow transplantation therapy or gene replacement therapy. On the contrary, the lessons learned from ERT can help make bone marrow transplantation and gene replacement therapy more effective, once the considerable problems of the latter two approaches have been resolved.

PRINCIPLES AND PROBLEMS

The general principle of ERT is straightforward: replace the missing or inactive enzyme with active enzyme. The enzyme deficiency may be localized in, and the normal exogenous enzyme will have to be delivered to, one of several different compartments, making it necessary to cross one or more natural barriers. When extracellular spaces such as the gastrointestinal tract or blood plasma are the target sites, access is easy. Pancreatic enzyme supplementation in cystic fibrosis and ERT in adenosine deaminase (ADA) deficiency are pertinent examples (see Chs. 1 and 10). Traversing the cellular plasma membrane to gain access to cytoplasm is necessary for replacement of a soluble, cytoplasmic enzyme. Although less explored, this approach is feasible in principle, using liposomes that fuse with the plasma membrane and that deliver their contents to the cytoplasm[9] or combination vesicles that exploit the properties of viral coat proteins (e.g. Sendai/influenza virus) to adhere to the plasma membrane, and cause internalization of viral packages across the cellular membrane.[10] Bacterial toxins (tetanus, botulinus) or plant toxins (ricin) that create pores in the cell membrane and allow hydrophilic proteins to cross a hydrophobic barrier may be exploited as carriers of enzymes into the cytoplasm. These mechanisms are being studied for therapeutic drug delivery,[11] but they have not yet been exploited for cytoplasmic ERT. Enzyme replacement aimed at cytoplasmic organelles (e.g., mitochondria) requires traversing the plasma membrane and yet another membrane, that of the organelle itself. The mechanisms by which proteins are normally translocated across mitochondrial membranes are beginning to be understood,[12] and it is not too far-fetched to think that appropriate protein packages could be developed to replace defective mitochondrial enzymes. By contrast, an organelle that does not require traversing the plasma membrane is the lysosome-vacuole system, since it communicates with the extracellular spaces through endocytic processes.[13] This is also the site at which undegraded substrates accumulate in lysosomal storage diseases; delivery of exogenous endocytosed enzyme to these storage sites is possible, owing to organelle fusion processes that result in mixing the contents of "old" storage vesicles and newly formed, endocytic vesicles.[13] This, in essence, is the principle of ERT in lysosomal storage diseases.[14] This approach has been tested successfully in simple cell culture systems,[15,16] but its application to in vivo situations is impeded by a number of problems (Table 8-1). This chapter briefly reviews the problems that have been or that may be encountered and outlines possible solutions and their advantages and disadvantages. The problems are listed in Table 8-1 in an operational, temporal sequence, but they are all interrelated, and it is difficult to point to the most important one. The last problem listed, however, is almost entirely unexplored and may turn out to be of greater importance than all the others, as its clarification may radically change our approach to the treatment of lysosomal storage disease, when the precise cellular function that needs to be restored is known.

Availability of Enzyme

For small-scale experiments, the choice of tissue is not really of concern, provided that (1) there is easy access to the tissue, (2) it yields adequate quantities of the desired enzyme, and most important, (3) the enzyme extracted from

Table 8-1. Problems in Lysosomal Enzyme Replacement Therapy

Availability of enzyme from
    Tissue extracts
    Cell cultures
    Microorganisms

Rapid plasma clearance and
    Effects on circulating substrate
    Nontarget cell uptake
    Target cell uptake

Targeting to
    Specific visceral organs
    Central nervous system
    All cells
    Specific cell types

Cellular uptake by
    Fluid-phase endocytosis
    Adsorptive, nonspecific endocytosis
    Acceptor- or receptor-mediated endocytosis

Evaluation of effectiveness in
    Cell cultures
    Animal models
    Patients

Complications and side effects
    Immunogenicity
    Interference with receptor function
    Dumping of catabolytes

Unclear pathogenesis
    Storage versus cellular pathology
    Developmental aspects
    Reversibility

it carries the appropriate recognition signals for cellular uptake. In terms of ERT, however, most lysosomal enzymes extracted from tissue available in large quantity, such as placenta, have glycosyl chains of the high mannose type[17] and are therefore recognized mostly by the mannose/N-acetylglucosamine-specific receptor present almost exclusively on activated macrophages and reticular endothelial system (RES) cells.[18,19] Ironically, in the case of the one storage disorder—Gaucher disease—in which the RES cells are the target of ERT, the pivotal enzyme, glucocerebrosidase, has a glycosyl chain of the "complex" type and is not efficiently recognized by the target cells. However, sequential deglycosylation resulting in exposure of glycosyl residues recognized by RES cell receptors[20] has made possible clinically effective ERT trials in this disease (see Ch. 9).

In addition to the type of enzyme that can be obtained, and the not always simple extraction procedures involved, one very real problem today is acquired immunodeficiency sydrome (AIDS); human immunodeficiency virus (HIV) in principle taints every available human tissue or body fluid and presents a risk both for those involved in the preparation of the enzyme and for the patient. Can one then turn to normal cells in culture? Even with scaled-up procedures, the problem is that the amount of enzyme that can be recovered from conventional cultures falls far short of what is needed for a sustained ERT regimen. In addition, we face again the same problem of glycosyl chain determinants that may not be the most appropriate for uptake by target cells. There are workable approaches using cell culture, based on molecular genetics technology, to produce large quantities of recombinant proteins[21] (see Ch. 14). This approach, however, is not as simple in the case of heteromultimeric enzymes, such as Hex A.

Microorganisms offer a relatively inexpensive and effective way to produce large amounts of the desired enzyme by recombinant DNA technologies. The drug industry is fully geared up for this, as shown by the production of recombinant insulin, human growth hormone, and plasminogen activator. The potential market for recombinant lysosomal enzymes is far smaller than that for these compounds, however, and even the Orphan Drug Act may not be enough of an incentive for their development. Furthermore, the microorganisms likely to be employed may yield nonglycosylated enzyme, or enzyme with glycosyl chains of inappropriate composition. A solution to this problem is to develop biochemical derivatization procedures that will endow these mass-produced proteins with the appropriate (glycosyl) recognition markers.

Plasma Clearance

It is now well recognized that the structure of the glycosyl chain is a key determinant of the plasma half-life of lysosomal enzymes and of their cellular disposition.[18,19] Obviously, short plasma half-life is not a problem if rapid clearance means that the enzyme is taken up mainly by the target cells. Conversely, a long half-life is not an advantage, unless uptake by target cells also takes place. In special cases, such as Fabry disease, in which undegraded trihexosylceramide is present in plasma, cell uptake may not be essential. Plasma-derived, sialylated, α-galactosidase infused intravenously into patients with Fabry disease proved effective in clearing circulating substrate.[22] In most cases, however, the outcome is less felicitous, as for instance in the case of human Hex A infused into GM2 gangliosidosis cats.[4] The rapid clearance attributable to removal by the RES cells by the mannose-specific receptor could be competitively blocked by intravenous mannans injection. As a consequence of the longer plasma half-life, enzyme uptake could be obtained in several tissues and organs; however, it represented only a small fraction of the normal enzyme content. Thus, it is clearly not enough to prevent preferential RES uptake of enzyme and prolong its plasma half-life. Active efficient uptake by target cells must also be obtained.

Targeting and Cell Uptake

This is a central topic in ERT and should be considered at two levels: organ targeting and cell targeting. So far, ERT trials have relied on intravenous administration of enzyme that results in its dilution in the general circulation.

If liver is the target organ, dilution is not a major concern because of the high percent perfusion rate of this organ, which will rapidly clear enzyme-containing venous blood, but it becomes relevant for other target organs. As in tumor chemotherapy,[23] it may be appropriate in some cases to consider direct organ access by arterial catheterization, if the appropriate pharmacokinetics criteria are met. This would not only permit delivery of highly concentrated enzyme to the organ but would also avoid first-pass hepatic RES sequestration. A special organ targeting case is that of the central nervous system (CNS), to which we devote special attention in a separate section.

More important than the macroscopic organ targeting problem, however, is that of targeting exogenous enzyme to specific cell types. The presence of normal enzyme in the pericellular space may not necessarily result in sufficient levels of internalization. It is therefore important to understand some aspects of endocytotic processess[24,25] crucial to effective enzyme replacement.

Fluid-Phase Endocytosis

In principle, uptake by all cells may be obtained by nonadsorptive, fluid-phase endocytosis, but this route has some drawbacks. The amount of protein internalized is directly proportional to its concentration in the pericellular space; thus, effective use of this route may require very high levels of exogenous enzyme in the interstitial fluid spaces of tissues, not easily attainable in vivo. This is especially relevant in the case of cells with low levels of constitutive endocytosis. Our observations on human placental Hex A uptake by cat and rat neural cells in mixed cell type cultures[7] are pertinent here. Enzyme uptake, detected by indirect immunofluorescence, differs significantly among the cell types present and parallels the relative pattern of uptake of Lucifer yellow, a marker for fluid phase endocytosis.[26] At enzyme concentrations that render vacuolized, phagocytic cells highly fluorescent, uptake by mature, neuron-specific enolase-positive neurons is barely detectable. Between these extremes, uptake is variable both within and among cell types but is generally higher in multipolar type I astrocytes and small process-bearing cells than in large undifferentiated epithelial-like cells. In addition, even high endocytic activity does not necessarily result in internalization to the lysosome. For instance, macrophages, which have a high rate of constitutive endocytosis, internalize large amounts of Lucifer yellow but return to the surface and release into the medium a large part of the endocytosed tracer, with only a small amount being retained and routed to lysosomes.[27]

Adsorptive Endocytosis

Another approach to nonspecific cell targeting is through charge-induced adsorptive endocytosis obtainable by cationization of proteins.[28] For instance, covalently bound low-molecular-weight poly-L-lysine imparts a strong positive charge to proteins, which are then adsorbed to the plasma membrane and subsequently internalized[29] with endocytic vacuoles.

We explored this approach by conjugating poly-L-lysine to Hex A by the carbodiimide method[6] and by reductive amination,[7] the advantage of the latter being disruption of the glycosyl chain and removal of the signal that would result in preferential RES uptake in vivo. We observed a remarkable enhanced uptake of poly-L-lysine Hex A over native Hex A both quantitatively in cultured fibroblasts[6,7] and qualitatively in cultured neural cells,[7] as well as in hepatocytes after intravenous infusion in cats.[30] Enzyme cationization for ERT, however, still presents problems. First, the relative increase in ligand uptake by cells results from its affinity for plasma membrane moieties that increases its concentration at the plasma membrane level; thus, the rate of ligand internalization increases, but in the ultimate analysis it reflects the rate of constitutive endocytosis.[24,29] Cells in which this rate is low would still internalize relatively low amounts of ligand. In fact, we observed that although the overall uptake of poly-L-lysine Hex A by neural cells in culture was greater than that of Hex A, the differences in uptake among cell types described above remained, with neurons showing the least uptake.[7,31] Second, intravenous injection of polycations in the amounts predictably needed for ERT may trigger an anaphylactoid reaction caused by platelet and mast cell degranulation.[32] Indeed, we encountered this complication in several cats infused with poly-L-lysine Hex A (Rattazzi MC: unpublished observations). Is the cationization approach useless for ERT? It may not be, if the target cells can be approached by a route other than the general circulation, and have a relatively high rate of constitutive endocytosis.

Acceptor-Mediated Endocytosis

Specific cell targeting depends on knowledge of the existence of specific acceptors or receptors on a given target cell type. This knowledge is not always available, owing either to difficulties in isolating that cell type and testing it in culture with a variety of ligands, or as in the case of the CNS, to anatomic barriers that complicate in vivo experiments. The advantage of acceptor- or receptor-mediated endocytosis for ERT, however, in terms of selectivity and efficiency, should motivate efforts toward their identification and utilization. *Acceptor* is defined as a moiety present in the plasma membrane, the physiologic function of which may be unknown, but that recognizes with high affinity a ligand exploitable for targeting and, within the context of ERT, is internalized by endocytosis. *Receptor* is defined as a moiety that recognizes with high affinity a specific physiologic ligand, the function of which comprises ligand internalization. The distinction is important because the rate of internalization of acceptor-bound ligands may be low, being solely a function of constitutive endocytosis, whereas that of receptor-bound ligands may be much higher, as a consequence of ligand binding-induced endocytosis and receptor recycling to the cell surface.[24]

Our studies aimed at increasing Hex A uptake by neurons[7] represent a pertinent example of cell-specific targeting and internalization through acceptor-mediated endocytosis. Neurons have high affinity for tetanus toxin (TT) that binds avidly (K_d about 10^{-8} M) to their plasma membrane because of its high content in gangliosides GD1a, GD1b, and GT.[33,34] The ganglioside-binding portion of the heavy chain of TT, termed fragment C (TTC), can be cleaved off by proteolysis and isolated[35] and has recently been obtained by recombinant DNA methodologies.[36] The 50-kd fragment, TTC, retains neuron-binding properties but is not toxic. We bound fragment C to Hex A by thiolation and mixed disulfide formation. The adduct, TTC Hex A, retained activity against GM2 ganglioside. When we exposed cultured rat or cat neurons to TTC Hex A, we were able to detect strong binding of the derivatized enzyme to the cell membrane. By contrast, native Hex A, as expected, was not bound. A key question, when dealing with acceptors rather than bona fide receptors, is whether binding of a ligand is followed by internalization to the lysosomes. In our experiments, internalization of TTC Hex A could be inferred morphologically from the granular distribution pattern of enzyme immunoreactivity, consistent with endosomal-lysosomal localization. We recently obtained more cogent evidence of lysosomal location using cultured neurons from GM2 gangliosidosis kittens, as well as a monoclonal antibody[37] to assess GM2 ganglioside by immunofluorescence microscopy. Data from these recent experiments[38] show clearly that incubation of these cells with TTC Hex A decreases GM2 ganglioside in most neurons, to levels below the limits of sensitivity of our assay system. By contrast, native Hex A at the same concentration has very little degradative effect. This finding indicates that TTC-mediated uptake results in lysosomal delivery of active Hex A.

What proportion of membrane-bound TTC Hex A is internalized and does it reach the lysosomes as a complex? These are questions to be answered in future experiments. Nevertheless, our data indicate clearly that TTC (or its ganglioside-binding domain) has a definite potential for specific, effective delivery of enzyme to neuronal lysosomes.

Receptor-Mediated Endocytosis

TTC has not been available commercially for the past 2½ years. This has prompted us to explore receptor-mediated endocytosis for Hex A targeting using the cation-independent mannose 6-phosphate-specific receptor (M6PR). The main endocellular function of this well-studied receptor system[39,40] appears to be the routing of newly synthesized lysosomal enzymes from the Golgi apparatus to their final destination, the lysosomes—at least in some cell types. In addition to this endocellular function, however, it also has an "external" cycle and can take up ligands at the plasma membrane and deliver them to the endosome-lysosome complex,[40] meeting the criteria for its use in an enzyme replacement setting. Furthermore, because of the finding that M6PR is identical with the insulin-like growth factor II receptor (IGFIIR), the extensive literature on organ, tissue, and cell localization of IGFIIR makes it clear that it is not confined to fibroblasts and mesenchymal cells, widening its potential use for cell targeting. Finally, its affinity for newly synthesized (M6P-rich "high uptake") lysosomal enzymes is high. The problem is that enzymes obtained from tissue extracts have resided in lysosomes for some time and have little, if any, 6-phosphate residues or their terminal mannosyl residues.[41,42] Although there are natural sources of high-uptake lysosomal enzymes, such as sperm fluid,[43] their value in preparing the large amounts of material needed for ERT is questionable.

To test the properties of the M6PR in an ERT setting, we recently devised a scheme to reintroduce M6P on Hex A using pentamannosyl 6-phosphate (PMP) from mutant yeast o-phosphomannans.[44] This compound had been used by others to derivatize proteins[45] for uptake by cells by the M6PR, but the derivatization conditions are not compatible with retention of enzymatic activity. Thus, we used p-aminophenylethylamine (PAPEA) to aminate PMP,[46] obtained by mild acid hydrolysis[44] of o-phosphomannans. We coupled PMP-PAPEA to Hex A by conventional carbodiimide methodology to obtain PMP Hex A. In initial immunofluorescence microscopy experiments,[47] we were able to show that PMP Hex A was endocytosed more effectively than native Hex A by cultured fibroblasts, astrocytes, and neurons. Although in some quantitative experiments PMP Hex A uptake by fibroblasts was as high as 20-fold that of native enzyme,[48] it now appears that the efficiency of PMP amination using PAPEA[46] is usually low, and uptake enhancement is more modest. These initial experiments are sufficiently encouraging, however, to warrant further work. We are currently exploring alternative PMP derivatization and coupling strategies that should permit more detailed study of the applicability of the M6PR system to ERT. Of particular interest in this respect is the reported existence of the M6PR-IGFII on neural cells,[49–52] consistent with our observations with PMP Hex A as ligand,[47,48] that may provide a targeting method complementary to that employing TTC. Also of interest is the reported existence of M6P-IGFIIR in muscle[53] and vascular endothelial cells[54] relevant to ERT in Pompe and Fabry disease, respectively. The report of endocytosis of high uptake α-glucosidase by cultured myocytes from a patient with Pompe disease, resulting in catabolism of stored glycogen,[55] is encouraging and suggests that PMP α-glucosidase may be useful in this case. Contrasting reports on internalization of M6P ligands by vascular endothelial cells[54,56] indicate the need for additional experiments, preferably in vivo, to establish to which segment of the vascular tree an ERT approach based on M6P-rich ligands may be applicable. It must be pointed out that although M6P-IGFIIR internalization with delivery of ligands to the lysosomal compartment has been established in a number of different cell types, for several others the available data do not permit the conclusion that the receptor either is present on the cell surface or is involved in endocytic processes. Furthermore, the receptor appears to be developmentally down-regulated[52,53] and may not be present postnatally on the plasma membrane of a given cell type in amounts sufficient to obtain therapeutically useful enzyme internalization, although it may undergo redistribution to the cell surface on ligand exposure.[40]

M6P-IGFIIR thus exemplifies the advantages and disadvantages of receptor-directed cell targeting. Although highly efficient in principle because of its high affinity for the appropriate ligand and because of recycling, a receptor system may be ill suited to practical ERT applications because (1) it may not be present on the cells of interest, (2) it is not always internalized, or (3) it recognizes as ligand a moiety that is difficult to obtain from natural sources. Nevertheless, as our knowledge of receptor-mediated endocytosis expands, it should be possible to assemble an armamentarium of receptor systems and related ligands to permit enzyme targeting to a large number of different cell types.

Effectiveness of Enzyme Replacement Therapy

Biochemical Evaluation

Experiments aimed at testing the first level of targeting strategies (i.e., specificity of binding, efficiency of cellular uptake, and lysosomal delivery of ligands) are not difficult in principle, provided that the system used (cells in culture) permits generalizations (e.g., M6P-IGFIIR in cultured fibroblasts). By contrast, the assessment of catabolic effectiveness of such strategies presents considerable difficulties. Ideally, one would prefer not only to maintain in culture the very target cells to which ERT will ultimately be directed but also to obtain cells that exhibit lysosomal storage of the relevant metabolite. This fortunate combination is very rarely encountered.[1,15,16,55-57] Researchers interested in ERT for neurodegenerative lysosomal disorders, in particular, are faced with these difficulties, for it is indeed possible to bypass cell cultures and carry out ERT experiments in non-neural organs and tissues in appropriate animal models, but it is much more complicated and expensive to devise in vivo experiments involving the CNS. Not only does the impermeability of the blood-brain barrier to proteins make organ targeting very difficult, but also the heterogeneity of cell types in neural tissue makes cell targeting and evaluation of results much more complicated than in visceral tissues. Our early experience with GM2 gangliosidosis cats, in which apparent effective delivery of Hex A by experimental blood-brain barrier permeabilization contrasted with minimal GM2 ganglioside degradation,[5] is a case in point, hence our efforts at developing a methodology for long-term cultures of neural cells derived from affected kittens.[58] The procedure (Dobrenis K, Rattazzi MC: unpublished data) uses cerebral cortex from postnatal kittens (1–3 days of age) that is enzymatically digested with a hyperosmotic solution of trypsin, hyaluronidase, chondroitinase ABC, and DNAase I, followed by mechanical dissociation. The cells are plated on a polylysine-collagen substratum and kept in media that permit enrichment in neurons or astrocytes or maintenance of mixed neural cell type cultures. We have monitored GM2 ganglioside storage in these cells by immunologic means, but it should not be difficult to apply more precise quantitative procedures. Numerous well-characterized animal models of neurodegenerative lysosomal diseases[59-60] are potentially amenable to the approach we have followed with GM2 gangliosidosis cats (see Ch. 11). The availability of neural cell cultures from affected animals would greatly simplify ERT assessment experiments.

Testing the biochemical effectiveness of ERT strategies in patients should be reserved, in our opinion, to instances in which appropriate animal models or culture systems are not available. Ethical and technical constraints make it difficult to obtain reliable data, although in rare cases a disease may lend itself to uncomplicated sampling, yielding data of direct relevance to treatment. The ERT trials in patients with Fabry disease by Desnick's group[22,61] provide the best, if not the only, example. Although sampling in these early studies was limited to blood plasma, the effectiveness of enzyme infusion in decreasing plasma trihexosylceramide concentration strongly suggested that this might be sufficient to prevent, or possibly reverse, vascular endothelial cell storage.

Functional Evaluation

Although useful for biochemical studies, cell cultures are of little value in assessing the ultimate effectiveness of ERT in lysosomal diseases, as the precise relationship between lysosomal storage and clinically evident functional impairment is not clear. Until this relationship is better understood, we are left, essentially, with one course of action: to pour enzyme in the cellular pathology black box, hoping to see a change in the histologic picture of a diseased tissue, accompanied by a change in the clinical picture. It is in this respect that animal models are particularly useful, and it is gratifying to see that their value is increasingly recognized. In the case of neurodegenerative disorders, the black box of lysosomal storage and cellular pathology is contained, so to speak, in the larger black box of neural function, which makes it even more imperative to test ERT modalities in an appropriate animal model. By this we mean a model in which abnormalities are manifest not only at the neural cell level but, most important, at the functional level. To borrow an example from the field of bone marrow transplantation, demonstration of decreased glycosaminoglycan storage in neural cells in mucopolysaccharidosis type I (MPS I) dogs[62] is very encouraging but does not necessarily permit the conclusion that the approach is therapeutically valid, because objective neurologic signs are not present in this model. Although readily available experimental animals are ill suited to the assessment of higher neural functions, regression or resolution of symptomatology reflecting neurodegenerative processes affecting simpler functions (e.g., motor system) in animal models would provide at least reasonable presumption of therapeutic potential with respect to more complex systems. Thus, despite the growing opposition to animal experimentation, ERT trials in animal models of neurodegenerative disorders are, in our opinion, a necessary stage that must precede possible application to patients. The concept that "there is nothing to lose" in carrying out therapeutic trials in patients affected by fatal diseases in the absence of supportive evidence from animal experiments does not easily apply to neurodegenerative disorders, in which partial restoration of function or mere extension of survival are not necessarily desirable outcomes.

In disorders not affecting the CNS, however, it may be appropriate to carry out trials in patients in the absence of data from experiments in animal models, provided that (1) basic enzyme-targeting mechanisms are understood and taken into account in the choice of enzyme or adduct; (2) the pathology of the disease in question is reasonably well understood, if not at the subcellular level, at least at the cell and tissue level; and (3) signs and symptoms of the disease closely reflect its pathology and can be monitored by relatively uncomplicated and minimally invasive techniques. Gaucher disease type I now fulfills nearly all these requirements, and the recent ERT trials in this disease by Brady's group[63] can be regarded as a culmination of these researchers' efforts at making it such (see Ch. 9). The very encouraging results of these trials should not only be a stimulus to pursue this therapeutic approach in other disorders but also an inducement to reflect on the steps that led to the successful outcome.

Complications and Side Effects

An insufficient number of extended in vivo experiments or trials in patients have been conducted to determine the immunologic consequences of lysosomal enzyme infusion, but the available data suggest that this may be less of a problem than anticipated. No immunologic reactions were detected in ERT for patients with Fabry, Gaucher, and Tay-Sachs disease[22,61,63,64] (see Ch. 9). In the first two cases, it is possible to explain this outcome by postulating that residual enzymatic activity or enzymatically inactive, immunologic cross-reacting material is sufficient for the organism not to recognize the exogenous enzyme as a foreign antigen.[65] In the case of Tay-Sachs disease, however, in which as a rule the α chain of Hex A is not synthesized, this argument does not seem to apply. Although in these trials infused Hex A was a mixture of native enzyme and presumably less immunogenic, polyvinylpyrrolidone (PVP)-derivatized enzyme, and enzyme infusion was intrathecal or intraventricular, sufficient enzyme was present in the general circulation to be a potential antigenic stimulus.[64] It can be hypothesized that lysosomal enzymes, by virtue of their well-known resistance to proteolysis, are less easily "processed" by antigen-presenting cells.[66] The antigenicity of exogenous lysosomal enzyme, however, might be increased by the modifications designed to improve its targeting and uptake, such as deglycosylation, which may expose epitopes sterically hindered in the native enzyme, or covalent linkage of an immunogenic peptide moiety like TTC. In general, however, immunosuppressive strategies developed for organ transplantation may be applicable to the control of immunologic complications in ERT.[67]

Nonimmunologic side effects may also be anticipated, at least in theory, in the case of receptor-mediated targeting and uptake, depending on the nature and characteristics of the receptor in question:

1. The (modified) exogenous enzyme may compete with the natural ligand for receptor binding, possibly preventing action of the latter at a critical moment.
2. Through receptor binding and activation, the exogenous enzyme may exert the same activity as the natural ligand, but at an inappropriate time or in excessive amount.
3. By up- or down-regulating the receptor by affecting its affinity or its availability at the cell surface, the exogenous enzyme may modify subsequent activity of the natural ligand.

Examples of these potentially inappropriate interactions can be found in recent work on the M6P/IGFIIR system: decreased affinity of the receptor for IGFII on binding of β-galactosidase[68]; activation of phospholipase C in renal proximal tubule cell membranes by both M6P-containing ligands and IGFII[69]; and receptor recruitment to the cell surface, with increased binding affinity, on cell exposure to M6P or IGFII.[70] The consequences of these phenomena at the cell or organ level are unknown but using an otherwise convenient receptor system for ERT may be complicated by them.

The effect on cellular metabolism of the sudden release of the products of enzymatic degradation of stored compounds may also be of concern. For example, observations on the exchange of membrane-bound gangliosides between the lysosomal compartment and the plasma membrane[71] suggest that GM3 ganglioside, resulting from catabolism of stored GM2 ganglioside by exogenous Hex A, could reach the plasma membrane in amounts sufficient to elicit effects of the type observed on exposure of cells to exogenous GM3 ganglioside, that affects receptor kinase activity.[72] Our knowledge of the physiologic role of glycosphingolipids and their breakdown products is fragmentary at best, but there is evidence that these compounds can exert profound influences on cell metabolism.[73] This makes it even more imperative to use appropriate model systems to investigate the effects of ERT at the cellular level.

Unresolved Basic Questions

The problems discussed above are mostly methodologic. They reflect a phenomenologic approach to the therapy of storage disorders and the underlying assumption of a causal relationship between storage in lysosomes and disease. With one possible exception,[74] however, both the precise cellular function(s) impaired by lysosomal storage and the mechanism(s) by which storage in a compartment teleologically regarded as terminal affects other cellular organelles or systems are unclear. In the case of neurodegenerative storage diseases, there are morphologic and biochemical observations[60,75–77] that suggest potentially fruitful lines of investigation. Although neural cells present experimental difficulties, the very complexity of their differentiation and function may be of help in dissecting the effects of impaired lysosomal degradation on cell metabolism. A clarification of these phenomena would not only be of value in terms of understanding normal and abnormal cell function but would also be of potential importance with respect to more rational approaches to therapy, based on simpler pharmacologic ways of intervention that ignore storage in lysosomes but are capable of correcting its metabolic effects.

A better understanding of pathogenesis would also help to predict whether intervention during the postnatal period might prevent or reverse the manifestation of the disease. The data from ERT trials in Gaucher disease[63] are encouraging with respect to reversibility of symptoms, but the RES cells affected by storage, in this instance, represent a highly dynamic cell system, quite different, for instance, from neural cells. In the latter, the metabolism of gangliosides is known to be developmentally regulated.[78,79] It is conceivable that impaired ganglioside catabolism might coincidentally have its greatest impact at some early critical stage of CNS development, irreversibly affecting differentiation-related cellular phenomena. Removal of accumulated ganglioside at a later stage may then be unlikely to be beneficial. Indications to this effect may be provided by ERT experiments in animal models. These complex organismic systems, however, do not permit simple approaches to the clarification of the underlying problem.

THE CHALLENGE OF CNS DELIVERY

We have left for last a problem that makes therapeutic attempts in neurodegenerative disorders difficult—that of delivering enzyme to the CNS across the blood-brain barrier. The anatomic substratum of the blood-brain barrier to proteins is the existence of tight junctions between cerebrovascular endothelial cells. To this one must add the scarcity of transendothelial vesicular transport, and pericytes and glial cell end-feet preventing diffusion of proteins that might penetrate the blood-brain barrier.[80]

Several years ago, Rapoport and his group[80] proposed a methodology that reversibly induces blood-brain barrier permeabilization, consisting of infusion of a bolus of hyperosmolar solution in the arterial blood supply to the brain. The resulting osmotic shrinkage of the endothelial cells causes temporary relaxation of tight junctions,[81] as well as extravasation of plasma and its content into the CNS spaces. This approach has been applied to the treatment of CNS tumors[82]; in this setting, it has generated controversy, mostly owing to the recognition that the normal nervous tissue surrounding the tumor is exposed to high concentrations of the chemotherapeutic agent, with consequent neurotoxic effects.[83,84] This very problem, however, suggests that, when applied to ERT, this method has the potential of effectiveness in delivering enzyme protein to the brain parenchyma. Both theoretically and in practice, however, blood-brain barrier permeabilization, even for a short time, may not be without consequences. Indeed, it can be argued that extravasation of blood-borne compounds in therapeutically significant amounts must be accompanied by extravasation of plasma proteins, enzymes, peptides, hormones, and other potentially harmful agents normally excluded from the internal milieu of the brain.[85,86] In fact, recent studies have shown neuronal damage coincident with plasma protein extravasation.[87] Finally, disruption of the blood-brain barrier may abrogate the immunologic isolation of the brain by permitting access to potential CNS antigens or entry of immunocompetent lymphocytes.[88,89] Thus, it seems important to explore alternative routes to deliver macromolecules to the brain parenchyma that are not based on forced entry by disruption of a physiologic barrier.

An alternative could be enzyme derivatization with a moiety that causes it to be transported by transcytosis across the vascular endothelium. Cationization is being explored for this purpose, with encouraging results.[90] Although it is not clear whether the amount of protein that crosses the blood-brain barrier with this method would be sufficient for effective ERT, a cationized enzyme would have the advantage of enhanced neural cell uptake, as shown by our experiments with poly-L-lysine Hex A. However, the effect of intravascular infusion of cationized protein on platelet and mast cell degranulation would be of concern. In addition, positively charged macromolecules can severely damage vascular endothelial cells causing prolonged disruption of the blood-brain barrier.[91]

Another potential alternative is to circumvent the blood-brain barrier altogether, by using the subarachnoid spaces and perivascular spaces as delivery routes.[92] Subarachnoidally injected horseradish peroxidase can spread from the subarachnoid spaces into the perivascular spaces and is detectable in the extracellular spaces.[93] Tight junctions in the pia mater and in the marginal glia are rare; even high-molecular-weight protein tracers introduced in the subarachnoid spaces can reach the extracellular space,[94,95] which represents about 20 percent of brain cortex mass.[96] Physiologically, however, this route appears to transport metabolites in the opposite direction. The interstitial fluid, secreted by the capillaries within the parenchyma, moves mainly by bulk flow from the pericellular space into the subarachnoid spaces,[97,98] and macromolecular markers microinjected into the parenchyma are transported centrifugally.[99] This outward flow may have been a contributing factor in the lack of effect of prolonged intraventricular and intrathecal infusion of Hex A in patients with Tay-Sachs disease.[64]

The direction of flow of interstitial fluid can be reversed, however, by increasing the osmolality of plasma. Plasma hyperosmolality induced by intravenous injection of mannitol (1.5 to 3 g/kg) in the dog results in bulk flow of CSF and interstitial fluid into the gray matter.[98] Similar results have been obtained in the rat by increasing plasma osmolality from 300 (baseline) to 360 mOsm/kg, with concomitant increased clearance of albumin from the CSF into brain tissue.[99]

In summary, then, there is an anatomic route between the subarachnoid and pericellular spaces that can admit proteins and, although the flow of interstitial fluid normally prevents protein penetration of extracellular space, this flow can be inverted by increasing plasma osmolarity. It is our contention that this phenomenon can be exploited for transporting proteins into the extracellular space. This novel approach is worth exploring for several reasons: (1) enzyme would be delivered directly to the CNS, necessitating smaller quantities for ERT; (2) the CNS would not be exposed to potentially harmful plasma components, as enzyme would be the only exogenous protein administered in a salt solution matching the CSF composition; (3) intrathecal injection of drugs and intravenous mannitol are both accepted clinical procedures,[100,101] and a combination of the two should be acceptable as well; and (4) the procedure may be applicable to CNS parenchymal delivery of other macromolecules (e.g., neuropeptides, growth factors, and antibodies). We have begun to test this hypothesis by developing and applying a modification of the subarachnoid perfusion technique in the cat.[96] In our experiments, a catheter is introduced in the cisterna magna, and a thin Teflon cannula is introduced in the interhemispheric space to reach the subarachnoid supracallosal cistern. Both are affixed in place. After recovery from this surgical procedure, the animal is perfused subarachnoidally with artificial CSF containing the usual dye-protein marker, Evans blue-albumin complex (0.1 to 0.2 percent) for sufficient time to permit even distribution of the marker over the brain surface. We have also visualized the process by using magnetic resonance imaging (MRI) and an appropriate contrast-enhancing agent.[102]

A bolus of mannitol (4 g/kg) is then infused intravenously, whereupon the osmolality of plasma rises to 360 to 370 mOsm/kg. The subarachnoid perfusion continues, while plasma osmolality decreases gradually, eventually to return to baseline (300 mOsm/kg). After about 2 hours from mannitol infusion, the perfusion is stopped, urinary losses of water and electrolytes are compensated by subcutaneous rehydration with Ringer's lactate, and the cat is allowed to recover. Cats recover from anesthesia in about 6 to 8 hours and exhibit normal behavior 12 to 18 hours after the procedure. At sacrifice, 18 to 24 hours later, macroscopic widespread blue staining of the cortex indicates penetration of the dye-protein complex into the parenchyma.

In animals subjected to the fully developed procedure, plasma osmolality reached 365 to 375 mOsm/kg. At sacrifice, the saline-perfused blood-free brain of the animals showed blue staining of the cortex that was more intense in the anterior one-half or two-thirds of the cerebral hemispheres. The medial aspect of the hemispheres adjacent to the tip of the inlet catheter was usually the most intensely stained area. In coronal sections, the blue staining of the tissue extended for a variable distance into the parenchyma, reaching the white matter in areas corresponding to the more intense surface stain. The basal ganglia were never stained, and the cerebellum was only occasionally stained. In animals in which plasma osmolality did not reach 365 to 370 mOsm/kg, only light staining was observed. In control animals, in which plasma osmolality remained at baseline, no staining of cortex was detectable.

In two cats subjected to the full procedure, the subarachnoid infusion solution also contained 0.5 and 1 mg/ml poly-L-lysine Hex B. In sections from cortical areas showing blue stain macroscopically, in addition to positive pial cells and subpial phagocytes, numerous cells positive for Hex B were visible by immunofluorescence microscopy at a distance of up to 1 mm from the cortical surface. Some cells were identifiable as microglia, but numerous positive cells were morphologically identifiable as astrocytes, and some were identifiable as neurons. The enzyme signal was localized in cytoplasmic granules consistent with lysosomes-endosomes. Areas devoid of blue stain were also negative for Hex B-containing cells. We present these preliminary observations not as the solution to the problem of brain delivery of proteins but as an initial set of data suggesting that this is an approach worth exploring. We are actively pursuing this line of investigation, with the aim of determining the minimum level of plasma hyperosmolality that results in parenchymal inflow, as well as possible adverse effects of this procedure. We have hopes that this method may prove useful in an ERT setting. Obviously, we will pursue this aspect with systematic enzyme infusion experiments in GM2 gangliosidosis kittens.

CONCLUSIONS

We have tried to present arguments in support of our initial statement that enzyme replacement is worthy of renewed efforts at determining its therapeutic value. The problems that discouraged clinical investigators from pursuing this approach have, for the most part, been identified. Some have been solved, and most appear soluble in the light of our increasing understanding of endocytic processes and their determinants. New technologies can provide sufficient amounts of enzyme for extended trials, with the potential of tailoring enzyme structure for optimal cell targeting. Finally, there are recent data from trials in patients indicating therapeutic effectiveness. ERT looks much more promising now than it did 15 years ago. To be sure, basic pathogenetic questions are still unanswered, but these can be addressed, given the availability of new biochemical and cell biology insights, improvements in cell isolation and culture methods, and well-characterized animal models. Neurodegenerative diseases are still a major challenge, but techniques are available that enable us to carry out animal experiments aimed at exploring both methodologic and conceptual aspects of ERT directed to the nervous system.

The results in Gaucher disease notwithstanding, we are still dealing with the therapeutic *potential* of ERT, with no clearly predictable outcome. The question then arises: Is it worth pursuing this line of investigation, which will at best provide a form of treatment, rather than bone marrow transplantation (see Ch. 12) or gene replacement (see Chs. 15 through 17), which could provide a permanent cure? Indeed, bone marrow transplantation and gene replacement therapy have the potential of being much more effective than ERT, but ERT still seems to present some advantages over the other two approaches. The concepts underlying ERT are accepted and time tested. By contrast, those underlying bone marrow transplantation and gene replacement therapy are still a matter of debate. Within the lysosomal storage disorders, the applicability of ERT appears wider than that of bone marrow transplantation, for which there is little evidence of neurologic effectiveness, or that of gene replacement therapy, which is similarly limited because of problems of CNS delivery and neural cell transfection (see Ch. 17). Furthermore, the methodologies of ERT are relatively uncomplicated, those of bone marrow transplantation are more complex and clinically demanding, and those of gene replacement therapy still in the process of being developed. Thus, it can be argued that, on balance, the medical-pharmacologic, somewhat simplistic, ERT approach has an edge over the more daring surgical approach of bone marrow transplantation and the scientifically more elegant and potentially definitive approach of gene replacement therapy. These pragmatic arguments should not be construed as advocating the former to the exclusion of the latter two. In fact, experimental or clinical applications of ERT can provide data useful for the development or further refinement of gene replacement therapy and bone marrow transplantation. With the exception of replacement of RES cells when these are the site of pathology, and of stable gene transfection in most cells in a given organ, respectively, the therapeutic effects of bone marrow transplantation and gene replacement therapy would depend on excretion of enzyme from normal or gene-normalized cells and its uptake by affected cells. At the cellular level, these approaches share with ERT the problems of enzyme recognition and uptake, assessment of biochemical and functional effectiveness, and possible complications discussed above. Thus, the results of biochemical, cell, or animal experiments and clinical trials of ERT will most probably contribute to a more rational and effective application of bone marrow transplantation and gene replacement therapy.

ACKNOWLEDGMENTS

The work in our laboratory was supported by grants NS 13667, NS 21404, and RR 05924 from the National Institutes of Health. Experiments on MRI localization of subarachnoidally perfused proteins were done in collaboration with Dr. P. Cahill and Dr. T. Vullo, of the MRI unit, Department of Radiology, New York Hospital, New York. We thank Dr. M.E. Slodki, of the USDA, Peoria, Illinois, for providing o-phosphomannans, and Dr. P. Livingston, of Memorial Sloan-Kettering Cancer Center, New York, for providing anti-GM2 ganglioside antibody. We also thank Mrs. Marion Feeney for typing the manuscript.

REFERENCES

1. Tager JM, Hamers MN, Schram AW, et al: An appraisal of human trials in enzyme replacement therapy of genetic diseases. p. 343. In Desnick RJ (ed): Enzyme Therapy in Genetic Diseases. Vol. 2. Alan R Liss, New York, 1980
2. Desnick RJ, Grabowski GA: Advances in the treatment of inherited metabolic diseases. Adv Hum Genet 11:281, 1981
3. Rattazzi MC, Lanse SB, McCullough RA, et al: Towards enzyme therapy in GM2 gangliosidosis: Organ disposition and induced central nervous system uptake of human beta-hexosaminidase in the cat. p. 179. In Desnick RJ (ed): Enzyme Therapy in Genetic Disease. Vol. 2. Alan R Liss, New York, 1980
4. Rattazzi MC, Appel AM, Baker HJ, et al: Toward enzyme replacement in GM2 gangliosidosis: Inhibition of hepatic uptake and induction of CNS uptake of human beta-hexosaminidase in the cat. p. 405. In Callahan JW, Lowden JA (eds): Lysosomes and Lysosomal Storage Diseases. Raven Press, New York, 1981
5. Rattazzi MC, Appel AM, Baker HJ: Enzyme replacement in feline GM2 gangliosidosis: Catabolic effects of human beta-hexosaminidase A. p. 213. In Desnick RJ, Patterson DF, Scarpelli DG (eds): Animal Models of Inherited Metabolic Diseases. Alan R Liss, New York, 1982
6. Rattazzi MC: Beta-hexosaminidase isozymes and replacement therapy in GM2 gangliosidosis. Isozymes Curr Top Biol Med Res 11:65, 1983
7. Rattazzi MC, Dobrenis K, Joseph A, et al: Modified β-d-N-acetylhexosaminidase isozymes for enzyme replacement in GM2 gangliosidosis. Isozymes Curr Top Biol Med Res 16:49, 1987
8. Cork LC, Munnell JF, Lorenz MD, et al: GM2 ganglioside storage disease in cats with beta-hexosaminidase deficiency. Science 196:1014, 1977
9. Pagano RE, Huang L: Interaction of phospholipid vesicles with cultured mammalian cells. II. Studies of mechanisms. J Cell Biol 67:49, 1975
10. Gitman AG, Graesman A, Loyter A: Targeting of loaded Sendai virus envelopes by covalently attached insulin molecules to virus receptor-depleted cells: Fusion-mediated microinjection of ricin A and simian virus 40 DNA. Proc Natl Acad Sci USA 82:7309, 1985
11. Simpson LL: Targeting drugs and toxins to the brain: The magic bullets. Int Rev Neurobiol 30:123, 1988
12. Nakai M, Hase T, Matsubara H: Precise determination of the mitochondrial import signal contained in a 70 KDa protein of yeast mitochondrial outer membrane. J Biochem (Tokyo) 105:513, 1989

13. Holtzman E: Endocytosis and heterophagy. p. 25. In Holtzman E (ed): Lysosomes. Plenum Press, New York, 1989
14. DeDuve C: From cytases to lysosomes. Fed Proc 23:1045, 1964
15. Fratantoni JC, Hall CW, Neufeld EF: Hurler and Hunter syndrome: Mutual correction of the defect in cultured fibroblasts. Science 162:570, 1968
16. O'Brien JS, Miller AL, Loverde AW, et al: Sanfilippo disease type B: Enzyme replacement and metabolic correction in cultured fibroblasts. Science 181:7563, 1973
17. Kornfeld R, Kornfeld S: Assembly of asparagine-linked oligosaccharides. Annu Rev Biochem 54:631, 1985
18. Stahl S, Schlesinger PH, Rodman JS, et al: Recognition of lysosomal glycosidases in vivo inhibited by modified glycoproteins. Nature 264:86, 1976
19. Schlesinger PH, Rodman JS, Doebber TW, et al: The role of extra hepatic tissue in the receptor-mediated clearance of glycoproteins terminated by mannose or N-acetylglucosamine. Biochem J 192:596, 1980
20. Furbish FC, Steer CS, Krett NL, et al: Uptake and distribution of placental glucocerebrosidase by rat hepatic cells and effects of sequential deglycosylation. Biochim Biophys Acta 673:425, 1981
21. Powell PP, Kyle JW, Miller RD, et al: Rat liver beta-glucuronidase. cDNA cloning, sequence comparisons and expression of a chimeric protein in COS cells. Biochem J 250:547, 1988
22. Desnick RJ, Dean KJ, Grabowski GA, et al: Enzyme therapy in Fabry's disease: Differential in vivo plasma clearance and metabolic effectiveness of plasma and splenic alpha-galactosidase-A isozymes. Proc Natl Acad Sci USA 76:5326, 1979
23. Collins JM: Pharmacologic rationale for regional drug delivery. J Clin Oncol 2:498, 1984
24. Silverstein SC, Steinman RM, Cohn ZA: Endocytosis. Annu Rev Biochem 46:669, 1977
25. Goldstein JL, Brown MS, Anderson RGW, et al: Receptor-mediated endocytosis. Annu Rev Cell Biol 1:1, 1985
26. Miller DK, Griffiths E, Lenard J, et al: Cell killing by lysosomotropic detergents. J Cell Biol 97:1841, 1983
27. Swanson JA, Silverstein SC: Pinocytic flow through macrophages. p. 15. In Pernis B, Silverstein SC, Vogel HJ (eds): Processing and Presentation of Antigens. Academic Press, San Diego, 1988
28. Schen WC, Ryser HJP: Conjugation of poly-L-lysine to albumin and horseradish peroxidase: A novel method of enhancing the cellular uptake of proteins. Proc Natl Acad Sci USA 75:1872, 1978
29. Ryser JHP, Drummond J, Schen WC: The cellular uptake of horseradish peroxidase and poly (lysine) conjugate by cultured fibroblasts are qualitatively similar despite a 900-fold difference in rate. J Cell Physiol 113:167, 1982
30. Rattazzi MC, O'Neil DC, Vladutiu GD: Towards enzyme replacement in GM2 gangliosidosis: Properties of poly-L-lysine-conjugated beta-hexosaminidase. Pediatr Res 17:217A, 1983
31. Rattazzi MC, Dobrenis K: Endocytosis of native and modified human beta-hexosaminidase by cultured rat and cat central nervous system cells. Soc Neurosci Abs 12:37, 1986
32. Ennis M, Pearce FL, Weston PM: Some studies on the release of histamine from mast cells stimulated with polylysine. Br J Pharmacol 70:329, 1980
33. Wellhoner NH: Tetanus neurotoxin. Rev Physiol Biochem Pharmacol 93:1, 1982

34. Morris NP, Consiglio E, Kohn LD, et al: Interaction of fragment B and C of tetanus toxin with neural and thyroid membranes and with gangliosides. J Biol Chem 255:6071, 1980
35. Bizzini B, Akert K, Glicksman M, et al: Preparation of conjugates using two tetanus toxin derived fragments. Toxicon 18:561, 1980
36. Makoff AJ, Oxer MD, Romanos MA, et al: Expression of tetanus toxin fragment C in *E. coli*: High level expression by removing rare codons. Nucleic Acid Res 17:10191, 1989
37. Natoli EJ, Livingston PO, Pukel CS, et al: A murine monoclonal antibody detecting N-acetyl- and N-glycolyl-GM2: Characterization of cell surface reactivity. Cancer Res 46:4116, 1986
38. Dobrenis K, Joseph A, Rattazzi MC: Neuronal lysosomal enzyme replacement using fragment C of tatanus toxin. (submitted)
39. Kornfeld S, Mellman I: The biogenesis of lysosomes. Annu Rev Cell Biol 5:483, 1989
40. Sahagian GG: The mannose-6-phosphate receptor: Function biosynthesis and translocation. Biol Cell 51:207, 1984
41. Glaser J, Rosen KJ, Brut FE, et al: Multiple isoelectric and recognition forms of human beta glucuronidase activity. Arch Biochem Biophys 166:536, 1975
42. O'Dowd BF, Cumming DA, Gravel RA, et al: Oligosaccharide structure and amino acid sequence of the major glycopeptides of mature human beta-hexosaminidase. Biochemistry 27:5216, 1988
43. Vladutiu GD, Rattazzi MC: I-cell disease: Desialylation of β-hexosaminidase and its effect on uptake by fibroblasts. Biochim Biophys Acta 539:31, 1978
44. Slodki ME, Ward RM, Boundy JA: Concanavalin A as a probe of phosphomannan molecular structure. Biochim Biophys Acta 304:449, 1973
45. Leichtner AM, Krieger M: Addition of mannose-6-phosphate containing oligosaccharide alters cellular processing of low density lipoprotein by parental and LDL-receptor-defective Chinese hamster ovary cells. J Cell Sci 68:183, 1984
46. Jeffrey AM, Zopf DA, Ginsburg V: Affinity chromatography of carbohydrate-specific immunoglobulins: Coupling of oligosaccharides to Sepharose. Biochem Biophys Res Commun 62:608, 1975
47. Rattazzi MC, Calabro A, Dobrenis K, et al: Enzyme replacement in GM2 gangliosidosis: Pentamannosyl-6-phosphate-beta hexosaminidase A for delivery to cells via the mannose-6-phosphate receptor. Am J Hum Genet 45:A10, 1989
48. Rattazzi MC, Calabro A, Dobrenis K, et al: Enzyme replacement: Pentamannosyl-6-phosphate beta hexosaminidase for delivery to cells via the mannose-6-phosphate receptor. Pediatr Res 27:136A, 1990
49. Lesniak MA, Hill JM, Kiess W, et al: Receptors for insulin-like growth factors I and II: Autoradiographic localization in rat brain and comparison to receptors for insulin. Endocrinology 123:2089, 1988
50. Nielsen FC, Gammeltoft S: Mannose-6-phosphate stimulates proliferation of neuronal precursor cells. FEBS Lett 262:142, 1990
51. Ocrant I, Valentino KL, Eng LF, et al: Structural and immunohistochemical characterization of insulin-like growth factor I and II receptors in the murine central nervous system. Endocrinology 123:1023, 1988
52. Sklar MM, Kiess W, Thomas CL et al: Developmental expression of the tissue insulin-like growth factor II/mannose-6-phosphate receptor in the rat. J Biol Chem 264:16733, 1989
53. Alexandrides T, Moses A, Smith RJ: Developmental expression of receptors for

insulin, insulin-like growth factor I (IGFI) and IGFII in rat skeletal muscle. Endocrinology 124:1064, 1989
54. Willemsen R, Wisselaar HA, van der Ploeg AT: Plasmalemmal vesicles are involved in transendothelial transport of albumin, lysosomal enzymes and mannose-6-phosphate receptor fragments in capillary endothelium. Eur J Cell Biol 51:235, 1990
55. Van der Ploeg AT, Loonen MC, Bolhuis PA, et al: Receptor-mediated uptake of acid alpha-glucosidase corrects lysosomal glycogen storage in cultured skeletal muscle. Pediatr Res 24:90, 1988
56. Hasholt L, Wandall A, Sorensen SA: Enzyme replacement in Fabry endothelial cells and fibroblasts: Uptake experiments and electron microscopical studies. Clin Genet 33:360, 1988
57. Brooks SE, Amsterdam D, Hoffman LM, et al: Cytology, growth characteristics and cellular alterations following SV40-induced transformation of human foetal brain cells derived from GM2 gangliosidosis and control. J Cell Sci 38:211, 1979
58. Dobrenis K, Becker J, Rattazzi MC: Cerebral cortex cultures from postnatal GM2 gangliosidosis cats. Soc Neurosci Abs 15:933, 1989
59. Desnick RJ, Patterson DF, Scarpelli DG: Animal Models of Inherited Metabolic Diseases. Alan R Liss, New York, 1982
60. Walkley SV: Pathobiology of neuronal storage diseases. Int Rev Neurobiol 29:191, 1988
61. Desnick RJ, Dean KJ, Grabowski GA, et al: Enzyme therapy XVII: Metabolic and immunologic evaluation of alpha galactosidase A replacement in Fabry's disease. p. 393. In Desnick RJ (ed): Enzyme Therapy in Genetic Diseases. Vol. 2. Alan R Liss, New York, 1989
62. Shull RM, Breider MA, Costantopoulos GC: Long-term neurologic effects of bone marrow transplantation in canine lysosomal storage disease. Pediatr Res 24:347, 1988
63. Barton NW, Brady RO, Dambrosia JM, et al: Replacement therapy for inherited enzyme deficiency—macrophage-targeted glucocerebrosidase for Gaucher's disease. N Engl J Med 324:1464, 1991
64. von Specht BV, Geiger B, Arnon R, et al: Enzyme replacement in Tay-Sachs disease. Neurology 29:848, 1979
65. Boyer SH, Siggers DC, Kreuger LJ: Caveat to protein replacement therapy for genetic disease: Immunologic implications of accurate molecular diagnosis. Lancet 2:654, 1973
66. Blum JS, Diaz R, Diment S, et al: Proteolytic processing in endosomal vesicles. Cold Spring Harbor Symp Quant Biol 54:287, 1989
67. Bennett VJ, Chang PL: Suppression of immunological response against a novel gene product delivered by implants of genetically modified fibroblasts. Mol Biol Med 8:471, 1990
68. Kiess W, Thomas CL, Sklar MM, et al: Beta-galactosidase decreases the binding affinity of the insulin-like growth factor II/mannose-6-phosphate receptor for insulin-like growth factor II. Eur J Biochem 190:71, 1990
69. Rogers SA, Purchio AF, Hammerman MR: Mannose-6-phosphate containing peptides activate phospholipase C in proximal tubular basolateral membranes from canine kidney. J Biol Chem 265:9722, 1990
70. Braulke T, Tippmer S, Chao HJ, et al: Regulation of mannose-6-phosphate/insulin-like growth factor II receptor distribution by activators and inhibitors of protein kinase C. Eur J Biochem 189:609, 1990
71. Sonderfeld S, Conzelmann E, Schwarzmann G, et al: Incorporation and

metabolism of ganglioside GM2 in skin fibroblasts from normal and GM2 gangliosidosis subjects. Eur J Biochem 149:247, 1985
72. Hanai N, Nores GA, MacLeod C, et al: Ganglioside mediated modulation of cell growth. Specific effect of GM3 and lyso-GM3 in tyrosine phosphorylation of the epidermal growth factor receptor. J Biol Chem 263:10915, 1988
73. Hannun YA, Bell RM: Functions of sphingolipids and sphingolipid breakdown products in cellular regulation. Science 243:500, 1989
74. Igisu H, Suzuki K: Progressive accumulation of toxic metabolite in a genetic leukodystrophy. Science 224:753, 1984
75. Walkley SV, Baker HJ, Rattazzi MC: Initiation and growth of ectopic neurites and meganeurites during postnatal development in ganglioside storage disease. Dev Brain Res 51:167, 1990
76. Koenig ML, Jope RS, Baker HJ, et al: Reduced Ca^{2+} flux in synaptosomes from cats with GM1 gangliosidosis. Brain Res 424:169, 1987
77. Walkley SU: Vulnerability of GABA-ergic neurons to neuroaxonal dystrophy in neuronal storage disorders. J Neuropathol Exp Neurol 48:350, 1989
78. Vanier MT, Holm M, Ohman R, et al: Developmental profiles of gangliosides in human and rat brain. J Neurochem 18:581, 1971
79. Yu RC, Macala LJ, Taki T, et al: Developmental changes in ganglioside composition and synthesis in embryonic rat brain. J Neurochem 50:1825, 1988
80. Rapoport SI: Blood-Brain Barrier in Physiology and Medicine. Raven Press, New York, 1976
81. Rapoport SI, Robinson PJ: Tight-junctional modifications at the basis of osmotic opening of the blood-brain barrier. Ann NY Acad Sci 481:250, 1986
82. Neuwelt EA, Dahlborg SA: Blood-brain barrier disruption in the treatment of brain tumors. Clinical implications. p. 195. In Neuwelt E (ed): The Clinical Impact of the Blood-Brain Barrier and Its Manipulations. Vol. 2. Plenum Press, New York, 1989
83. Fishman RA: Is there a therapeutic role for osmotic breaching of the blood-brain barrier? (Editorial.) Ann Neurol 22:298, 1987
84. Groothuis DR, Warnke PC, Molnar P, et al: Effect of hyperosmotic blood-brain barrier disruption on transcapillary transport in canine brain tumors. J Neurosurg 72:441, 1990
85. Maier-Hauff K, Baethmann AJ, Lange M, et al: The kallikrein-kinin system as mediator of vasogenic brain edema. Part 2. Studies on kinin formation in focal and perifocal brain tissue. J Neurosurg 61:97, 1984
86. Jezova D, Johansson BB, Oprsalova Z, et al: Changes in blood-brain barrier function modify the neuroendocrine response to circulating substances. Neuroendocrinology 49:428, 1989
87. Salahuddin TS, Johansson BB, Kalimo H, et al: Structural changes in the rat brain after carotid infusions of hyperosmolar solutions. Acta Neuropathol (Berl) 77:5, 1988
88. Pollack IF, Lund RD: The blood-brain barrier protects foreign antigens in the brain from immune attack. Exp Neurol 108:114, 1990
89. Kajivara K, Ito H, Fukumoto T: Lymphocyte infiltration into normal rat brain following hyperosmotic blood-brain barrier opening. J Neuroimmunol 27:233, 1990
90. Triguero D, Buciak JB, Yang J, et al: Blood-brain barrier transport of cationized immunoglobulin G: enhanced delivery compared to native protein. Proc Natl Acad Sci USA 86:4761, 1989

91. Nagy Z, Peters H, Huttner I: Charge-related alterations of the cerebral endothelium. Lab Invest 49:662, 1983
92. Jones EG: On the mode of entry of blood vessels into the cerebral cortex. J Anat 106:507, 1970
93. Rennels ML, Gregory TF, Blaumarnis TF, et al: Evidence for a paravascular fluid circulation in the mammalian central nervous system, provided by the rapid distribution of tracer protein throughout the brain from the subarachnoid space. Brain Res 326:47, 1985
94. Brightman MW: The distribution within the brain of ferritin injected into the cerebrospinal fluid compartments. Am J Anat 117:193, 1965
95. Brightman MW, Reese TS: Junctions between intimately apposed cell membranes in the vertebrate brain. J Cell Biol 40: 618, 1969
96. Levin Va, Fenstermacher JO, Patlak CS: Sucrose and inulin space measurements of cerebral cortex in four mammalian species. Am J Physiol 219:1528, 1970
97. Wood JM: Physiology, pharmacology and dynamics of cerebrospinal fluid. p. 1. In Wood JM (ed): Neurobiology of Cerebrospinal Fluid. Plenum Press, New York, 1980
98. Rosenberg GA, Kyner WT, Estrada E: Bulk flow of brain interstitial fluid under normal and hyperosmolar conditions. Am J Physiol 238:F42, 1980
99. Cserr HF: Role of secretion and bulk flow of brain interstitial fluid in brain volume regulation. Ann NY Acad Sci 529:9, 1988
100. Ommaya AK: Implantable devices for chronic access and drug delivery to the central nervous system. Cancer Drug Deliv 1:169, 1984
101. Ropper AH, Rockoff HA: Treatment of intracranial hypertension. p. 23. In Ropper AH, Kennedy SF (eds): Neurological and Neurosurgical Intensive Care. Aspen Publications, Rockville, MD, 1988
102. Ogan MD, Schmiedl V, Moseley ME, et al: Albumin labeled with Gd-DTPA, an intravascular contrast enhancing agent for magnetic resonance blood pool imaging: Preparation and characterization. Invest Radiol 22:665, 1987

9

ENZYME REPLACEMENT THERAPY FOR TYPE I GAUCHER DISEASE

Roscoe O. Brady · Norman W. Barton

INTRODUCTION

Gaucher disease is an inherited metabolic disorder caused by deficient activity of glucocerebroside-β-glucosidase in the organs and tissues of patients with this condition.[1,2] A number of mutations in the gene coding for glucocerebrosidase that result in diminished catalytic effectiveness have been identified.[3-6] The pathophysiologic consequences of diminished enzymatic activity include hepatosplenomegaly, anemia, thrombocytopenia, leukopenia, abnormalities of liver function, and skeletal damage. In some patients, pulmonary and renal function may also be compromised. If there is no central nervous system (CNS) involvement, patients are categorized as having type I Gaucher disease. This is the most frequently encountered lipid storage disorder, and it is the most prevalent genetic disease in persons of Ashkenazic Jewish ancestry. There are two less frequently encountered phenotypes. One of them is designated type II (acute neuronopathic) Gaucher disease. This condition is found in affected infants, who rarely survive beyond the second year of life. These patients have extensive brain damage, with loss of neurons and infiltration of phagocytic cells. The other comparatively rare phenotype is type III (chronic neuronopathic) Gaucher disease. These patients usually begin to manifest neurologic symptoms during childhood. The condition is characterized by horizontal supranuclear gaze paresis, progressive cognitive decline, and frequent myoclonic seizures. Horizontal supranuclear gaze paresis is the sole neurologic abnormality in a subset of patients with type III Gaucher disease.[7] These latter patients have severe systemic disease with marked hepatic and pulmonary dysfunction. Since all three forms of Gaucher disease are characterized by reduction of glucocerebrosidase activity, assays based on determinations of the activity of this enzyme in extracts of leukocytes or cultured skin fibroblasts are widely used to diagnose homozygotes,[8] to identify heterozygotes,[9] and, when desired, to monitor pregnancies at risk of this disorder, either through amniocentesis in mid-second trimester[10] or by chorionic villus biopsy in the first trimester.[11] Genetic counseling employing these techniques is in wide use throughout the world.

THERAPEUTIC CONSIDERATIONS

Soon after the enzymatic defect was identified in Gaucher disease, several therapeutic strategies were suggested for these patients. Chief among them were organ and bone marrow transplantation, enzyme supplementation, and gene replacement.[12] Organ transplantation trials with kidney[13] and spleen[14] grafts were totally unsatisfactory. Bone marrow transplantation has been successful in a few cases,[15] but considerable morbidity and mortality are associated with this procedure. Gene replacement is under active investigation using in vitro[16,17] and in vivo systems,[18–20] but no clinical trials have been carried out to date. Extensive study in lower animals and primates will be required before this approach can be considered for clinical trial in patients with Gaucher disease. These limitations have led to the conclusion that enzyme replacement is currently the therapeutic procedure of choice. The development of this technology and its successful clinical application are reviewed here.

EARLY INVESTIGATIONS

Sandhoff Variant of Tay-Sachs Disease

In order to determine whether an exogenous enzyme could reduce the quantity of accumulated material in a patient with a metabolic storage disorder, it was first necessary to obtain a sufficiently pure preparation of the enzyme so that its metabolic potential could be investigated without harm to the recipient. A further desideratum was that it be obtained from a human source to reduce the risk of sensitization of the patient to the exogenous protein. The first enzyme of this type that was available for clinical investigation was hexosaminidase A. When a preparation of this enzyme, isolated from human urine, was injected intravenously into a patient with the Sandhoff variant of Tay-Sachs disease, none of the enzyme reached the CNS. This was not an unexpected observation, since the blood-brain barrier prevents passage of macromolecules from the circulation to the brain. However, a 43 percent reduction in the quantity of a neutral sphingoglycolipid called globoside was observed in the plasma of the recipient.[21] The concentration of globoside is increased in the circulation of these patients because its catabolism is impaired. Reduction of the plasma globoside concentration in this patient was the first indication that an exogenous enzyme could affect the level of an accumulating lipid in a heritable metabolic disorder.

Fabry Disease

Further experiments examining the effect of exogenous enzyme in a sphingolipid storage disorder were performed in patients with Fabry disease. The requisite ceramidetrihexosidase (α-galactosidase A) was isolated from human placental tissue.[22] When two males with this disorder were given small amounts of the enzyme intravenously, there was a rapid reduction in the quantity of the accumulating lipid (ceramidetrihexoside) in the circulation of each recipient.[23] Thereafter, the plasma ceramidetrihexoside level gradually returned to the preinfusion value within 72 hours of enzyme administration. Although this effect has been confirmed with α-galactosidase A derived from human plasma and splenic tissue,[24] no additional studies on enzyme replacement in patients with Fabry disease have been reported. Several apparent obstacles contribute to this situation. The first is the question of selective delivery of enzyme to cells damaged by accumulating ceramidetrihexoside. In principle, one would like to target the exogenous enzyme to endothelial cells of the vascular system and glomerular and tubular cells in the kidney, where pathologic quantities of ceramidetrihexoside accumulate. Second, in view of the large amounts of enzyme required for successful treatment of Gaucher disease, one might anticipate a similar requirement of α-galactosidase A in patients with Fabry disease. It is hoped that the availability of clones of the α-galactosidase A gene[25,26] will permit production of sufficient quantities of enzyme by recombinant DNA technology so that replacement trials in this disorder can be extended.

Initial Studies on Gaucher Disease

Work began at the National Institutes of Health during the late 1960s on the isolation of glucocerebrosidase. Once again, we chose human placental tissue as the source material, to minimize the chance of sensitization of the recipients to exogenous enzyme. We encountered a great deal of difficulty in purifying this enzyme, as it turned out to be an extraordinarily hydrophobic protein. Eventually, a procedure was developed in which detergent was included in all the fractionation steps. This technique provided a small amount of glucocerebrosidase of sufficient purity that it was considered safe for intravenous administration to patients with Gaucher disease. In order to assess the effect of the enzyme, we performed needle biopsies of the liver before and 24 hours after enzyme infusion. Two patients received this preparation in the initial study.[27] The first recipient was a 15½-year-old splenectomized boy with type III Gaucher disease. He received a total of 25 IU of glucocerebrosidase. The amount of glucocerebroside in his liver decreased 26 percent after enzyme infusion. The second patient was a 51-year-old splenectomized woman with type I Gaucher disease. She received 55 IU of glucocerebrosidase intravenously. Again, a 26 percent reduction in liver glucocerebroside was observed. Her initial hepatic glucocerebroside level was 2.3 times that of the first recipient.

Of considerable interest was an apparent redistribution of glucocerebroside among erythrocytes, plasma, and tissue depots after enzyme infusion. In both patients, the amount of glucocerebroside associated with red blood cells gradually fell over a 72-hour period from a threefold elevation to the normal range. The reduction of erythrocyte-associated glucocerebroside persisted in both recipients over a lengthy period with a gradual return to preinfusion levels after a single injection of enzyme.[28] It was presumed that this slow reappearance of glucocerebroside in the blood was a reflection of its comparatively slow reaccumulation in major storage organs such as the liver. However, no direct information is available concerning potential equilibria between the storage compartments. Previous calculations suggested that only about 5 percent of the glucocerebroside turned over each day actually accumulates in the typical patient with type I Gaucher disease.[29] The protracted reduction of erythrocyte-associated glucocerebroside seemed to augur well for the success of enzyme replacement therapy.

This encouraging concept was, however, short-lived. After months of work, we obtained 155 IU of glucocerebrosidase by Pentchev's procedure and infused it into a 22-year-old woman with type I Gaucher disease. We observed only an 8 percent reduction of hepatic glucocerebroside in this patient.[30] She had accumulated 25 times more glucocerebroside in her liver than the first patient and 11 times more than the second recipient. There was no reduction in erythrocyte-associated glucocerebroside in the third recipient. This finding suggested that insufficient material had been removed from storage organs such as the liver, thus preventing a detectable reduction of erythrocyte glucocerebroside as a result of redistribution from the circulation.

Despite the limited effect of exogenous enzyme in the third recipient, the amount of glucocerebroside cleared from the liver of each of these patients was a function of the quantity of enzyme infused. In each case, 25 µmol of hepatic glucocerebroside was eliminated for each international unit of glucocerebrosidase administered.[30] On the basis of these observations, we estimated that more than 1900 IU of enzyme would theoretically be required to effect the complete removal of hepatic glucocerebroside in the third patient. Since it had taken nearly 1 year to obtain the 155 IU administered to the third patient, it was mandatory that a more efficient enzyme isolation procedure be developed. An entirely new method was devised that took advantage of the exceptionally strong hydrophobicity of glucocerebrosidase.[31] Two hydrophobic column chromatography steps were used in the final phases of the new purification procedure. The method provided highly purified glucocerebrosidase in excellent yield. However, for the purification procedure to be effective, lipid associated with the placental proteins had to be removed by butanol extraction before hydrophobic chromatography. When the extraction step was not included, glucocerebrosidase did not adhere to the hydrophobic columns.

Clinical investigations were then undertaken with hydrophobic column-purified glucocerebrosidase. When 267 IU of this enzyme preparation was injected into a 5-year-old boy with type I Gaucher disease, there was only a 12 percent reduction of liver glucocerebroside. Next an 8-year-old boy was given 668 IU of glucocerebrosidase. In the latter case, a 15 percent reduction of hepatic glucocerebroside occurred. Seven additional patients with type I Gaucher disease were given 833 IU of glucocerebrosidase. Three of them showed a reduction of hepatic glucocerebroside within a range of 16 to 80 percent. However, in four of the recipients, there was no detectable clearance of glucocerebroside from the liver (Table 9-1). On the basis of investigations with glucocerebrosidase purified by our original procedure, these results were intensely disquieting. At this point, several critical questions had to be answered. The first was whether glucocerebrosidase prepared by the new hydrophobic chromatographic method would catalyze the catabolism of glucocerebroside in situ. We knew from enzymatic assays with labeled glucocerebroside that the enzyme was catalytically effective in vitro. Using this preparation, we found that in vitro addition of the enzyme to liver biopsy specimens obtained from patients with Gaucher disease resulted in the catabolism of all the glucocerebroside in the specimens.[32] This was a key observation, since it was considered possible that we had removed lipid activators such as phosphatidylserine[33] during the butanol extraction step of the purification procedure. Demonstration of the enzymatic degradation of glucocerebroside in liver biopsy specimens in vitro left us with the possibility that removal of lipids during purification might have had other consequences. Since targeting of liposomes[34,35] to macrophages of the reticuloendothelial system could be increased by the addition of phosphatidylserine, delipidation may have diminished delivery of the exogenous enzyme to storage cells. In point of fact, the reduced effectiveness of exogenous glucocerebrosidase in patients with Gaucher disease hinges on its lack of delivery to the storage cells. Resolution of this limitation is described in the next section.

Table 9-1. Inconsistent Effect of Unmodified Glucocerebrosidase on Hepatic Glucocerebroside Levels in Patients With Type I Gaucher Disease[a]

Patient	Age	Sex	Hepatic Glucocerebroside (mg of glucocerebroside/g wet wt)		Change (%)
			Preinfusion	Postinfusion	
6	30	F	1.5	0.3	−80
7	38	M	3.1	3.1	0
8	56	F	3.2	3.2	0
9	33	M	14.	14.	0
10	32	F	15.	15.	0
11	35	F	25.	21.	−16
12	37	F	26.	13.	−50

[a] Each patient received 833 IU of hydrophobic column-purified placental enzyme. The normal value of glucocerebroside in human liver is approximately 0.04 mg/g.

MAKING EXOGENOUS GLUCOCEREBROSIDASE EFFECTIVE

While these experiments were in progress, several important pieces of information became available. The first was the demonstration that lysosomal enzymes are glycoproteins.[36] The second was that pulmonary macrophages were shown to have a lectin on their surface that specifically interacts with the

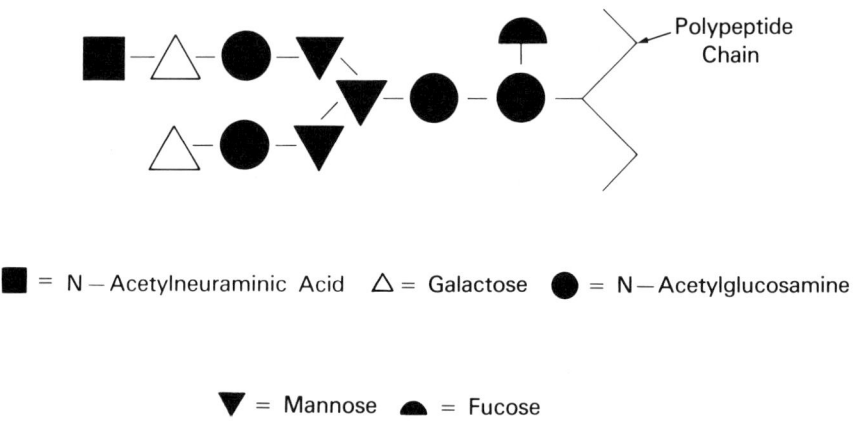

Fig. 9-1. Major carbohydrate units of native human placental glucocerebrosidase.

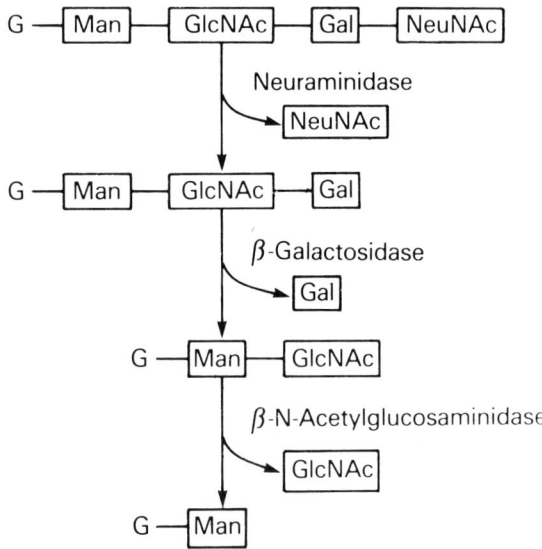

Fig. 9-2. Schematic representation of procedure for sequential enzymatic deglycosylation of human placental glucocerebrosidase. G, glucocerebrosidase containing N-asparagine linked N-acetylglucosamine (±)fucose-N-acetylglucosamine-mannose; Man, mannose; GlcNAc, N-acetylglucosamine; Gal, galactose; NeuNAc, N-acetylneuraminic acid.

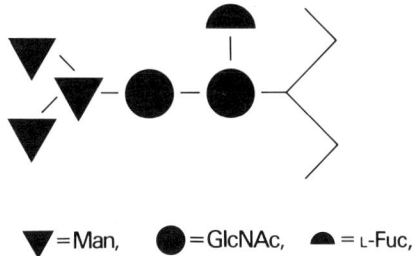

▼ = Man, ● = GlcNAc, ▲ = L-Fuc,

Fig. 9-3. Carbohydrate units of mannose-terminated glucocerebrosidase.

sugar mannose.[37] Glucocerebrosidase was known to be a lysosomal enzyme.[38] The purified human placental enzyme was demonstrated to be a glycoprotein[39] containing approximately 7 percent carbohydrate. Most of the oligosaccharide chains of the placental enzyme are of the complex type (Fig. 9-1). Most of these chains are terminated with N-acetylneuraminic acid. However, some of the branches have terminal galactose residues. The work of Ashwell and Morell established that glycoproteins that terminate with galactose are largely endocytosed by hepatocytes. We observed that intravenously injected native placental glucocerebrosidase was largely cleared from the blood by this route in experimental animals.[40] Glucocerebroside does not accumulate in hepatocytes, presumably because they are able to excrete at least some of this material through the bile.[41] We began to examine ways to increase the number of mannose-terminated chains on placental glucocerebrosidase to direct it preferentially to macrophages (Kupffer cells) in the liver. The first attempt to target the enzyme in this fashion consisted of covalently linking linear pentamannosyl side chains to the protein. This procedure did not increase the delivery of glucocerebrosidase to Kupffer cells in experimental animals. The second approach was to link trimannosyldilysyl residues covalently to the enzyme. This resulted in an approximately fourfold increase in delivery of the enzyme to nonparenchymal cells in the liver.[42] In the belief that even greater delivery to Kupffer cells would be required, we elected to produce mannose-terminated glucocerebrosidase by sequential enzymatic cleavage of the three external carbohydrate residues from the enzyme, that is, N-acetylneuraminic acid, galactose, and N-acetylglucosamine[40,43–48] (Fig. 9-2). The delivery of mannose-terminated glucocerebrosidase (Fig. 9-3) to Kupffer cells in experimental animals was greatly increased when compared with unmodified enzyme (Table 9-2).

EARLY TRIAL WITH MANNOSE-TERMINATED GLUCOCEREBROSIDASE

Emboldened by the prospect of enhanced delivery of intravenously injected glucocerebrosidase to the storage cells, we infused 167 IU into eight patients with type I Gaucher disease on a weekly basis, over a 6-month period. There

Table 9-2. Uptake of Hydrophobic Column-Purified Native and Mannose-Terminated Glucocerebrosidase by Hepatocytes and Nonparenchymal Cells in Rats

Enzyme Form (Dose)	Cell Type	Enzymatic Activity		Enrichment
		Control	Infused	-fold
Unmodified (15 IU)	Hepatocytes	104^a	256^a	2.5
	Nonparenchymal cells	7^a	48^a	6.9
Mannose-terminated (50 IU)	Hepatocytes	117^b	117^b	0
	Nonparenchymal cells	48^b	2312^b	48.0
Mannose-terminated (50 IU)	Hepatocytes	79^b	126^b	1.6
	Nonparenchymal cells	30^b	2119^b	70.0

^{a} Nanomoles of glucocerebroside hydrolyzed per 10^6 cells per hour.
^{b} Nanomoles of glucocerebroside hydrolyzed per mg of protein per hour.
(Data from Furbish et al.[46])

was no evidence of clinical improvement in seven of the recipients. However, one of the recipients was a 4-year-old boy with severe anemia and visceromegaly (Fig. 9-4). Fourteen infusions of enzyme were administered to this patient over a period of 26 weeks. He subsequently received weekly injections of enzyme for the next 79 weeks. The amount of enzyme initially administered was equivalent to 12.3 IU/kg/wk.

Between weeks 26 and 45, there was a clear rise in the patient's hemoglobin from the pretreatment level of 6.8 g/dl to a plateau value of 10.2 g/dl[49] (Fig. 9-5). During this time, the platelet count rose from 30,000/mm^3 to 54,000/mm^3. Phagocytic activity of the spleen decreased. Increased mineralization of the long bones along with decreased endosteal scalloping and the appearance of a

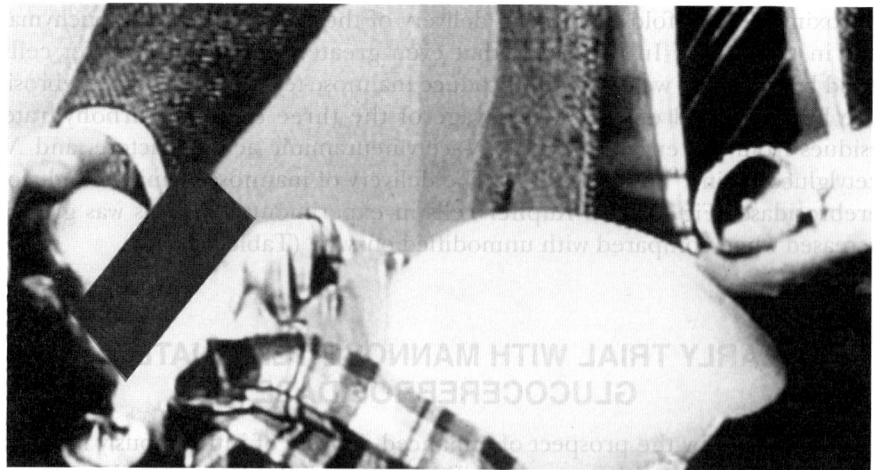

Fig. 9-4. Hepatosplenomegaly in a 4-year-old patient with type I Gaucher disease before enzyme replacement.

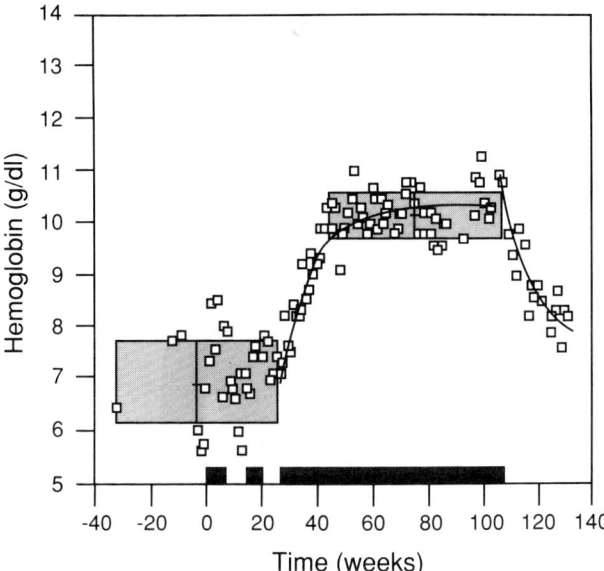

Fig. 9-5. Effect of weekly intravenous infusions of mannose-terminated glucocerebrosidase on the hemoglobin concentration in a child with type I Gaucher disease. Solid black bars along the abscissa represent periods of enzyme administration. Shaded areas represent the mean pretreatment and steady-state values observed during enzyme supplementation (±1 SD). Logistic growth functions (illustrated by the curved lines) closely approximate the changes in hemoglobin concentration observed during enzyme treatment and withdrawal.

finer trabecular pattern first became apparent after approximately 18 months of treatment. Enzyme infusions were suspended over a 25-week period. During this interval, the patient's hemoglobin and platelet count returned to their pretreatment values. Enzyme infusions were resumed at age 6 years, 7 months and have been continued. At the time of this writing, his hemoglobin was within the normal range, and the platelet counts varied between 85,000 and 90,000/mm^3. His liver and spleen size are greatly reduced (Fig. 9-6). The possible production of antimannose-terminated glucocerebrosidase antibody has been carefully monitored by a highly sensitive enzyme-linked immunosorbent assay (ELISA) procedure. No antibody has been detected, nor has there been any clinical evidence to suggest sensitization to the exogenous enzyme.

DOSE-RESPONSE STUDY

Since the single clinically responsive patient in the preceding trial was the smallest subject in the study, he received the greatest amount of enzyme for body weight of all the recipients. Accordingly, a critical investigation was conducted to determine the threshold for detectable responses to mannose-terminated glucocerebrosidase. A single infusion of enzyme over the dosage range of 1 to 200 IU/kg of body weight was administered to 23 patients with type I Gaucher disease. The quantity of glucocerebroside was evaluated in liver biop-

162   Treatment of Genetic Diseases

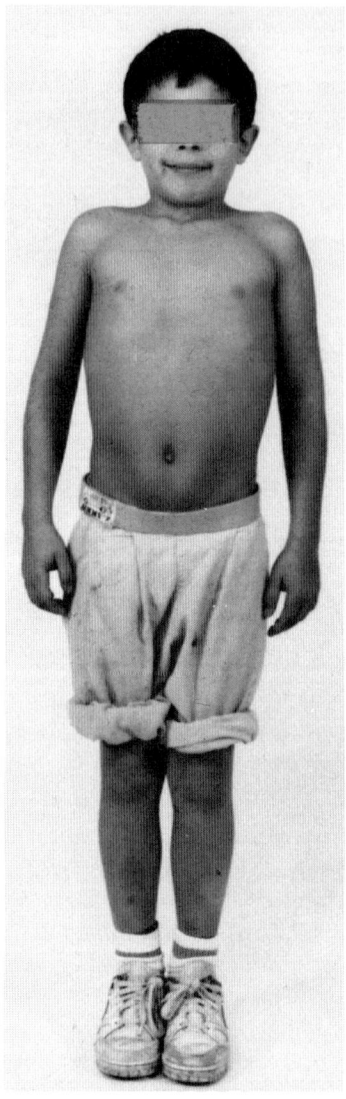

Fig. 9-6. Abdominal contour of the patient (DOB, 11/15/79) illustrated in Figure 9-4 at age 10, after administration of mannose-terminated glucocerebrosidase over a 3½-year period.

sy specimens obtained before enzyme infusion and compared with the amount of glucocerebroside in similar specimens obtained 44 hours after the enzyme infusion. Other parameters monitored before and after enzyme infusion included (1) histologic and electron microscopic analyses of biopsy tissue and (2) histochemical staining for acid phosphatase and lysozyme in the biopsy specimens. Consistent improvement in these parameters was observed when patients received 30 or more IU of mannose-terminated glucocerebrosidase per kilogram of body weight.

CLINICAL EFFICACY OF MACROPHAGE-TARGETED GLUCOCEREBROSIDASE

On the basis of the beneficial responses observed in the child who received 12.3 IU/kg/wk of mannose-terminated glucocerebrosidase and the consistency of the responses observed in the dose-response study at enzyme dosages of 30 or more IU/kg, we designed a formal efficacy trial to measure clinical responses to mannose-terminated glucocerebrosidase. Twelve nonsplenectomized patients with type I Gaucher disease received 60 IU/kg of mannose-terminated human placental glucocerebrosidase every other week over a 6-month period. The enzyme was intravenously infused over 2 hours. Hematologic changes, blood chemistries, organ size, and skeletal manifestations were carefully monitored. These parameters were also followed over a 6-month period in 12 other nonsplenectomized patients who did not receive enzyme. In addition, serum samples from patients in both groups were analyzed frequently with a highly sensitive ELISA procedure to monitor for the development of antibody to mannose-terminated glucocerebrosidase.

Hematologic Effects of Enzyme Replacment

A significant increase in hemoglobin concentration became apparent in all patients who received the enzyme between the third and fourth months.[50] These values continued to rise throughout the trial period, plateauing in the normal range in most recipients. In addition, there was an increase in platelet count, but this effect was not as pronounced as the rise in hemoglobin concentration. These hematologic parameters either did not change or progressively worsened in the patients who did not receive enzyme.

Effect of Enzyme Replacement on Hepatosplenomegaly

All patients in the efficacy trial had enlarged livers and spleens. The volume of these organs was quantitated by magnetic resonance imaging (MRI). There were significant reductions in splenic size in all recipients of the enzyme, ranging from a minimum of 14 percent to a maximum of 75 percent. The degree of hepatomegaly also decreased, although this effect was not so dramatic as the change in splenic size. Reduction in liver size ranged from 8 to 22 percent. By contrast, liver and spleen size in the noninfusion group remained stable or became slightly larger over the course of the study.

Effect of Enzyme Replacement on the Skeleton

The effect of mannose-terminated glucocerebrosidase on the extent of skeletal involvement in patients with Gaucher disease was evaluated by several procedures. In addition to conventional radiographs, three novel techniques were used to assess skeletal changes. These tests included (1) quantitative MRI for measurement of the fat content of the bone marrow, (2) dual-energy quantitative computed tomography (CT) of the skeleton for measurement of fat content and bone density, and (3) quantitative triple xenon scanning to determine the amount of stored glucocerebroside in the bone without necessitating bone biopsy. In addition, bone marrow biopsies were obtained for several of the recipients.

Although skeletal changes characteristically occurred later than other parameters, such as improvement in hemoglobin, the above procedures consistently showed evidence of skeletal improvement. The most dramatic finding was clearing of the Gaucher cell infiltrate as seen in histologic sections of the bone marrow. It is apparent from these data that administration of mannose-terminated glucocerebrosidase can induce substantial improvement in the skeleton in patients with Gaucher disease.

Effect of Enzyme Replacement on the Quality of Life

In addition to the biochemical and clinical responses to enzyme replacement, we have consistently observed major improvement in the life-style of patients with Gaucher disease. Several of the severely affected children have been able to return to school and resume normal social activities that they were unable to realize before receiving the enzyme. Adults who were unable to perform their customary work and other activities have resumed their employment and are no longer dependent on their families or other forms of support. Thus, their aspirations have been fulfilled as a consequence of receiving exogenous glucocerebrosidase.

CONCLUSIONS

Many man-years of intensive labor have been invested in bringing enzyme replacement therapy to fruition for Gaucher disease. Early in the course of these investigations, the nonphysiologic nature of injecting an exogenous enzyme into the bloodstream of a patient with the anticipation that it would be physiologically effective was clearly recognized.[51] Each organ in the body produces its own complement of catabolic enzymes whose activity is a function of the respective metabolic load.[52] That an intravenous injection of such an agent would produce a detectable response was completely uncertain. However, it was known that some exogenous enzymes were preferentially taken up by cells of the reticuloendothelial system.[53] It was therefore suggested that this situation constituted a rational basis to undertake enzyme replacement.[12] What was not apparent at the time these studies were begun was that lysosomal enzymes are glycoproteins[36] and that macrophages have a high-affinity lectin on their surface for mannose-terminated glycoproteins.[37] Practical application of this information required years of trial and dedicated effort to bring it to a successful conclusion. It is apparent that we are entering a new era of medical advancement. It seems likely that the information obtained in the clinical efficacy trial in Gaucher disease will be rapidly extrapolated to other heritable metabolic disorders, such as Fabry disease and type B Niemann-Pick disease. Because of the difficulties in delivering enzymes across the blood-brain barrier,[54] it remains to be determined how soon this strategy can be extended to conditions in which the CNS is involved. Furthermore, it seems likely that enzyme replacement will prove useful in other metabolic disorders, such as the mucopolysaccharidoses and the glycogenoses, using similar tissue-specific targeting strategies that have proved so effective in Gaucher disease. We look forward with confidence to extensions of this approach to the treatment of an increasing number of human metabolic problems in the coming years.

REFERENCES

1. Brady RO, Kanfer JN, Shapiro D: The metabolism of glucocerebrosides. II. Evidence of an enzymatic deficiency in Gaucher's disease. Biochem Biophys Res Commun 18:221, 1965
2. Brady RO, Kanfer JN, Bradley RM, et al: Demonstration of a deficiency of glucocerebroside-cleaving enzyme in Gaucher's disease. J Clin Invest 45:1112, 1966
3. Tsugi S, Choudary PV, Martin BM, et al: A mutation in the human glucocerebrosidase gene in neuronopathic Gaucher's disease. N Engl J Med 316:570, 1987
4. Tsugi S, Martin BM, Barranger JA, et al: Genetic heterogeneity in type 1 Gaucher disease: Multiple genotypes in Ashkenazic and non-Ashkenazic individuals. Proc Natl Acad Sci USA 85:2349, 1988
5. Graves PN, Grabowsi GA, Eisner R, et al: Gaucher disease type 1: Cloning and characterization of a cDNA encoding acid β-glucosidase from an Ashkenazic Jewish patient. DNA 7:521, 1988
6. Firon N, Eyan N, Kolodny EH, et al: Genotype assignment in Gaucher disease by selective amplification of the active glucocerebrosidase gene. Am J Hum Genet 46:527, 1990
7. Yu K-T, Merrick HFW, Verderese C, et al: Horizontal supranuclear gaze palsy: A marker for severe systemic involvement in Type III Gaucher's disease. Neurology 40(suppl 1):S357, 1990
8. Kampine JP, Brady RO, Kanfer JN, et al: The diagnosis of Gaucher's disease and Niemann-Pick disease using small samples of venous blood. Science 155:86, 1967
9. Brady RO, Johnson WG, Uhlendorf BW: Identification of heterozygous carriers of lipid storage diseases. Am J Med 51:423, 1971
10. Schneider EL, Ellis WG, Brady RO, et al: Infantile (Type II) Gaucher's disease: In utero diagnosis and fetal pathology. J Pediatr 81:1134, 1972
11. Evans MI, Moore C, Kolodny EH, et al: Lysosomal enzymes in chorionic villi, cultured amniocytes, and cultured skin fibroblasts. Clin Chem Acta 157:109, 1986
12. Brady RO: Sphingolipidoses. N Engl J Med 275:312, 1966
13. Desnick SJ, Desnick RJ, Brady RO, et al: Renal transplantation in type 2 Gaucher's disease. Birth Defects 9:109, 1973
14. Groth CG, Berstrom K, Collste L, et al: Immunologic and plasma protein studies in a splenic homograft recipient. Clin Exp Immunol 10:359, 1972
15. Starer F, Sargent JD, Hobbs JR: Regression of the radiological changes of Gaucher's disease following bone marrow transplantation. Br J Radiol 60:1189, 1987

16. Sorge J, Kuhl W, West C, et al: Complete correction of the enzymatic defect of type 1 Gaucher disease fibroblasts by retroviral-mediated gene transfer. Proc Natl Acad Sci USA 84:906, 1987
17. Nolta JA, Sender LS, Barranger JA, et al: Expression of human glucocerebrosidase in murine long-term bone marrow cultures after retroviral vector-mediated transfer. Blood 75:787, 1990
18. Correll PH, Fink JK, Brady RO, et al: Production of human glucocerebrosidase in mice following retroviral gene transfer into multipotential hematopoietic progenitor cells. Proc Natl Acad Sci USA 86:8912, 1989
19. Fink JK, Correll PH, Perry LK, et al: Correction of glucocerebrosidase deficiency following retroviral-mediated gene transfer into hematopoietic progenitor cells from patients with Gaucher disease. Proc Natl Acad Sci USA 87:2334, 1990
20. Correll PH, Kew Y, Perry LK, et al: Expression of human glucocerebrosidase in long-term reconstituted mice following retroviral-mediated gene transfer into hematopoietic stem cells. Hum Gene Ther 1:277, 1990
21. Johnson WG, Desnick RJ, Long DM, et al: Intravenous injection of purified hexosaminidase A into a patient with Tay-Sachs disease. Birth Defects 9:120, 1973
22. Johnson WG, Brady RO: Ceramidetrihexosidase from human placenta. Methods Enzymol 28:849, 1972
23. Brady RO, Tallman JF, Johnson WG, et al: Replacement therapy for inherited enzyme deficiency: Use of purified ceramidetrihexosidaase in Fabry's disease. N Engl J Med 289:9, 1973
24. Desnick RJ, Dean KJ, Grabowski GA, et al: Enzyme therapy in Fabry's disease: Differential in vivo plasma clearance and metabolic effectiveness of plasma and splenic α-galactosidase A isozymes. Proc Natl Acad Sci USA 76:5326, 1979
25. Bishop DF, Colhoun DH, Bernstein HS, et al: Human α-galactosidase A: Nucleotide sequence of a cDNA clone encoding the mature enzyme. Proc Natl Acad Sci USA 83:4859, 1986
26. Tsuji S, Martin BM, Kaslow DC, et al: Signal sequence and DNA mediated expression of human lysosomal α-galactosidase A. Eur J Biochem 165:275, 1987
27. Brady RO, Pentchev PG, Gal AE, et al: Replacement therapy for inherited enzyme deficiency: Use of purified glucocerebrosidase in Gaucher's disease. N Engl J Med 291:989, 1974
28. Pentchev PG, Brady RO, Gal AE, et al: Replacement therapy for inherited enzyme deficiency: Sustained clearance of accumulated glucocerebroside in Gaucher's disease following infusion of purified glucocerebrosidase. J Mol Med 1:73, 1975
29. Kattlove HE, Williams JC, Gaynor E, et al: Gaucher cells in chronic myelogenous leukemia: An acquired abnormality. Blood 33:379, 1969
30. Brady RO, Pentchev PG, Gal AE, et al: Enzyme replacement therapy for the sphingolipidoses. p. 523. In Volk BD, Schneck L (eds): Current Trends in Sphingolipidoses and Allied Disorders. Plenum Press, New York, 1976
31. Furbish FS, Blair HE, Shiloach J, et al: Enzyme replacement therapy in Gaucher's disease: Large-scale purification of glucocerebrosidase suitable for human administration. Proc Natl Acad Sci USA 74:3560, 1977
32. Pentchev PG, Barranger JA, Gal AE, et al: Incorporation of exogenous enzymes into lysosomes. A theoretical and practical means for correcting lysosomal blockage. p. 150. In Walborg EF Jr (ed): Glycoproteins and Glycolipids in Disease Processes. ACS Symposium Series 80. American, Washington DC, 1978
33. Dale GL, Villacorte D, Beutler E: Solubilization of glucocerebrosidase from human placenta and demonstration of a phospholipid requirement for it catalytic

activity. Biochem Biophys Res Commun 71:1048, 1976
34. Fidler IJ, Sone S, Fogler WE, et al: Eradication of spontaneous metastases and activation of alveolar macrophages by intravenous injection of liposomes containing muramyl dipeptide. Proc Natl Acad Sci USA 78:1680, 1981
35. Allen TM, Williamson P, Schlegel RA: Phosphatidylserine as a determinant of reticuloendothelial recognition of liposome models of the erythrocyte surface. Proc Natl Acad Sci USA 85:8067, 1988
36. Goldstone A, Koenig H: Lysosomal hydrolases as glycoproteins. Life Sci 9:1341, 1970
37. Stahl PD, Rodman JS, Miller MJ, et al: Evidence for receptor-mediated binding of glycoproteins, glycoconjugatges, and lysosomal glycosidases by alveolar macrophages. Proc Natl Acad Sci USA 75:1399, 1978
38. Weinreb NJ, Brady RO, Tappel AL: The lysosomal localization of sphingolipid hydrolases. Biochim Biophys Acta 159:141, 1968
39. Takasaki S, Murray GJ, Furbish FS, et al: Structure of N-asparagine-linked oligosaccharide units of human placental β-glucocerebrosidase. J Biol Chem 259:10112, 1984
40. Furbish FS, Steer CJ, Barranger JA, et al: The uptake of native and desialylated glucocerebrosidase by rat hepatocytes and Kupffer cells. Biochem Biophys Res Commun 81:1047, 1978
41. Tokoro T, Gal AE, Gallo LL, et al: Studies of the pathogenesis of Gaucher's disease: Tissue distribution and biliary excretion of 14C-l-glucosylceramide in rats. J Lipid Res 28:968, 1987
42. Doebber TW, Wu MS, Bugianesi RL, et al: Enhanced macrophage uptake of synthetically glycosylated human placental β-glucocerebrosidase. J Biol Chem 257:2193, 1982
43. Steer CJ, Furbish FS, Barranger JA, et al: The uptake of agalacto-glucocerebrosidase by rat hepatocytes and Kupffer cells. Fed Eur Biochem Soc Lett 91:202, 1978
44. Furbish FS, Krett NL, Barranger JA, et al: Fucose plays a role in the clearance and uptake of glucocerebrosidase by liver cells. Biochem Biophys Res Commun 95:1768, 1980
45. Brady RO, Furbish FS: Enzyme replacement therapy: Specific targeting of exogenous enzymes to storage cells. p. 587. In Martonosi AN (ed): Membranes and Transport. Vol. 2. Plenum Press, New York, 1982
46. Furbish FS, Oliver KL, Zirzow GC, et al: Interaction of human placental glucocerebrosidase with hepatic lectins. p. 219. In Barranger JA, Brady RO (eds): The Molecular Basis of Lysosomal Storage Disorders. Academic Press, San Diego, 1984
47. Furbish FS, Steer CJ, Krett NL, et al: Uptake and distribution of placental glucocerebrosidase in rat hepatic cells and effects of sequential deglycosylation. Biochim Biophys Acta 673:425, 1981
48. Murray GJ: Lectin-specific targeting of lysosomal enzymes to reticuloendothelial cells. Methods Enzymol 149:25, 1987
49. Barton NW, Furbish FS, Murray GJ, et al: Therapeutic response to intravenous infusions of glucocerebrosidase in a patient with Gaucher disease. Proc Natl Acad Sci USA 87:1913, 1990
50. Barton NW, Brady RO, Doppelt SH, et al: Clinical effectiveness of enzyme replacement in Gaucher's disease. Clin Res 38:457A, 1990
51. Brady RO: Enzyme replacement in the sphingolipidoses. p. 461. In Barranger JA, Brady RO (eds): The Molecular Basis of Lysosomal Storage Disorders. Academic Press, San Diego, 1984

52. Kampine JP, Kanfer JN, Gal AE, et al: Response of sphingolipid hydrolases in spleen and liver to increased erythrocytorrhexis. Biochim Biophys Acta 137:135, 1967
53. Wakim KG, Fleisher GA: Fate of enzymes in body fluids—Experimental study. IV. Relationship of reticulo-endothelial system to activities and disappearance rates of various enzymes. J Lab Clin Med 61:107, 1963
54. Barranger JA, Rapoport SI, Fredericks WR, et al: Modification of the blood-brain barrier: Increased concentration and fate of enzymes entering the brain. Proc Natl Acad Sci USA 76:481, 1979

10

PEG-ENZYME REPLACEMENT THERAPY FOR ADENOSINE DEAMINASE DEFICIENCY

Michael S. Hershfield · Sara Chaffee

INTRODUCTION

The goal of using purified enzymes to treat inborn errors of metabolism was actively pursued during the 1970s and largely ignored during the 1980s. This reflects major difficulties encountered in attempts to treat lysosomal storage diseases (see Ch. 8) and the appearance of recombinant DNA techniques that have made *gene therapy* a feasible objective. During the past 5 years, severe combined immunodeficiency disease (SCID) caused by deficiency of adenosine deaminase (ADA) has been at the focal point of both therapeutic strategies. SCID has features that make it an attractive candidate for initial attempts at stem cell gene replacement; it is also well suited to an innovative technology that uses covalent attachment of polyethylene glycol (PEG) to enhance the therapeutic characteristics of enzymes. Moreover, like other forms of SCID, ADA deficiency can be treated, and potentially cured, by bone marrow transplantation. This chapter summarizes the clinical experience to date with PEG-modified bovine ADA, which in March 1990 became the first enzyme product to be approved by the U.S. Food and Drug Administration (FDA) as parenteral replacement therapy for an inherited disease. The discussion focuses on the role of PEG-ADA among alternative treatments for ADA deficiency and speculates on other genetic disorders that might be treatable with PEG-modified proteins.

In patients with ADA deficiency, estimated to be 20 to 25 percent of cases of SCID,[1-3] immune development is "poisoned" by exposure to adenosine (Ado) and 2'-deoxyadenosine (dAdo) (reviewed by Kredich and Hershfield[3]). dAdo, the more toxic nucleoside, is normally not salvaged but, in the absence of ADA, T lymphocytes with relatively high levels of dAdo-phosphorylating activity and slow rates of dATP turnover can undergo expansion of the dATP pool. dATP accumulation inhibits ribonucleotide reductase in dividing cells, blocking DNA replication, and can also cause ATP depletion by more than one mechanism. In addition to dATP-mediated toxicity, Ado inhibits and dAdo inactivates S-adenosylhomocysteine (AdoHcy) hydrolase, which can lead to the buildup of AdoHcy, a potent inhibitor of transmethylation. Cumulatively, these actions affect both immature and mature T cells and B cells, resulting in SCID. The primary stage of development affected by ADA deficiency is not known, but the very high level of ADA activity in immature thymocytes may be required to protect cells destined to reach maturity from large amounts of dAdo generated by the dissolution of nuclear DNA of "negatively selected" thymocytes that undergo massive programmed death in the fetal thymus. Besides direct toxicity, ADA deficiency may alter the differentiation of lymphohematopoietic stem cells. High levels of dATP might alter the product of terminal deoxynucleotidyl transferase (TdT) or might otherwise interfere with DNA rearrangements critical to lymphopoiesis.

Most patients are susceptible from infancy to recurrent opportunistic infections of the skin and respiratory and digestive tracts with fungal, viral, and bacterial organisms. Later onset of serious infections at 6 to 18 months of age is not uncommon and, in a few patients, illness has begun after 2 years of age.[2,3] Fatal graft-versus-host disease (GVHD) can result from transfusion of unirradiated blood and from disseminated infection from vaccination with live viruses. At diagnosis, T-cell depletion is profound, and blood mononuclear cells do not respond to mitogens or antigens in vitro. Immunoglobulin levels are low for age but may be near-normal; except in the mildest cases, responses to immunization are defective. Erythrocytes invariably show markedly elevated dATP and less than 10 percent of normal AdoHcyase activity; ATP is often decreased.[3,4] These clinical, immunologic, and biochemical hallmarks of untreated ADA deficiency are easily monitored and are useful in judging the effectiveness of therapy.

THERAPY FOR ADA DEFICIENCY

Supportive care consisting of varying degrees of isolation, combined with vigorous antibiotic and immunoglobulin replacement therapy, is aimed at controlling or slowing the consequences of infection. Without therapy directed specifically at restoring immune function, however, either by transplantation of marrow stem cells or by replacing the missing enzyme, most children with ADA deficiency will either die of overwhelming sepsis or suffer progressive and debilitating consequences of recurrent infection.

Bone Marrow Transplantation

Transplantation of bone marrow from an HLA-identical sibling donor can reconstitute normal immune function, can be done without prior immunosuppression, and carries a low risk of GVHD; unfortunately, matched donors are available for only a few patients. Several approaches have been developed to permit the use of marrow from an HLA-haploidentical (parental) donor, depleted of mature T cells to reduce the risk of GVHD. After haploidentical transplantation, engraftment may be delayed for many months and has been characterized by chimerism for circulating lymphocytes, with T cells of donor and B cells of host origin; defective humoral immunity may persist, requiring long-term therapy with IVIg. ADA-deficient children have not responded as well as SCID patients with normal ADA. Pretransplant cytoablation of ADA-deficient patients, used to increase the rate of engraftment, may also increase morbidity.[4-11] With engraftment, biochemical abnormalities may be completely corrected but are more often partially corrected. Erythrocyte dATP and plasma Ado levels have ranged from normal to higher than those found pretransplantation and AdoHcyase activity in red blood cells remains less than 20 percent of normal[12-14] (also unpublished results of this laboratory).

Partial Exchange Transfusion

Replacement therapy for ADA deficiency, unlike that for lysosomal storage disorders, may not require that exogenous enzyme enter affected cells to reverse pathogenic metabolic abnormalities. First, intracellular ADA substrates can equilibrate rapidly with plasma by facilitated diffusion mediated by the nonconcentrative nucleoside transport system of the cell membrane.[15] Second, since DNA degradation is a limited process in dividing cells, most of the dAdo encountered by lymphocytes may arise externally, released into the circulation by macrophages involved in degrading nuclei of negatively selected thymocytes and red blood cell progenitors.[16,17] Thus, if sufficient ADA activity could be maintained in plasma, either as the enzyme itself or as a population of ADA-containing cells, catabolism of circulating dAdo could prevent toxic metabolites from accumulating in enzyme-deficient cells (Fig. 10-1).

Because of the rapid clearance of purified ADA, the approach pursued from 1975 to about 1980 was repeated partial exchange transfusion of red blood cells from normal donors, irradiated to prevent GVHD.[18] With aggressive transfusion, blood ADA activity can temporarily approach normal levels, but in practice this is not achieved. Metabolic abnormalities were not completely corrected and, while immune function and clinical status improved in some children,[18-21] the response was inadequate or unsustained in most.[22-27] However, experience with red blood cell transfusions suggested that enzyme therapy might be more effective if higher circulating levels of ADA could be maintained and the dangers of iron overload and transmission of viral infections avoided.

PEG-ADA

Polyethylene glycols are flexible, hydrophilic, inert polymers of the unit, $-(O-CH_2-CH_2)_n-$ used extensively as additives to enhance solubility or to stabilize topical and parenteral drug preparations. Among the latter are some forms of immunoglobulin and factor VIII. Preclinical animal studies showed

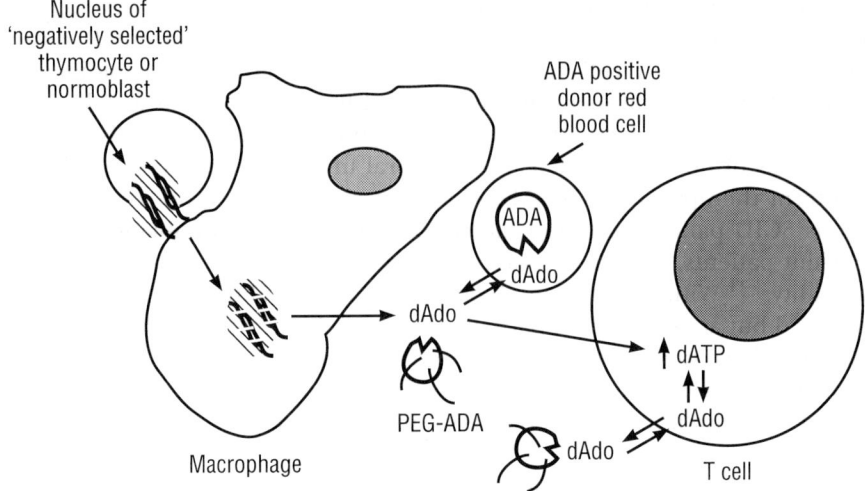

Fig. 10-1. Basis for effectiveness of circulating adenosine deaminase (ADA) activity in correcting metabolic abnormalities in ADA deficiency. The bold, unidirectional arrows indicate the route by which dAdo, produced in and secreted from ADA-deficient macrophages, is taken up by ADA-deficient T cells, resulting in expansion of the dATP pool (and other toxic effects not shown). PEG-ADA or ADA-containing transfused red blood cells can interrupt this chain of events by deaminating circulating dAdo.

that PEGs are quantitatively excreted by the kidney. Except for the shortest forms ($n = 2$ to 10), which share some of the properties of ethylene glycol, and some very large forms, PEGs are nontoxic. Abuchowski and Davis and colleagues[28–30] first investigated the effects of covalently attaching PEG, after chemical activation of a terminal OH, to primary amino groups of bovine serum albumin (BSA) and a series of enzymes, including bovine ADA.[31] Modification with PEG was often achieved without major loss of enzymatic activity. It inhibited proteolysis and the binding of antibodies in vitro; in animals, it enhanced circulating life and reduced immunogenicity compared with the unmodified protein. The latter effect was attributed to interference by PEG strands with epitope recognition and uptake by antigen-processing cells. Other studies showed that PEG-modified proteins could be tolerigenic.[32,33]

Attachment of monomethoxy-PEG of average molecular weight 5000 to approximately 60 percent of the lysine residues of bovine ADA diminished its immunogenicity and increased its circulating half-life from less than 30 minutes to about 30 hours in mice.[31] No toxicity was seen in mice after injection of 10,000 U/kg of PEG-ADA. In 1984, Enzon, Inc. obtained permission for clinical investigation of PEG-ADA as an Orphan Drug for treatment of ADA deficiency. In April 1986, in collaboration with pediatric immunologists at Duke University, we initiated a trial of PEG-ADA (modified with disuccinyl-monomethoxy-PEG$_{5000}$[34]) in two patients.[35] Measurement of the circulating life of PEG-ADA and the effects on levels of red blood cell dATP and AdoHcy hydrolase activity established a weekly intramuscular injection schedule.

Metabolic abnormalities were corrected, followed by improvement in immune function and clinical status; no toxicity was encountered.[35] Similar results have been reported in two other children.[36,37]

Since 1986, this laboratory has led an international trial of PEG-ADA, monitoring plasma ADA levels, metabolic response, and development of antibody to PEG-ADA in all patients. Collaborating investigators at eight institutions in the United States and at three centers in Europe are following immune function and clinical status in the 16 patients treated through August 1990. None of these patients has had an HLA-matched donor and, aside from the first patient, none was judged by their physicians and parents to be a candidate for haploidentical bone marrow transplantation after consideration of the risks and potential benefits of that procedure. A report of the experience with PEG-ADA in 14 patients who have been followed for more than 3 months is in preparation and is summarized here.

All 14 patients have had combined immunodeficiency and have spanned the spectrum of severity associated with ADA deficiency. Six of the first eight patients were older children ranging in age from 3.8 to 15 years. They included the first patient, who had failed two attempts at haploidentical bone marrow transplantation and had not responded to transfusion therapy[10,35]; three patients who had been receiving monthly transfusion therapy for 7 to 9 years[19,35] (the only long-term survivors of this approach); a child with late-onset of serious infections after 3 years of age[36]; and another who, despite recurrent pneumonia and upper respiratory infections from infancy, was not diagnosed until 8 years of age.[38] The remaining patients were started on therapy at 6 weeks to 2 years of age. Six patients have now been treated for 3 to more than 4 years, another six patients for 14 to 24 months, and two patients for 4 and 6 months.

PEG-ADA, which has been given the generic name of pegademase (Adagen, Enzon, Inc.), is currently supplied at 250 U/ml. Eleven patients are currently receiving weekly intramuscular injections of 20 or 30 U/kg; three have been receiving 25 to 30 U/kg twice weekly for up to 7 months. Peak plasma ADA activity occurs 24 to 72 hours after injection and decays with a half-life of approximately 3 to more than 6 days. Preinjection plasma ADA activity is maintained at one to more than four times the normal level of total blood ADA activity. The PEG-enzyme is not taken up in significant amounts by hematopoietic cells. By 4 to 8 weeks of treatment, the dAdo nucleotide (dAXP) content of erythrocytes has declined from 20 to 70 percent of total adenine nucleotides to less than 1 percent. Over the past 6 months, red blood cell dAXP levels of all patients have averaged 6.4 ± 2 nmol/ml packed cells, compared with pretreatment values of 200 to 1200 in untransfused, and 50 to 70 nmol/ml packed cells in transfused patients. AdoHcy hydrolase activity has also normalized.

At current doses, circulating mature T lymphocytes that respond to mitogens in vitro have usually appeared within a few weeks to 3 months, but this has taken more than 6 months. As with haploidentical transplantation, the immune system then matures over a longer time. In several patients, lymphocyte counts have risen for 6 to 12 months and then fallen to a lower plateau during the second year of treatment. Patients can roughly be placed in three groups. About 40 percent have recovered substantial, if not completely normal, immune function, as judged by in vitro lymphocyte responses to mitogens, antigens, and allogeneic cells and by specific antibody responses to immunization. A similar percentage have developed normal or near-normal proliferative responses to mitogens and significant in vitro responses to some antigens and allogeneic cells and have shown specific antibody responses after immunization. However, these patients remain more T lymphopenic than the first group and have diminished responses to some antigens. Two patients have shown slower and more limited recovery of T cells and lymphocyte proliferative responses after 37 and 19 months of treatment.

All patients have shown clinical improvement, including those with limited recovery of immune function and persisting T lymphopenia. During the first year of treatment, chronic diarrhea, dermal and gastrointestinal candidiasis, and recurrent pneumonia have resolved roughly in relationship to the time course of return of immune function. Beyond the first year, life-threatening opportunistic infections have ceased to occur, and upper respiratory infections have been similar in severity, duration, and frequency to those of healthy children of the same age. Thus far, one patient has contracted chickenpox and another mumps during their second and third years of treatment; both patients had uneventful courses. In patients with preexisting chronic lung disease, chest x-rays and oxygen requirements have stabilized or improved. Growth has been maintained and in some cases has improved markedly. Restrictions on social interaction have been minimized, and older patients are attending school.

The persisting lymphopenia and diminished in vitro response of circulating lymphocytes to antigens in some patients warrant concern. In view of the clinical benefit observed in these patients, however, and the degree to which their course has changed from the natural history of SCID, no guidelines for adjunctive therapy have been established. Some patients have continued to receive prophylaxis for *Pneumocystis carinii* and IVIg; in others with similar or greater T lymphopenia, all prophylaxis has been discontinued without clinical deterioration. One view is that residual abnormalities have the same prognostic significance as similar degrees of lymphopenia or diminished in vitro responses in patients with partial combined immunodeficiency, acquired immunodeficiency syndrome (AIDS), or Wiskott-Aldrich syndrome. This analogy fails to take into account differences in etiology of immunodeficiency and the fact that pathogenic mechanisms continue to operate and suppress immune function in the latter conditions, whereas PEG-ADA substantially eliminates the consequences of ADA deficiency that cause immunodeficiency. Although the repertoire of circulating lymphocytes in PEG-ADA-treated patients may be restricted compared with normal persons, this may not reflect the status of lymphoid tissues. The clinical response thus far suggests more reason for optimism than for pessimism regarding long-term protective immune function. Only continued monitoring of every patient receiving PEG-ADA will establish whether protection will be sustained.

The basis for variation in recovery of lymphocyte counts and immune function is unclear. It is possible that all patients have the same capacity for immune reconstitution, but in some patients, extracellular PEG-ADA fails to achieve adequate correction of intracellular metabolic abnormalities. This is difficult to reconcile with the uniform correction of erythrocyte biochemical abnormalities observed among patients. Studies of metabolites in lymphoid cells, however, although more difficult to perform routinely, might be more informative. Alternatively, despite adequate enzyme replacement, the capacity for complete reconstitution may be variable. It is important to note that ADA-deficient fetuses[39] and newborns[40,41] are lymphopenic, and their red blood cells show high levels of dATP. Thus, by the time replacement therapy is begun, some irreversible damage to stem cells or the thymic microenvironment may have occurred in some patients, possibly related to the degree of residual ADA activity. Viral infection before treatment may also influence the extent of recovery of immune function.

Immune Response to PEG-ADA

Since PEG-ADA is the first PEG-modified protein to be used therapeutically, the possibility that efficacy might be limited during chronic use by antibody to the bovine enzyme has been an important concern. We have developed an enzyme-linked immunosorbent assay (ELISA) with which to monitor antibodies directed against ADA in the first 12 patients.[42] Since PEG interferes with binding to plastic, unmodified bovine ADA was used as the immobilized antigen. The ability of unmodified bovine ADA or PEG-ADA preincubated with plasma to block the ELISA was used to establish specificity. Before treatment, none of the patients showed specific antibody to ADA. Some who now show specific antibody responses to other foreign antigens have not developed anti-ADA antibody after 2 to 4 years of therapy. In eight patients, ELISA-detectable IgG anti-ADA antibodies have appeared after 3 to 9 months of treatment, usually coincident with appearance of circulating mature T cells. In 6 of these patients (10 of 12 in all), there has been no correlation between anti-ADA ELISA levels and the circulating life or activity of PEG-ADA: preinjection plasma ADA levels have remained above the normal total red blood cell ADA activity.

In PEG-ADA patient #8, preinjection plasma ADA fell to baseline levels at about 5 months of treatment, associated with a fall in half-life of plasma ADA activity from more than 6 days at the start of therapy to less than 2 days. This finding coincided with the appearance of ELISA-detectable anti-ADA IgG and of direct ADA inhibitory activity in IgG prepared from plasma. ADA inhibitory activity was not present in control plasma, in IVIg preparations, or in IgG prepared from the plasma of other PEG-ADA patients.[42] We were able to induce tolerance to ADA by modifying an empirical regimen developed for patients with hemophilia, about 15 percent of whom develop high titer antibodies to factor VIII.[43] After therapy was interrupted for 8 weeks, PEG-ADA was administered twice weekly, along with weekly IVIg infusion and a tapering course of prednisone.[44] After an anamnestic increase, ADA-inhibitory antibody disappeared over 2 to 3 months, and plasma ADA activity rose to a range consistent with the dose of PEG-ADA. After 4 months, this regimen was discontinued, and a weekly injection schedule of PEG-ADA was resumed. The half-life of PEG-ADA returned to approximately 6 days, and plasma ADA levels have been maintained for the past 12 months. This same regimen has been partially successful in a second patient (followed by Dr. A. Fischer and Dr. D. Girault in Paris), in whom enhanced clearance of PEG-ADA and ADA inhibitory activity developed after 5 months of therapy. Both patients, now in their 24th and 14th months of therapy, have shown excellent recovery of immune function.

ROLE OF PEG-ADA IN THE TREATMENT OF ADA DEFICIENCY

PEG-ADA is certainly safer and more effective than red blood cell transfusion, but its place among other alternative therapies for ADA deficiency is yet to be defined (Fig. 10-2). Transplantation of bone marrow from an HLA-matched sibling remains the treatment of choice. In the absence of such a donor, the choice between PEG-ADA and haploidentical transplantation is a matter of judgment for the physicians and family of each new patient. The potential for curing the disease must be weighed against the potential morbidity and even mortality associated with transplantation versus the more limited experience with PEG-ADA and the need for lifelong treatment. Both approaches have produced slow or partial immune reconstitution in some patients. An attempt was recently made (Ochs HD, et al: unpublished observations) to compare recovery of humoral immunity in three transplanted and three PEG-ADA-treated patients, based on immunization with bacteriophage φX174, a well-characterized test of integrated immune response to a T-cell-dependent antigen. The PEG-ADA-treated patients showed essentially normal responses, while responses in the transplanted patients were abnormal in some respects. This is a small sample, and no single test can predict the adequacy of future responses to specific pathogens. Moreover, both approaches are in evolution,

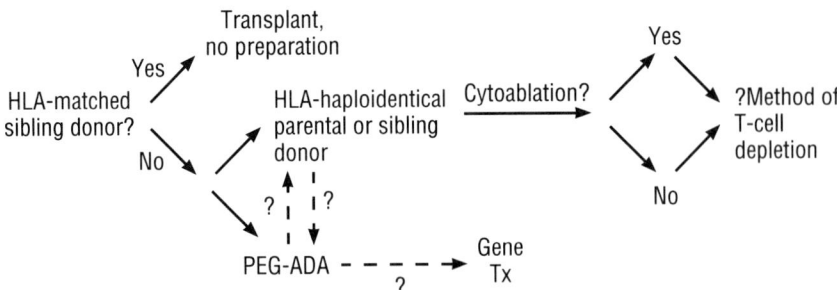

Fig. 10-2. Options for treatment of ADA deficiency.

and their relative merits may change with experience. Several quite different techniques for haploidentical transplantation are being evaluated. With PEG-ADA, larger doses and combined therapy with PEG-ADA and PEG-interleukin-2 (IL-2) are being studied in selected patients.

Thus far, only one patient has been placed on PEG-ADA after unsuccessful haploidentical transplantation,[35] and no patient, after starting on PEG-ADA, has undergone transplantation. Both situations may arise, and some notes of caution are in order. The effectiveness of subsequent treatment with PEG-ADA in restoring immune function might be limited in failed transplant patients who had undergone pretransplant cytoablation. Conversely, PEG-ADA should probably be discontinued before transplantation to decrease the chance of graft rejection by functional host lymphocytes. Experience with a patient in whom PEG-ADA was temporarily withheld for several weeks (see above) indicated that a return to baseline immune dysfunction should occur within 2 to 4 weeks of discontinuing PEG-ADA. During this period, such precautions as relative isolation and prophylaxis with IVIg and antibiotics could be used to prevent clinical deterioration.

A protocol for a retrovirus-mediated ADA gene supplementation experiment was approved in July 1990 by the Recombinant DNA Advisory Committee of the National Institutes of Health. The study would be open to patients who had received PEG-ADA for at least 9 months and who manifest recurrent thrush or abnormalities in any 3 of 12 tests of immune function. There is no requirement that a patient shall have either failed to benefit clinically from PEG-ADA therapy or shown clinical deterioration on PEG-ADA. Patients would continue to receive PEG-ADA throughout the study. According to the protocol, at approximately monthly intervals over a 2-year period, samples of mature blood T cells, obtained by leukapheresis, would be cultured in vitro with IL-2, transfected with an ADA vector, and then reinfused (without IL-2). Since mature T cells, not stem cells, are the target, this approach is, like PEG-ADA, a treatment rather than a potentially curative procedure. This non-stem cell strategy appears to have two aims: (1) to prolong the survival, and possibly the function, of mature T cells (which presumably had arisen in response to PEG-ADA) and (2) to provide circulating ADA-expressing cells that may, like transfused red blood cells (Fig. 10-1), help eliminate circulating dAdo and allow ADA-negative T cells to survive.

In regard to the first aim, little is known about the repertoire diversity, life span, or functional capacity (after reinfusion) of IL-2-dependent mature T cells grown for varying periods in vitro. In regard to the second aim, it should be noted that 2.5 ml of PEG-ADA provides the ADA activity of approximately 10^{12} normal T cells. Whether sufficient transfected ADA-expressing T cells could be given to supplement this activity substantially is unclear. Moreover, dAdo arising extracellularly (see above) must be prevented from entering ADA-deficient T cells if they are to survive and function. In the case of PEG-ADA, protection from extracellular dAdo is a direct function of the level of ADA activity maintained in plasma. In the case of transfected T cells, however, a cell membrane is interposed between the circulating dAdo and intracellular

ADA. The probability of dAdo entering an ADA-negative versus an ADA-positive T cell would depend on the relative proportions of the two cell types, not the ADA activity of the latter. Thus, a given amount of ADA in transfected T cells might not protect ADA-negative T cells as effectively as the same amount of ADA that is free in plasma. It is noteworthy that, in contrast to the reconstitution of functional ADA-negative T cells that occurs with PEG-ADA therapy, such cells do not reappear after haploidentical transplantation for ADA deficiency (performed without pretransplant cytoablation).[8,9] This may reflect inadequate metabolic correction by even normal numbers of ADA-expressing donor T cells in the transplanted patients. Finally, a concern is that, since patients will continue to be treated with PEG-ADA, evaluation of the clinical and metabolic effectiveness of gene introduction will be difficult in some patients who are apparently considered candidates for the gene supplementation protocol.

To summarize, PEG-ADA has been well tolerated and of clinical benefit to all patients; no adverse reactions have occurred in any patient and none has had to discontinue therapy. The convenience of weekly intramuscular injections compares favorably with replacement therapy for diabetes and hemophilia; as with therapy for the latter, PEG-ADA will be expensive because the number of patients is very small. Continued monitoring will be necessary to evaluate the long-term benefit and any consequences of receiving as much as 1 g/yr of PEG contained in the enzyme conjugate.

FUTURE APPLICATIONS

PEG-modified enzymes might be used to treat other enzyme deficiency diseases in which substrates and necessary cofactors equilibrate with plasma and in which the enzyme retains activity in plasma. Products of the enzymatic reaction must also be nontoxic and able to undergo additional required metabolism after formation in plasma. Any enzyme deficiency disease in which erythrocyte transfusion partially corrects metabolic abnormalities is a potential candidate, but this criterion requires that the enzyme normally occurs in erythrocytes and that substrate efficiently crosses the red blood cell membrane. Limited metabolic improvement after transfusion therapy has been observed in patients with deficiencies of purine nucleoside phosphorylase,[45] arginase,[46,47] and fumarylacetoacetase (tyrosinemia type I).[48] Argininosuccinate lyase deficiency[49] is a potential candidate for PEG-enzyme therapy that we are now investigating. It has been suggested that Gaucher disease[50] and Fabry disease[51] might be treated by maintaining high levels of circulating enzyme. It is also possible that benefit could be achieved in these disorders if PEG-modified glucocerebrosidase or galactosidase were slowly taken up by the affected cell populations, that is, the circulating PEG-enzyme acts as a long-circulating depot. In each case, it will have to be determined that the enzyme can be PEG modified successfully, with the desired effects on circulating life and immunogenicity in animals and assurance that it is nontoxic. The availability of animal models of the deficiency disease would be of obvious value.

ACKNOWLEDGMENTS

This work was supported in part by grant RO1 DK 20902 from the National Institutes of Health. Sara Chaffee was a Fellow of the Leukemia Society of America. The clinical trial of PEG-ADA was also supported in part by Enzon, Inc. Many collaborators were involved in the care and investigation of the patients who participated in the PEG-ADA clinical trial. We acknowledge the important contribution of these investigators, the patients, and their families, and of many dedicated employees of Enzon, Inc., with whom we have worked closely. We appreciate the helpful comments of Dr. Ricardo U. Sorensen.

REFERENCES

1. Giblett ER, Anderson JE, Cohen F, et al: Adenosine deaminase deficiency in two patients with severely impaired cellular immunity. Lancet 2:1067, 1972
2. Hirschhorn R, Vawter GE, Kirkpatrick JA Jr, Rosen FS: Adenosine deaminase deficiency: Frequency and comparative pathology in autosomally recessive severe combined immunodeficiency. Clin Immunol Immunopathol 14:107, 1979
3. Kredich NM, Hershfield MS: Immunodeficiency diseases caused by adenosine deaminase deficiency and purine nucleoside phophorylase deficiency. p. 1045. In Scriver CR, Beaudet AL, Sly WS, Valle D (eds): The Metabolic Basis of Inherited Disease. 6th Ed. McGraw-Hill, New York, 1989
4. Morgan C, Levinsky RJ, Hugh-Jones K, et al: Heterogeneity of biochemical, clinical and immunological parameters in severe combined immunodeficiency due to adenosine deaminase deficiency. Clin Exp Immunol 70:491, 1987
5. Reisner Y, Kapoor N, Kirkpatrick D, et al: Transplantation for severe combined immunodeficiency with HLA-A,B,D,DR incompatible parental marrow cells fractionated by soybean agglutinin and sheep red blood-cells. Blood 61:341, 1983
6. Friedrich W, Goldmann SF, Ebell W, et al: Severe combined immunodeficiency: Treatment by bone marrow transplantation in 15 infants using HLA-haploidentical donors. Eur J Pediatr 144:125, 1985
7. Fischer A, Friedrich W, Levinsky R, et al: Bone-marrow transplantation for immunodeficiencies and osteopetrosis: European survey, 1968–1985. Lancet 2:1080, 1986
8. Buckley RH, Schiff SE, Sampson HA, et al: Development of immunity in human severe primary T cell deficiency following haploidentical bone marrow stem cell transplantation. J Immunol 136:2398, 1986
9. Markert ML, Hershfield MS, Schiff RI, Buckley RH: Adenosine deaminase and purine nucleoside phosphorylase deficiencies: Evaluation of therapeutic interventions in eight patients. J Clin Immunol 7:389, 1987
10. Vossen JM: Bone marrow transplantation in the treatment of primary immunodeficiencies. Ann Clin Res 19:285, 1987
11. Moen RC, Horowitz SD, Sondel PM, et al: Immunologic reconstitution after haploidentical bone marrow transplantation for immune deficiency disorders: Treatment of bone marrow cells with monclonal antibody CT-2 and complement. Blood 70:664, 1987
12. Chen SH, Ochs HD, Scott CR, et al: Adenosine deaminase deficiency: Disappearance of adenine deoxynucleotides from a patient's erythrocytes after successful marrow transplantation. J Clin Invest 62:138, 1978

13. Rich KC, Richman CM, Mejias E, Daddona P: Immunoreconstitution by peripheral blood leukocytes in adenosine deaminase-deficient severe combined immunodeficiency. J Clin Invest 66:389, 1980
14. Hirschhorn R, Roegner-Maniscalco V, Kuritsky L, Rosen FS: Bone marrow transplantation only partially restores purine metabolites to normal in adenosine deaminase-deficient patients. J Clin Invest 68:1387, 1981
15. Jarvis SM: Kinetics and molecular properties of nucleoside transporters in animal cells. p. 102. In Gerlach E, Becker BF (eds): Topics and Perspectives in Adenosine Research. Springer-Verlag, Berlin, 1987
16. Chan T-S: Purine excretion by mouse peritoneal macrophages lacking adenosine deaminase activity. Proc Natl Acad Sci USA 76:925, 1979
17. Henderson JF, Smith CM: Mechanisms of deoxycoformycin toxicity in vivo. p. 208. In Tattersal MHN, Fox RM (eds): Nucleosides in Cancer Treatment. Academic Press, Sydney, 1981
18. Polmar SH, Stern SC, Schwartz AL, et al: Enzyme replacement therapy for adenosine deaminase deficiency and severe combined immunodeficiency. N Engl J Med 295:1337, 1976
19. Rubinstein A, Hirschhorn R, Sicklick M, Murphy RA: In vivo and in vitro effects of thymosin and adenosine deaminase on adenosine-deaminase-deficient lymphocytes. N Engl J Med 300:387, 1979
20. Dyminski JW, Daoud A, Lampkin BC, et al: Immunological and biochemical profiles in response to transfusion therapy in an adenosine deaminase-deficient patient with severe combined immunodeficiency disease. Clin Immunol Immunopathol 14:307, 1979
21. Hirschhorn R, Papageorgiou P, Kesariwala HH, Taft LT: Amelioration of neurologic abnormalities after "enzyme replacement" in adenosine deaminase deficiency. N Engl J Med 303:377, 1980
22. Schmalsteig FC, Mills GC, Nelson JA, et al: Limited effect of erythrocyte and plasma infusions in adenosine deaminase deficiency. J Pediatr 93:597, 1978
23. Ziegler JB, Lee CL, van der Weyden MB, et al: Severe combined immunodeficiency and adenosine deaminase deficiency: Failure of enzyme replacement therapy. Arch Dis Child 55:452, 1980.
24. Hutton JJ, Wiginton DA, Coleman MS, et al: Biochemical and functional abnormalities in lymphocytes from an adenosine deaminase deficient patient during enzyme replacement therapy. J Clin Invest 68:413, 1981
25. Davies EG, Levinsky RJ, Webster DR, et al: Effect of red cell transfusion, thymic hormone and deoxycytidine in severe combined immunodeficiency due to adenosine deaminase deficiency. Clin Exp Immunol 50:303, 1982
26. Polmar SH: Metabolic aspects of immunodeficiency. Semin Hematol 17:30, 1980
27. Hirschhorn R: Inherited enzyme deficiencies and immunodeficiency: Adenosine deaminase (ADA) and purine nucleoside phosphorylase (PNP) deficiencies. Clin Immunol Immunopathol 40:157, 1986
28. Abuchowski A, McCoy JR, Palczuk NC, et al: Effect of attachment of polyethylene glycol on immunogenicity and circulating life of bovine liver catalase. J Biol Chem 252:3582, 1977
29. Abuchowski A, van Es T, Palczuk NC, Davis FF: Alteration of immunological properties of bovine serum albumin by covalent attachment of polyethylene glycol. J Biol Chem 252:3578, 1977
30. Abuchowski A, Davis FF: Soluble polymer-enzyme adducts. p. 367. In Holcenberg JS, Robert J (eds): Enzymes as Drugs. John Wiley & Sons, New York, 1981

31. Davis S, Abuchowski A, Park YK, Davis FF: Alteration of the circulating life and antigenic properties of bovine adenosine deaminase in mice by attachment of polyethylene glycol. Clin Exp Immunol 46:649, 1981
32. Wie SI, Wie CW, Lee WY, et al: Suppression of reaginic antibodies with modified allergens. III. Int Arch Allergy Appl Immunol 64:84, 1981
33. Lee WY, Sehon AH, Akerblom E: Suppression of reaginic antibodies with modified antigens. IV. Int Arch Allergy Appl Immunol 64:100, 1981
34. Abuchowski A, Kazo GM, Verhoest CR Jr, et al: Cancer therapy with modified enzymes. I. Antitumor properties of polyethylene glycol-asparaginase conjugates. Cancer Biochem Biophys 7:175, 1984
35. Hershfield MS, Buckley RH, Greenberg ML, et al: Treatment of adenosine deaminase deficiency with polyethylene glycol-modified adenosine deaminase. N Engl J Med 316: 589, 1987
36. Levy Y, Hershfield MS, Fernandez-Mejia C, et al: Adenosine deaminase deficiency with late onset of recurrent infections: Response to treatment with polyethylene glycol-modified adenosine deaminase (PEG-ADA). J Pediatr 113:312, 1988
37. Bory C, Boulieu R, Souillet G, et al: Comparison of red cell transfusion and polyethylene glycol-modified adenosine deaminase therapy in an adenosine deaminase-deficient child. Pediatr Res 28:127, 1990
38. Geffner ME, Stiehm ER, Stephure D, Cowan MJ: Probable autoimmune thyroid disease and combined immunodeficiency disease. Am J Dis Child 140:1194, 1986
39. Linch DC, Levinsky RJ, Rodeck CH, et al: Prenatal diagnosis of three cases of severe combined immunodeficiency: Severe T cell deficiency during the first half of gestation in fetuses with adenosine deaminase deficiency. Clin Exp Immunol 56:223, 1984
40. Hirschhorn R, Roegner V, Rubinstein A, Papageorgiou P: Plasma deoxyadenosine, adenosine and erythrocyte deoxyATP are elevated at birth in an adenosine deaminase deficient child. J Clin Invest 65:768, 1980
41. Hershfield MS, Kurtzberg J, Aiyar VN, et al: Abnormalities in S-adenosylhomocysteine hydrolysis, ATP catabolism, and lymphoid differentiation in adenosine deaminase deficiency. Ann NY Acad Sci 451:78, 1985
42. Chaffee S, Hershfield MS: Immune response to polyethylene glycol-modified bovine adenosine deaminase (PEG-ADA). Pediatr Res 27:155A, 1990 (abst)
43. Nilsson IM, Berntorp E, Zettervall O: Induction of immune tolerance in patients with hemophilia and antibodies to factor VIII by combined treatment with intravenous IgG, cyclophosphamide, and factor VIII. N Engl J Med 318:947, 1988
44. Lee N, Kobayashi RH, Chaffee S, et al: Suppression of an inhibitory antibody to bovine adenosine-deaminase (ADA) and improved cellular immunity following intravenous immunoglobulin and polyethylene-glycol (PEG-ADA). Pediatr Res 27:155A, 1990 (abst)
45. Rich KC, Mejias E, Fox IH: Purine nucleoside phosphorylase deficiency: Improved metabolic and immunologic fuction with erythrocyte transfusions. N Engl J Med 303:973, 1980
46. Takeshi S, Nakabayashi H, Shimizu H, et al: A successful trial of enzyme replacement therapy in a case of argininemia. Tohoku J Exp Med 142:239, 1984
47. Mizutani N, Hayakawa C, Maehara M, et al: Enzyme replacement therapy in a patient with hyperargininemia. Tohoku J Exp Med 151:301, 1987
48. Linblad B, Friden J, Greter J, et al: Treatment of hereditary tyrosinaemia (fumarylacetoacetase deficiency) by enzyme substitution. J Inher Metab Dis 9:257, 1986
49. Brusilow AW, Horwich AL: Urea cycle enzymes. p. 629. In Scriver CR, Beaudet AL,

Sly WS, Valle D (eds): The Metabolic Basis of Inherited Disease. 6th Ed. McGraw-Hill, New York, 1989
50. Dawson G, Sweeley CC: In vivo studies of glycosphingolipid metabolism in porcine blood. J Biol Chem 245:410, 1970
51. Desnick RJ, Dean KJ, Grabowski G, et al: Enzyme therapy. XVII. Metabolic and immunologic evaluation of α-galactosidase A replacement in Fabry disease. p. 393. In Desnick RJ (ed): Enzyme Therapy in genetic diseases. Vol. 2. Alan R Liss, New York, 1980

11

TRANSPLANTATION IN ANIMAL MODEL SYSTEMS

Mark Haskins · Henry J. Baker ·
Edward Birkenmeier · Peter M. Hoogerbrugge ·
Ben J.H.M. Poorthuis · Takeshi Sakiyama ·
Robert M. Shull · Rosanne M. Taylor ·
Mary Anna Thrall · Steven U. Walkley

INTRODUCTION

Naturally occurring diseases of animals can provide insights into the causal mechanisms and pathogenesis of, and therapy for, human disease. Animal models are particularly useful in the study of genetic disorders, in which evaluation of therapy requires a whole mammalian organism. Accurate models of specific genetic disorders cannot be readily produced by experimental manipulation; therefore, the rather large group of naturally occurring animal models of human genetic disease forms the foundation for testing therapeutic strategies that cannot be assessed adequately in clinical trials, owing to the limitations of human experimentation.

Transplantation as a form of therapy for genetic disease serves primarily either to replace or supplement abnormal cells with their normal equivalent or to act as a source of a normal gene product. The use of animal models of genetic disease provides the opportunity to evaluate efficacy, to determine methods to overcome problems associated with therapy, and to test strategies on uniform mutations, either on a constant genetic background using inbred strains or on outbred subjects that more closely resemble the human patient, using larger domestic species. The use of animal models also permits evaluation of positive and negative controls, which is not possible in human subjects.

TRANSPLANTATION TO REPLACE OR SUPPLEMENT DEFECTIVE CELLS, TISSUES, OR ORGANS

Several animal model systems have been used to evaluate the effect of transplanting normal cells into subjects with genetic defects in those cell types to be transplanted. In hypogonadal mice, the gonads are infantile, secondary sex organ development is impaired, there are decreased levels of luteinizing hormone (LH) and follicle-stimulating hormone (FSH) in the pituitary gland and plasma, and females are infertile.[1] This syndrome is caused by a genetic deficiency of hypothalamic gonadotropin-releasing hormone (GnRH), which in normal animals is produced by the cell bodies of the preoptic area of the brain. Transplantation of fetal preoptic tissue into the third ventricle of adult female hypogonadal mice produced vaginal opening, which requires estrogen produced by stimulated ovaries, in 16 to 40 days. The animals went into constant vaginal estrus and were mated. Seven of ten animals became pregnant, with six females delivering live litters. Microscopic examination showed that GnRH-containing processes extended into the lateral median eminence.

The *mdx* mouse is a model of Duchenne muscular dystrophy. Myoblasts, which are muscle precursor cells, were transplanted into 70 mice, aged 5 to 27 days.[2] Nucleated myoblasts were injected into the muscles of affected mice, which were then examined 20 to 99 days later. Fusion of transplanted cells into myocytes occurred in 39 animals. Dystrophin, the gene product absent in *mdx* mice, was detected by fluorescent staining and immunoblot analysis, with two mice having 30 to 40 percent of normal dystrophin levels in selected areas. The dystrophin was present in its normal location in the sarcolemma in 10 to 40 percent of muscle fiber profiles.

Rats with a hereditary deficiency of hepatic bilirubin UDP-glucuronyltransferase have been transplanted by intraportal or intrasplenic injection of isolated hepatocytes from normal rats.[3–7] In these trials, total plasma bilirubin was decreased, and conjugated bilirubin was detected in the bile, correcting the metabolic abnormality. These effects were abolished when the transplanted cells were rejected.

Osteopetrosis is a hereditary disease caused by a deficiency or dysfunction of osteoclasts that results in generalized skeletal sclerosis. Several models exist in the rat and mouse and appear to be heterogeneous with respect to pathogenesis and treatment, with approximately one-half of the models cured by bone marrow transplantation.[8–16] In those responding to transplantation, providing donor stem cells that produce normal osteoclasts restores function, indicating that the defect results from the altered or decreased number of osteoclasts. In those animals that do not respond, the defect apparently resides external to the donor-derived osteoclast.

Hereditary cyclic hematopoiesis in the collie dog results in periodic alternation of peripheral blood cell numbers over a course of approximately every 12 days.[17] The syndrome also includes fever, gingivitis, diarrhea, and lameness. The successful treatment of this disorder by bone marrow transplantation helped establish, together with in vitro studies, that the defect is present in the primitive hematopoietic stem cell.[18]

Combined immunodeficiency in Arabian foals has been treated by transplantation of fetal liver cells, fetal liver and thymus cells, histocompatible bone marrow cells, equine lymphocyte antigen, and haploidentical bone marrow cells.[19-21] Transplantation of adequate numbers of histocompatible stem cells completely reconstituted the immune system of affected foals, supporting the hypothesis that combined immunodeficiency results from a defect of early lymphoid precursor cells.

Chédiak-Higashi syndrome (CHS) is characterized by the presence of very large lysosomes in a variety of cell types, including bone marrow-derived cells, hepatocytes, and kidney epithelium.[22-27] Hypopigmentation, phagocyte dysfunction, and a complete platelet storage pool deficiency that results in a bleeding tendency are also part of the syndrome. In beige mice with this syndrome, treatment with bone marrow transplantation has resulted in substantially reduced bleeding times and in restoration of serotonin concentrations in dense granules to normal.[28] Conversely, normal mice transplanted with affected donor marrow acquired the platelet storage pool deficiency. Of seven cats with this syndrome treated by bone marrow transplantation, aged 4 to 35 months, four survived to show a gradual restoration of neutrophil migration, whole blood platelet aggregation, and platelet ATP secretion.[29] However, abnormal lysosome distribution in the eye, skin, hepatocytes, and renal tubular epithelium remained unaltered at 8 months post-transplantation.

TRANSPLANTATION TO ACT AS A SOURCE OF A NORMAL GENE PRODUCT

In these approaches to therapy, normal enzyme produced by the transplanted cells may be secreted in a form that permits uptake by other, nontransplanted, diseased cells. Cattle with generalized glycogenosis type II have severe muscle weakness; intracytoplasmic accumulation of glycogen in skeletal muscle, cardiac muscle, and neurons; and a deficiency of lysosomal α-1,4-glucosidase.[30-31] Multiple transplantation of normal bovine amnion below the external oblique abdominal muscle in two affected calves did not raise blood or tissue enzyme levels, nor did it appear to alter the progress of the disease.[32]

Acatalasemic mice have an unstable catalase protein that turns over rapidly but is not associated with naturally occurring disease.[33] Affected animals have blood and brain catalase activities of 1 percent and 38 percent that of normal, respectively. At approximately 4 months after bone marrow transplantation, the 36 percent of acatalasemic mice that survived had blood catalase activity restored to normal, but brain catalase activity did not change significantly.

A series of animal models with deficient lysosomal hydrolase activities have been treated by bone marrow transplantation. Table 11-1 lists each disease, the respective hydrolase involved, and the references[34-100] describing the model

Table 11-1. Transplantation in Animal Models of Lysosomal Disorders

Disease	Enzymatic Defect	References
Fucosidosis	α-Fucosidase	34–39
GM1 gangliosidosis	β-Galactosidase	40–46
Krabbe disease	Galatosylceramidase	47–54
α-Mannosidosis	α-Mannosidase	55–56
MPS I	α-L-Iduronidase	Dog: 57–65 Cat: 66–70
MPS VI	Arylsulfatase B	71–82
MPS VII	β-Glucuronidase	Dog: 83–86 Mouse: 87–88 C3H/Rij mouse: 89–92
Niemann-Pick disease	Sphingomyelinase	93–100

Abbreviation: MPS, mucopolysaccharidosis.

and results of bone marrow transplantation. The mucopolysaccharidosis (MPS) VII–C3H/Rij mouse does not have pathology associated with the reduced activity of β-glucuronidase. Transplantation studies in cats with GM1 gangliosidosis, α-mannosidosis, and MPS I and in dogs with MPS VII have been performed recently and have not yet been published elsewhere.

The results of bone marrow transplantation in these models are summarized in Tables 11-2 through 11-17. For those models in which transplantation

Table 11-2. Experience With Bone Marrow Transplantation in Animal Models of Lysosomal Disorders

Disease	Species	No. Transplanted	Age at Transplantation
Fucosidosis	Dog	27	2–30 mo
GM1 gangliosidosis	Cat	4	2–6 mo
Krabbe disease	Mouse	>40	1–22 d
α-Mannosidosis	Cat	1	65 d
MPS I	Dog	6	20–26 wk
MPS I	Cat	1	11 wk
MPS VI	Cat	21	3–47 mo
MPS VII	Dog	2	6 wk
MPS VII	Mouse	>40	27–95 d
MPS VII	C3H/Rij mouse	50	8–12 wk
Niemann-Pick disease	Mouse	15	5–6 wk

Abbreviation: MPS, mucopolysaccharidosis.

Table 11-3. Bone Marrow Transplantation in Animal Models of Lysosomal Disorders: Genotype of the Donor

Disease	Donor Status
Fucosidosis	53% normal; 47% heterozygous
GM1 gangliosidosis	25% normal; 75% heterozygous
Krabbe disease	Normal
α-Mannosidosis	Heterozygous
MPS I dog	33% normal; 67% heterozygous
MPS I cat	Not determined
MPS VI	Not determined
MPS VII dog	Heterozygous
MPS VII mouse	Normal
MPS VII C3H/Rij mouse	Normal
Niemann-Pick disease	Normal

Abbreviation: MPS, mucopolysaccharidosis.

Table 11-4. Proportion of Animals Engrafted and Procedural Mortality

Disease	Engraftment (%)	Mortality (%)
Fucosidosis	83	33
GM1 gangliosidosis	100	50
Krabbe disease	>80	<20
α-Mannosidosis	100	0
MPS I dog	83	50
MPS I cat	100	100
MPS VI	74	28
MPS VII dog	100	50
MPS VII mouse	100	2
MPS VII C3H/Rij mouse	100	Negligible
Niemann-Pick disease	93	7

Abbreviation: MPS, mucopolysaccharidosis.

Table 11-5. Complications of Transplantation

Disease	Complications
Fucosidosis	Respiratory, pancreatitis, hepatopathy; GVHD
Krabbe disease	Rejection
GM1 gangliosidosis	Metabolic
α-Mannosidosis	Diarrhea, dehydration
MPS I dog	Fail to engraft, GVHD, encephalitis
MPS I cat	Fever, septicemia
MPS VI	Septicemia, failure to engraft
MPS VII dog	Pulmonary failure, septicemia, GVHD
MPS VII mouse	Tumors[a]
MPS VII C3H/Rij mouse	Liver tumors[a]
Niemann-Pick disease	Death during transplantation

[a] A relationship between the tumors and irradiation, aging, or other factors has not been established.

Abbreviations: GVHD, graft-versus-host disease; MPS, mucopolysaccharidosis.

Table 11-6. Proportion of Donor-Derived Peripheral Blood Cells in Transplanted Animals

Disease	Chimerism	Method
Fucosidosis	100% in 9; 65% in 1	Karyotype, RBC antigens, RFLP
GM1 gangliosidosis	ND	
Krabbe disease	100%	MHC difference
α-Mannosidosis	ND	
MPS I dog	100%	Enzyme levels
MPS I cat	ND	
MPS VI	95%	Karyotype, WBC inclusions
MPS VII dog	>90%	Karyotype, enzyme, WBC inclusions
MPS VII mouse	10–90%, proportional to radiation dose	WBC inclusions
MPS VII C3H/Rij mouse	100%	Monoclonal antibody
Niemann-Pick disease	ND	

Abbreviations: MPS, mucopolysaccharidosis; RBC, red blood cell; RFLP, restriction fragment length polymorphism; MHC, major histocompatibility complex; WBC, white blood cell; ND, not determined.

Table 11-7. Effect of Transplantation on Urinary Substrate Excretion

Disease	Urinary Substrate
Fucosidosis	Significant improvement
GM1 gangliosidosis	ND
α-Mannosidosis	ND
MPS I dog	Significant improvement
MPS I cat	ND
MPS VI	Significant improvement
MPS VII dog	Significant improvement
MPS VII mouse	ND

Abbreviations: MPS, mucopolysaccharidosis; ND, not determined.

Table 11-8. Effect of Transplantation on the Longevity of Affected Animals

Disease	Longevity
Fucosidosis	Age dependent—no change to significant improvement
GM1 gangliosidosis	Significant improvement
Krabbe disease	Significant improvement
α-Mannosidosis	Significant improvement
MPS I dog	Improved because of musculoskeletal improvement
MPS I cat	Died 3 wk post-transplantation
MPS VI	Worse to significant improvement
MPS VII dog	Increased because of musculoskeletal improvement
MPS VII mouse	Significant improvement
Niemann-Pick disease	No change

Abbreviation: MPS, mucopolysaccharidosis.

Table 11-9. Effect of Transplantation on Corneal Disease

Disease	Clinical Findings	Morphology
α-Mannosidosis	ND	ND
MPS I dog	Significant improvement	Significant improvement
MPS VI	Worse to significant improvement	No change
MPS VII dog	ND	ND
MPS VII mouse	Significant improvement	Significant improvement
Niemann-Pick disease		No change

Abbreviations: MPS, mucopolysaccharidosis; ND, not determined.

Table 11-10. Effect of Transplantation on Skeletal Disease

Disease	Bone Morphology
α-Mannosidosis	ND
MPS I dog	Significant improvement
MPS VI	Slight improvement
MPS VII dog	Significant clinical improvement
MPS VII mouse	Variable improvement

Abbreviations: MPS, mucopolysaccharidosis; ND, not determined.

Table 11-11. Effect of Transplantation on Abnormal Kidney Morphology

Disease	Renal Morphology
Krabbe disease	No change
MPS I dog	Significant improvement
MPS VII mouse	Variable improvement

Abbreviation: MPS, mucopolysaccharidosis.

Table 11-12. Effect of Transplantation on Liver Enzyme, Substrate, and Morphology

Disease	Enzyme	Substrate	Morphology
Fucosidosis	+++	+++	+++
Krabbe disease	+	ND	NR
GM1 gangliosidosis	+++	ND	+++
α-Mannosidosis	ND	ND	ND
MPS I dog	+++	+++	+++
MPS I cat	ND	ND	ND
MPS VI	ND	+++	+
MPS VII dog	ND	ND	ND
MPS VII mouse	+++	+	+++
MPS VII C3H/Rij mouse	+++	NR	NR
Niemann-Pick disease	+	+	+++

Abbreviations: MPS, mucopolysaccharidosis; ND, not determined; NR, not relevant; +, slight improvement; +++, significant improvement.

Table 11-13. Effect of Transplantation on the Peripheral Nervous System

Disease	Effect
Fucosidosis	+++
GM1 gangliosidosis	+++
Krabbe disease	+++ with time
α-Mannosidosis	ND
MPS I dog	ND
MPS I cat	ND
MPS VII dog	ND
MPS VII mouse	ND
Niemann-Pick disease	No change to +

Abbreviations: MPS, mucopolysaccharidosis; ND, not determined; +, slight improvement; +++, significant improvement.

Table 11-14. Effect of Transplantation on Enzymatic Activity in the Central Nervous System

Disease	CNS Enzyme Restoration
Fucosidosis	Gradual increase to 20% of normal by 6 mo
GM1 gangliosidosis	Regional improvement (cerebellum)
Krabbe disease	20% of donor at 100 d
α-Mannosidosis	ND
MPS I dog	2.4–5.3% of donor levels
MPS I cat	ND
MPS VI	No change
MPS VII dog	ND
MPS VII mouse	Slight improvement
MPS VII C3H/Rij mouse	No change
Niemann-Pick disease	No change

Abbreviations: MPS, mucopolysaccharidosis; ND, not determined.

Table 11-15. Effect of Transplantation on the Accumulated Substrate in the Central Nervous System

Disease	CNS Substrate
Fucosidosis	ND
GM1 gangliosidosis	ND
Krabbe disease	+++
α-Mannosidosis	ND
MPS I dog	+++
MPS I cat	ND
MPS VI	ND
MPS VII dog	ND
MPS VII mouse	No change to +
Niemann-Pick disease	No change

Abbreviations: MPS, mucopolysaccharidosis; ND, not determined; +, slight improvement; +++, significant improvement.

Table 11-16. Effect of Transplantation on the Morphology of the Central Nervous System Lesions

Disease	Light Microscopy	Electron Microscopy
Fucosidosis	Significant improvement if transplanted before onset of severe disease	Significant improvement if transplanted before onset of severe disease
GM1 gangliosidosis	No improvement	No improvement
Krabbe disease	Significant improvement in cerebellum; lesions not prevented	Remyelination; storage in Schwann cells and oligodendroglia present
α-Mannosidosis	ND	ND
MPS I dog	Significant improvement	Slight improvement in neurons; significant in vessels and glia
MPS I cat	ND	ND
MPS VI	ND	ND
MPS VII dog	ND	ND
MPS VII mouse	No change to slight improvement in glia	Neuronal storage remained; character of storage altered
Niemann-Pick disease	No change	No change to slight improvement

Abbreviation: MPS, mucopolysaccharidosis; ND, not determined.

Table 11-17. Overall Effect of Bone Marrow Transplantation on Disease

Disease	Result
Fucosidosis	Improved survival; improvement in CNS, PNS, and visceral tissues
GM1 gangliosidosis	Improved longevity and CNS function; improvement in liver and dorsal root ganglia
Krabbe disease	Some improvement in longevity and in CNS and PNS; not in kidney
α-Mannosidosis	Improved longevity; normal CNS function at >7 mo of age
MPS I dog	Improvement in most systems affected
MPS I cat	Died 3 wk post-transplantation
MPS VI	Some improvement in liver and in cardiovascular and skeletal systems; not ocular
MPS VII dog	Improved mobility
MPS VII mouse	Improved longevity; improvement in liver, spleen, and cornea; variable improvement in bone and kidney; little improvement in CNS
Niemann-Pick disease	Improvement in liver and spleen; little or no improvement in the CNS, PNS, and eye

Abbreviations: CNS, central nervous system; PNS, peripheral nervous system; MPS, mucopolysaccharidosis.

has been performed only during the past several months, limited data are available. Results are included only for those models in which abnormalities were expected in the system described.

CONCLUSIONS

The conclusions to be drawn from the above data, the published reports, and our observations of bone marrow transplantation therapy in these models include the following:

1. The age at transplantation appeared significant, with animals transplanted at younger ages showing greater improvement than if transplanted when older.
2. Although not critically examined, no obvious relationship was apparent between the genotype of the donor and the alteration of disease.
3. Mortality was low in mice but significant in dogs and cats. Engraftment was high in those animals that survive.
4. Complications were those typical for transplantation procedures including the consequences of immunosuppression and graft-versus-host disease. A relationship between the tumors seen in the MPS VII mice and irradiation, aging, or other factors has not yet been established.

5. The proportion of peripheral blood leukocytes that was donor derived after transplantation was high and was related to the dose of irradiation.
6. In all three diseases in which undegraded substrate is excreted in the urine, the levels were found to be significantly reduced.
7. In general, most affected animals that were successfully transplanted had a significantly increased life span compared with untreated, affected animals.
8. Corneal clearing occurred in some diseases treated by bone marrow transplantation, but not others. Consistent with bone marrow transplantation therapy in cats with MPS VI, three normal cats were recipients of MPS VI-affected cornea transplants. After 3 years with the abnormal corneas residing in genotypically normal cats, no clinical clearing could be demonstrated (Aguirre G, Raber I, Haskins M: manuscript in preparation).
9. Similar to cornea, the pathology of renal, bone, and peripheral nervous system was variably improved.
10. Of all the organ systems, with the exception of peripheral blood, which is virtually totally of donor origin, the liver showed the most consistent improvement. A proportion of this improvement logically arises from the bone marrow origin of Kupffer cells, which will be replaced over time by donor-derived cells. In the Krabbe and C3H/Rij mice, increased enzymatic activity in liver parenchymal cells was observed.
11. Central nervous system (CNS) enzyme levels remained unchanged or showed slight improvement in most models, with the exception of the fucosidosis dog and Krabbe mouse, in which the change was significant. There is a reasonable correlation between enzyme level and CNS substrate concentration, with the exception of the MPS I dog, in which a significant reduction in substrate occurred with only a slight increase in CNS enzymatic activity. This correlation cannot be evaluated for fucosidosis, as the substrate levels have not yet been evaluated. Variable alteration in intraneuronal storage was seen, with significant changes reported for fucosidosis and MPS I dogs. For most diseases, there does not appear to be a correlation between longevity and alteration in CNS enzymatic activity, substrate concentration, or pathology.

Our understanding of the pathogenic mechanisms of brain dysfunction in these diseases is limited. Consequently, it is difficult to determine whether treatment is effective. Most approaches evaluating the therapeutic effect of transplantation on CNS pathology have concentrated on the reduction of intralysosomal, stored material. However, other cytopathologic abnormalities have been observed in many of these disorders, including cell death, ectopic dendrite growth and new synapse formation, and axonal spheroid formation.[45,56,70,101–103] What significance these pathologic changes have in CNS dysfunction, how they relate to lysosomal storage per se, and how amenable they are to treatment are questions that are only beginning to be addressed.[104] Transplantation therapy should lead to improved understanding of the reversibility of these cytopathologic changes and may provide insight into their relative contributions to brain dysfunction.

SUMMARY

Transplantation as an approach to therapy in genetic diseases is well established for certain disorders and holds great promise for others. The availability of animals with genetic diseases that are pathophysiologically equivalent to those in humans has facilitated the evaluation of therapeutic strategies, from neuronal to bone marrow transplantation. The use of these animals to explore the advantages, problems, and limitations associated with therapy for specific genetic diseases should permit measured progress in the application to human patients.

Many genetic diseases have not yet been discovered or characterized in animals. However, several have been described in small laboratory species, particularly the mouse and rat, in which advantages include the availability of inbred strains in which subjects are immunologically compatible and in which short reproductive cycles can produce large numbers of affected subjects. Several diseases have only been discovered in larger domestic animals, particularly dogs and cats, because (1) the population structure of domestic animals is favorable for the detection of genetic influences in disease and (2) clinical specialization and the increasing sophistication of diagnostic methods in veterinary medicine have improved the probability of detection. These species also provide the ability to evaluate therapy in large outbred species, which are more similar in some ways to human patients. When the same disease occurs in several animal species, the opportunity exists to test how well the therapeutic theory and technology can be transferred from one species to another, increasing confidence that the procedure will work and be safe when performed in humans.

The hope is that the therapeutic trials in transplantation described above, and trials using transplantation modalities in conjunction with gene therapy, will prove useful to those who care for human patients—most frequently young children with devastating disease and their families. It is also hoped that the insights and technologies developed in these studies will be directly applicable to the improvement of animal health.

ACKNOWLEDGMENTS

This work was supported by grants DK-25759, RR-02512, DK-38857, DK-41082, AM-32126, AR-37095, NS-18804, and NS-10967 from the National Institutes of Health, by the Lucille P. Markey Charitable Trust, by the Mrs. Cheever Porter Foundation, by grant 28-1021 from Praeventie Fonds (The Netherlands), by an anonymous donor, by a project grant from the National Health and Medical Research Council of Australia, and by grants 6110911 and 62109006 from the Ministry of Education, Science, and Culture of Japan. We would like to acknowledge our co-workers in these studies: G.D. Aguirre, J.E. Barker, M.A. Breider, W. Bruyninckx, T. Byrne, C. Chieffo, S. Colgan, G. Constantopoulos, R.J. Desnick, B.R.H. Farrow, R. Fulton, P. Gasper, U. Giger,

T. Kitagawa, S. Miyawaki, E. Neufeld, R. Park, D. Patterson, I. Raber, S. Roberts, E. Schuchman, W.S. Sly, G.J. Stewart, Kinuko Susuki, Kiunihiko Susuki, M. Tadokoro, D.W. van Bekkum, C. Vogler, G. Wagemaker, M.A. Walker, D. Wenger, J. Wolfe, and S. Wurzelmann.

REFERENCES

1. Gibson MJ, Krieger DT, Charlton HM, et al: Mating and pregnancy can occur in genetically hypogonadal mice with preoptic area brain grafts. Science 225:949, 1984
2. Partridge TA, Morgan JE, Coulton GR, et al: Conversion of mdx myofibers from dystrophin-negative to -positive by injection of normal myoblasts. Nature 337:176, 1989
3. Matas AJ, Sutherland DER, Steffes MW, et al: Hepatocellular transplantation for metabolic deficiencies: Decrease in plasma bilirubin in Gunn rats. Science 192:892, 1976
4. Groth CG, Arborgh B, Bjorken C, et al: Correction of hyperbilirubinemia in the glucuronyl transferase-deficient rat by intraportal hepatocyte transplantation. Transplant Proc 9:313, 1976
5. Sutherland DER, Matas AJ, Steffes MW, et al: Transplantation of liver cells in an animal model of congenital enzyme deficiency: The Gunn rat. Transplant Proc 9:317, 1977
6. Vroemen JPAM, Buurman WA, Heirrwegh KPM, et al: Hepatocyte transplantation for enzyme deficiency disease in congenic rats. Transplantation 42:130, 1986
7. Vroemen JPAM, Blanckaert N, Buurman WA, et al: Treatment of enzyme deficiency by hepatocyte transplantation in rats. J Surg Res 39:267, 1985
8. Walker DG: Bone resorption in osteopetrotic mice by transplants of normal bone marrow and spleen cells. Science 190:784, 1975
9. Marks SC: Osteopetrosis in the *ia* rat cured by spleen cells from a normal littermate. Am J Anat 146:331, 1976
10. Milhaud G, Labat ML, Graf B, Thillard MJ: Guérison de l'osteopétrose congénitale du rat. CR Acad Sci Paris 283:531, 1976
11. Milhaud G, Labat ML, Parant M, et al: Immunological defect and its correction in the osteopetrotic mutant rat. Proc Natl Acad Sci USA 74:339, 1977
12. Marks SC: Osteopetrosis in the toothless rat (tl) rat: Presence of osteoclasts but failure to respond to parathyroid extract or to be cured by infusion of spleen or bone marrow cells from normal littermates. Am J Anat 149:289, 1977
13. Marks SC, Seifert MF, McGuire JL: Congenitally osteopetrotic (*op/op*) mice are not cured by transplants of spleen or bone marrow cells from normal littermates. Metab Bone Dis Relat Res 5:183, 1984
14. Seifert MF, Marks SC: Congenitally osteosclerotic (*oc/oc*) mice are resistant to cure by transplantation of bone marrow of spleen cells from normal littermates. Tissue Cell 19:29, 1987
15. Marks SC: Congenital osteopetrotic mutations as probes of the origin, structure, and function of osteoclasts. Clin Orthop 189:239, 1984
16. Milhaud G, Labat M-L, Parant M, et al: Immunological defect and its correction in the osteopetrotic mutant rat. Proc Natl Acad Sci USA 74:339, 1977
17. Lund JE, Padgett GA, Ott RL: Cyclic neutropenia in grey collie dogs. Blood 29:452, 1967

18. Dale DC, Graw RG: Transplantation of allogeneic bone marrow in canine cyclic neutropenia. Science 183:83, 1974
19. McGuire TC, Poppie MJ: Hypogammaglobulinemia and thymic hypoplasia in horses: A primary combined immunodeficiency disorder. Infect Immun 8:272, 1973
20. Bue CM, Davis WC, Magnuson NS, et al: Correction of equine severe combined immunodeficiency by bone marrow transplantation. Transplantation 42:14, 1986
21. Perryman LE, Bue CM, Magnuson NS, et al: Immunologic reconstitution of foals with combined immunodeficiency. Vet Immunol Immunopathol 17:495, 1987
22. Kramer JW, Davis WC, Prieur DJ: The Chédiak-Higashi syndrome of cats. Lab Invest 36:554, 1977
23. Hargis AM, Prieur DJ: Animal model: Light and electron microscopy of hepatocytes of cats with Chédiak-Higashi syndrome. Am J Med Genet 22:659, 1985
24. Hargis AM, Prieur DJ: Animal model: Renal lesions in cats with Chédiak-Higashi syndrome. Am J Med Genet 26:167, 1987
25. Colgan SP, Thrall MA, Gasper PW: Platelet aggregation and ATP secretion in whole blood of normal cats and cats homozygous and heterozygous for Chédiak-Higashi syndrome. Blood Cells 15:585, 1989
26. Holland JM: Serotonin deficiency and prolonged bleeding in beige mice. Proc Soc Exp Biol Med 151:32, 1976
27. Novak EK, Hui S-W, Swank RT: Platelet storage pool deficiency in mouse pigment mutations associated with seven distinct genetic loci. Blood 63:536, 1984
28. Novak EK, McGarry MP, Swank RT: Correction of symptoms of platelet storage pool deficiency in animal models for Chédiak-Higashi syndrome and Hermansky-Pudlak syndrome. Blood 66:1196, 1985
29. Colgan SP, Thrall MA, Gasper PW, et al: Reversal of neutrophil mobility defect and platelet pool deficiency in feline Chédiak-Higashi syndrome with bone marrow transplantation. Bone Marrow Transplant 7:110, 1991
30. Richards RB, Edwards JR, Cook RD, White RR: Bovine generalized glycogenosis. Neuropathol Appl Neurobiol 3:45, 1977
31. Cook RD, Dorling PR, Gawthorpe JM, et al: Bovine generalized glycogenosis. A model for human disease. J Neuropathol Exp Neurol 37:563, 1978
32. Howell JM, Dorling PR, Dimarco PN, Taylor EG: Multiple implantation of normal amnion into cattle with generalized glycogenosis type II. J Inher Metab Dis 10:3, 1987
33. Tuchman M, Blazar BR, Krivit W: Brain catalase activity following syngeneic bone marrow transplantation in acatalasemic mice. Birth Defects 22:165, 1986
34. Healy PJ, Farrow BRH, Nicholas FW: Canine fucosidosis: A biochemical and genetic investigation. Res Vet Sci 36:354, 1987
35. Taylor RM, Farrow BRH, Healy PJ: Canine fucosidosis: Clinical findings. J Small Anim Pract 28:291, 1987
36. Taylor RM, Farrow BRH: Animal models series: Fucosidosis. Comp Med Bull 20:2, 1988
37. Taylor RM, Stewart GJ, Farrow BRH, Healy PJ: Enzyme replacement in nervous tissue after allogeneic bone marrow transplantation for fucosidosis in dogs. Lancet 2:772, 1986
38. Taylor RM: Canine fucosidosis. Clinical and pathological studies of the disease and experimental treatment by allogeneic bone marrow transplantation. PhD thesis. University of Sydney, Sydney, Australia, 1988
39. Taylor RM, Farrow BRH, Stewart GJ, et al: The clinical effects of lysosomal enzyme

replacement by bone marrow transplantation after total lymphoid irradiation on neurologic disease in fucosidase deficient dogs. Transplant Proc 20:89, 1988
40. Baker HJ, Lindsey JR, McKhann GM, Farrell DF: Neuronal GM1 gangliosidosis in a Siamese cat with β-galactosidase deficiency. Science 174:838, 1971
41. Farrell DF, Baker HJ, Herdon RM, et al: Feline GM1 gangliosidosis: Biochemical and ultrastructural comparisons with the disease in man. J Neuropathol Exp Neurol 32:1, 1973
42. Baker HJ, Mole JA, Lindsey JR, et al: Animal models of human ganglioside storage diseases. Fed Proc 35:1193, 1976
43. Reynolds GD, Baker HJ, Reynolds RH: Enzyme replacement using liposome carriers in feline GM1 gangliosidosis fibroblasts. Nature 275:754, 1978
44. Baker HJ, Reynolds GD, Walkley SU, et al: The gangliosidoses: Comparative features and research applications. Vet Pathol 16:635, 1979
45. Walkley SU, Baker HJ, Purpura DP: Morphological changes in feline gangliosidosis: A Golgi study. p. 419. In Rose FC, Behon PO (eds): Animal Models of Neurological Diseases. Pitman Medical, London, 1980
46. Baker HJ, Walkley SU, Rattazzi MC, et al: Feline gangliosidoses as models of human lysosomal storage diseases. p. 203. In Desnick RJ, Patterson DF, Scarpelli DG (eds): Animal Models of Inherited Metabolic Diseases. Alan R Liss, New York, 1982
47. Duchen LW, Eicher EM, Jacobs JM, et al: Hereditary leukodystrophy in the mouse: The new mutant twitcher. Brain 103:695, 1980
48. Igisu H, Shimomura K, Kishimoto Y, Suzuli K: Lipids of developing brain of twitcher mouse. Brain 106:405, 1983
49. Igisu H, Suzuki K: Progressive accumulation of toxic metabolite in a genetic leukodystrophy. Science 224:753, 1984
50. Yeager AM, Brennan S, Tiffany C, et al: Prolonged survival and remyelination after hematopoietic cell transplantation in the twitcher mouse. Science 225:1052, 1984
51. Ichioka T, Kishimoto Y, Brennan S, et al: Hematopoietic cell transplantation in murine globoid cell leukodystrophy: Effects on levels of galactosylceramidase, psychosine and galactocerebrosides. Proc Natl Acad Sci USA 84:4259, 1987
52. Hoogerbrugge PM, Suzuki K, Suzuki K, et al: Donor-derived cells in the central nervous system of twitcher mice after bone marrow transplantation. Science 239:1035, 1988
53. Hoogerbrugge PM, Poorthuis BJHM, Romme AE, et al: Effect of bone marrow transplantation on enzyme levels and clinical course in the neurologically affected twitcher mouse. J Clin Invest 81:1790, 1988
54. Suzuki K, Hoogerbrugge PM, Poorthuis BJHM, et al: The twitcher mouse: CNS pathology following bone marrow transplantation. Lab Invest 58:302, 1988
55. Vandervelde M, Frankhauser R, Bichsel P, et al: Hereditary neurovisceral mannosidosis associated with α-mannosidase deficiency in a family of Persian cats. Acta Neuropathol (Berl) 58:64, 1982
56. Walkley SU, Siegel DA: Comparative studies of the CNS in swainsonine-induced and inherited feline α-mannosidosis. p. 57. In James LF, Elbein AD, Molyneux RJ, Warren CD (eds): Swainsonine and Related Glycosidase Inhibitors. Iowa State University Press, Ames, IA, 1989
57. Breider MA, Shull RM, Constantopoulos G: Long-term effects of bone marrow transplantation in dogs with mucopolysaccharidosis I. Am J Pathol 134:677, 1989
58. Constantopoulos G, Scott JA, Shull RM: Corneal opacity in canine MPS I. Invest Ophthalmol 30:1802, 1989

59. Constantopoulos G, Shull RM, Hastings N, Neufeld EF: Neurochemical characterization of canine α-L-iduronidase deficiency disease (a model of human mucopolysaccharidosis I). J Neurochem 45:1213, 1985
60. Shull RM, Breider MA, Constantopoulos G: Long-term neurological effects of bone marrow transplantation in a canine lysosomal storage disease. Pediatr Res 24:347, 1988
61. Shull RM, Hastings N, Selcer RR, et al: Bone marrow transplantation in canine mucopolysaccharidosis I: Effects within the central nervous system. J Clin Invest 79:435, 1987
62. Shull RM, Helman RG, Spellacy E, et al: Morphological and biochemical studies of canine mucopolysaccharidosis I. Am J Pathol 114:487, 1984
63. Shull RM, Munger RJ, Spellacy E, et al: Animal model of human disease: Canine α-L-iduronidase deficiency. A model of mucopolysaccharidosis I. Am J Pathol 109:224, 1982
64. Shull RM, Walker MA: Radiographic findings in a canine model of mucopolysaccharidosis I: Changes associated with bone marrow transplantation. Invest Radiol 23:124, 1988
65. Spellacy E, Shull RM, Constantopoulos G, Neufeld EF: A canine model of human α-L-iduronidase deficiency. Proc Natl Acad Sci USA 80:6091, 1983
66. Haskins ME, Jezyk PF, Desnick RJ, et al: Alpha-L-iduronidase deficiency in a cat: A model of mucopolysaccharidosis I. Pediatr Res 13:1294, 1979
67. Haskins ME, Jezyk P, Desnick RJ, et al: Mucopolysaccharidosis in a domestic short-haired cat: A disease distinct from that seen in the Siamese cat. J Am Vet Med Assoc 175:384, 1979
68. Haskins ME, Aguirre GD, Jezyk PF, et al: The pathology of feline α-L-iduronidase deficient mucopolysaccharidosis. Am J Pathol 112:27, 1983
69. Haskins ME, McGrath JT: Meningiomas in young cats with mucopolysaccharidosis I. J Neuropathol Exp Neurol 42:664, 1983
70. Walkley SU, Haskins ME, Shull R: Alterations in neuron morphology in mucopolysaccharidosis I: A Golgi study. Acta Neuropathol (Berl) 75:611, 1988
71. Cowell KR, Jezyk PF, Haskins ME, Patterson DF: Mucopolysaccharidosis in a cat. J Am Vet Med Assoc 169:334, 1976
72. Jezyk P, Haskins ME, Patterson DF, et al: Mucopolysaccharidosis in a cat with arylsulfatase B deficiency: A model of Maroteaux-Lamy syndrome. Science 198:834, 1977
73. Haskins ME, Jezyk PF, Patterson DF: Mucopolysaccharidosis VI in three families of cats with arylsulfatase B deficiency: Leukocyte studies and carrier identification. Pediatr Res 13:1203, 1979
74. Haskins ME, Aguirre GD, Jezyk PF, Patterson DF: The pathology of feline arylsulfatase B-deficient mucopolysaccharidosis. Am J Pathol 101:657, 1980
75. Haskins ME, Bingle SA, Northington JW, et al: Spinal cord compression and hindlimb paresis in cats with mucopolysaccharidosis VI. J Am Vet Med Assoc 182:983, 1983
76. Aguirre G, Stramm L, Haskins ME: Feline mucopolysaccharidosis VI: General ocular and pigment epithelial pathology. Invest Ophthalmol Vis Sci 24:991, 1983
77. McGovern MD, Mandell N, Haskins ME, Desnick RJ: Animal model studies of allelism: Characterization of arylsulfatase B mutations in homoallelic and heteroallelic (genetic compound) homozygotes with feline mucopolysaccharidosis VI. Genetics 110:733, 1985
78. Konde LJ, Thrall MA, Gasper P, et al: Radiographically visualized changes associated

with mucopolysaccharidosis VI in cats. Vet Radiol 28:223, 1987
79. Gasper PW, Thrall MA, Wenger DA, et al: Correction of feline arylsulfatase B deficiency (mucopolysaccharidosis VI) by bone marrow transplantation. Nature 312:467, 1984
80. Wenger DA, Gasper PW, Thrall MA, et al: Bone marrow transplantation in the feline model of arylsulfatase B deficiency. p. 177. In Krivit W, Paul N (eds): Bone Marrow Transplantation of Lysosomal Storage Diseases. March of Dimes, Birth Defects: Original Article Series 22, 1986
81. Thrall MA, Dial SM, Gasper PW, et al: Bone marrow transplantation as therapy for feline mucopolysaccharidosis VI. In Whitley CB (ed): Mucopolysaccharidosis and Mucolipidosis. Alan R Liss, New York (in press)
82. Dial SM, Byrne T, Thrall MA, et al: Urine glycosaminoglycan concentrations in mucopolysaccharidosis VI affected cats following bone marrow transplantation or leukocyte infusion. Clin Chem (submitted)
83. Haskins ME, Desnick RJ, DiFerrante N: Beta glucuronidase deficiency in a dog: A model of mucopolysaccharidosis VII. Pediatr Res 18:980, 1984
84. Schuchman EH, Tordyan TK, Haskins ME, Desnick RJ: Characterization of the defective β-glucuronidase activity in canine mucopolysaccharidosis type VII. Enzyme 42:174, 1989
85. Wolfe JH, Schuchman EH, Stramm LE, et al: Restoration of normal function in mucopolysaccharidosis type VII cells by retroviral vector-mediated gene transfer. Proc Natl Acad Sci USA 87:2877, 1990
86. Stramm LE, Wolfe JH, Schuchman EH, et al: β-Glucuronidase mediated pathway essential for retinal pigment degradation of glycosaminoglycans. Exp Eye Res 50:521, 1990
87. Birkenmeier EH, Mavisson MT, Beamer WG, et al: Murine mucopolysaccharidosis type VII: Characterization of a mouse with β-glucuronidase deficiency. J Clin Invest 83:1258, 1989
88. Vogler C, Birkenmeier EH, Sly WS, et al: A murine model of mucopolysaccharidosis VII: Gross and microscopic findings in beta-glucuronidase-deficient mice. Am J Pathol 136:207, 1990
89. Paigen K, Swank RT, Tomino S, Ganshow RE. The molecular genetics of mammalian glucuronidase. J Cell Physiol 85:379, 1975
90. Slavin S, Yatziv S: Correction of enzyme deficiency in mice by allogeneic bone marrow transplantation with total lymphoid irradiation. Science 210:150, 1980
91. Yatziv S, Wies L, Morecki S, et al: Long-term enzyme replacement therapy in beta-glucuronidase-deficient mice by allogeneic bone marrow transplantation. J Lab Clin Med 99:792, 1982
92. Hoogerbrugge PM, Poorthuis BJHM, Mulder AH, et al: Correction of lysosomal enzyme deficiency in various organs of beta-glucuronidase-deficient mice by allogeneic bone marrow transplantation. Transplantation 43:609, 1987
93. Sakiyama T, Tsuda M, Kitagawa T. A lysosomal disorder in mice: A model of Neiman-Pick disease. J Inher Metab Dis 5:239, 1982
94. Miyawaki S, Mitsuoka S, Sakiyama T, Kitagawa T: Sphingomyelinosis, a new mutation in the mouse: A model of Niemann-Pick disease in humans. J Hered 73:257, 1982
95. Miyawaki S, Mitsuoka S, Sakiyama T, Kitagawa T: Time course of hepatic lipids accumulation in a strain of mice with an inherited deficiency of sphingomyelinase. J Hered 74:465, 1983
96. Sakiyama T, Tsuda M, Owada M, et al: Bone marrow transplantation for Niemann-

Pick mice. Biochem Biophys Res Comm 113:605, 1983
97. Sakiyama T, Kitagawa T, Jhon H, Miyawaki S: Bone marrow transplantation for Niemann-Pick mice. J Inher Metab Dis 6:129, 1983
98. Sakiyama T, Tsuda M, Owada M, et al: Therapeutic bone marrow transplantation in Niemann-Pick mice. Acta Paediatr Jpn 27:64, 1985
99. Sakiyama T, Tsuda M, Owada M, et al: Bone marrow transplantation in Niemann-Pick mice. J Inher Metab Dis 9:305, 1986
100. Yasumizu R, Miyawaki S, Sugiura, et al: Allogeneic bone marrow plus liver transplantation in the C57BL Ksj Spm Spm Mouse, an animal model of Niemann-Pick disease. Transplantation 49:759, 1990
101. Walkley SU: Vulnerability of gabaergic neurons to neuroaxonal dystrophy in neuronal storage diseases. J Neuropathol Exp Neurol 48:350, 1989
102. Walkley SU, Baker HJ, Rattazzi MC: Initiation and growth of ectopic neurites and meganeurites during postnatal cortical development in ganglioside storage disease. Dev Brain Res 51:167, 1990
103. Walkley SU, Wurzelmann S, Rattazzi MC, Baker HJ: Distribution of ectopic neurite growth and other geometrical distortions of CNS neurons in feline GM2 gangliosidosis. Brain Res 510:63, 1990
104. Walkley SU, Wurzelmann S, Siegel DA: Ectopic axon hillock-associated neurite growth is maintained in metabolically reversed swainsonine-induced neuronal storage disease. Brain Res 410:89, 1987

12

BONE MARROW TRANSPLANTATION FOR STORAGE DISEASES

William Krivit · Elsa G. Shapiro

INTRODUCTION

More than 150 patients with storage diseases have undergone bone marrow transplantation during the past 10 years. Several general observations can be made:

1. Normal levels of leukocyte enzymatic activity derived from the newly engrafted donor marrow always occur after successful engraftment.
2. Enzyme replacement is life-long after 1 year of engraftment, and no further medications are required.
3. The abnormal substrates decrease to within normal limits after replacement of the deficient enzyme in these inborn errors of metabolism, on the basis of histologic and biochemical analyses of biopsied hepatocytes, tonsils, conjunctiva, and bone marrow, as well as urine and cerebrospinal fluid (CSF).
4. This metabolic correction has consistently improved the visceral lesions, such as hepatosplenomegaly and sleep apnea, in mucopolysaccharide syndromes.
5. Neuropsychological capabilities after bone marrow transplantation for storage diseases have varied widely. Some patients have maintained their rate of learning or intelligence quotient (IQ), while others have continued to deteriorate. This heterogeneity must result from the basic genetic differences between patients with the same enzymatic deficiency. Furthermore, the entrance of the enzyme into the central nervous system (CNS) by means of monocyte diapedesis only occurs over a period of months to years. The variations in the transplant process and the basic tempo of the primary disease also account for the major differences in CNS outcome.

6. The challenge now is to better define the limitations of, and the benefit from, the use of bone marrow transplantation for storage diseases.

In 20 institutions in the United States and Canada, we have formed a consortium to develop protocols to assess the efficacy of bone marrow transplantation in different storage diseases. Eligibility rules, biochemical and molecular designation of genetic heterogeneity, and specific protocols for neuroradiologic, neurophysiologic, and neuropsychological evaluation have been established. Criteria for failure or success with the use of specified end points have been identified. Patients will or will not receive transplants based on the presence or absence of histocompatible donors. This prospective comparative study will evaluate results by a rigorous statistical analysis.

BACKGROUND

Cross-correction In Vitro

The lysosome, an organelle identified by the presence of acid phosphatase inside the membrane-lined sac, was first described by DeDuve.[1] The first delineation of a lysosomal storage disease (i.e., acid α-glucosidase deficiency or Pompe disease) was made by Hers.[2] Neufeld[3] used abnormal fibroblasts to determine whether the pathophysiology of the "giant lysosomes" could be corrected by co-culturing with normal fibroblasts. This dramatic disappearance of the abnormal lysosomal accumulation in the diseased cells taken from patients with Hurler and Hunter syndromes was designated a *cross-correction* phenomenon. This is now an accepted, generalized concept that has been shown to be a phenomenon dependent on the uptake of secreted lysosomal enzymes from the media of cultured fibroblasts by the mannose 6-phosphate receptor-mediated pathway.[4,5] Lymphocytes also have served as a source of enzymatic correction of diseased fibroblasts.[6] In addition, glial cell cross-correction between normal and hypoxanthine-guanine phosphoribosyltransferase (HGPRT)-deficient cells has been demonstrated.[7]

Cross-correction In Vivo: The Hepatocyte and Kupffer Cell

Cross-correction has been investigated using bone marrow transplantation as an experimental tool. The bone marrow origin of hepatic macrophages (Kupffer cells) was demonstrated in humans by Y chromosome analysis, after transplantation.[8] Experimentally, the rapidity of this turnover is within 8 days after bone marrow transplantation.[9] The Kupffer cells permanently change from that of recipient to donor type. The Kupffer cells eventually are derived entirely from donor and, as such, will continuously provide a source of normal enzymatic activity. The intimate relationship between a Kupffer cell and its hepatocytes has served to demonstrate this phenomenon quite readily. In animals with glucuronidase deficiency, the replacement of enzymatically deficient cells by normal Kupffer cells is followed by enzymatic correction in the adjacent hepatocytes.[10] In patients who have received a bone marrow transplant, biopsies of the liver have shown that this cross-correction occurs in vivo. The evidence includes the decreased substrate concentration as measured biochemically and assessed morphologically.[11]

Microglia of the Central Nervous System

The microglia in the CNS are the cells analogous to the Kupffer cells. The total number of microglia in the mouse brain, 3.5×10^6, is comparable to the number of Kupffer cells in the liver on a weight-for-weight basis. However, microglia possess up to twice the membrane surface area of Kupffer cells (i.e., the perimeter of a microglia cell is approximately seven times that of a Kupffer cell[12]). The proportion of cells in brain that are microglia varies from 5 percent in the cortex and corpus callosum to 12 percent in the substantia nigra. In normal human white matter, the proportion of glial cells that are microglia is even greater, about 15 percent. Microglia are the major cell type expressing major histocompatibility complex (MHC) class II antigens in human white matter.

The microglia are derived from the bone marrow hematopoietic system. This finding is based on evidence accumulated in recent years based on several different animal models.[13-16] The localization of the hematopoietically derived microglia in close proximity to other cells has resulted in the clearance of accumulated substrates and the presence of increased enzymatic activity in the brain after bone marrow transplantation. Dogs with α-L-iduronidase or α-fucosidase deficiencies have been studied; in each, there has been morphologic and biochemical evidence of substrate degradation and return of the pathologic state to normal status after bone marrow transplantation.[17,18] In addition, in the α-fucosidase model, prevention of clinical disease has been observed by very early transplantation.[17] After bone marrow engraftment in humans, male (donor) cells have been demonstrated by Y chromosome in situ hybridization to be present in the brain parenchyma of female recipients (Unger E, Sung JH, Blazer B, Krivit W: unpublished results).

Krabbe disease (galactosylceramidase deficiency) in the mouse (twitcher mice) is an excellent model for the study of bone marrow transplantation. In this model, bone marrow transplantation has been reported to change the course of disease. The improvement in the neurologic outcome occurred with a decrease in psychosine, the "toxic" substrate. Concomitantly, the pathognomonic globoid cells disappeared. The clinical counterpart of this experience is presented in the accompanying chapter on bone marrow transplantation in human Krabbe disease (see Ch. 13).

The above experimental data are consonant with clinical observations that a period of at least 6 months to 1 year is needed to demonstrate changes in the neuropsychological and neurologic findings. For example, the fact that enzyme can be detected in the CNS several months after transplantation is presented in Chapter 13. In this disease, a 13-year-old child affected with the juvenile-onset form of Krabbe disease showed improvement in cognitive function during the year after bone marrow transplantation. Galactosylceramidase has been documented in the brain of another Krabbe disease patient who had been engrafted for 2 years. Conversely, the enzyme was absent in the brains of two patients who, despite being fully engrafted, died of graft-versus-host disease (GVHD) 2 months after transplantation. Taken together, the data support a tentative conclusion that the time required for enzyme penetration of the CNS is a minimum of several months.

Bone marrow transplantation for CNS correction in experimental lysosomal-deficient animal models has previously been reviewed.[19] The conclusion from all these data is that new bone marrow hematopoietic stem cell proliferation does provide a normal donor set of macrophages, lymphocytes, monocytes, and microglia. The entrance rate is variable and depends on the disease entity, the animal studied, the age of transplant, the age of onset and rapidity of disease process, the type of transplant, engraftment percentage, and the sensitivity of methods used to detect new enzyme in the CNS.

Blood-Brain Barrier—Not an Impediment to Leukocyte Diapedesis

The blood-brain barrier is a real and well-demonstrated physiologic mechanism that protects the brain from high-molecular-weight proteins that circulate in the plasma. No infringement on this concept need be made to understand how the CNS can receive new enzymatic activity. Circulating leukocytes (monocytes) enter the brain parenchyma by diapedesis through the capillaries, set up residence, and proliferate. Two clinical observations seen after bone marrow transplant support this concept.[20]

First, in metachromatic leukodystrophy, in which disease is limited to the CNS, with no impaired visceral involvement, we have noted seven patients who had bone marrow transplantation whose clinical course is disparate from that of either historical or family controls.[21,22] The only feasible explanation for the clinical changes is that new donor monocyte-macrophages enter the CNS and cause substrate (sulfatide) decrease and new enzyme utilization.

Second, enzymatic activity in the plasma is not present after bone marrow transplantation, in contrast to the normal amounts found in the leukocytes.[23] This discordance is to be expected, since hepatocytes are presumably a major source of plasma enzymes. Bone marrow is the site of leukocyte enzymatic repopulation. Thus, any amelioration of CNS signs and symptoms in patients after bone marrow transplantation must rely on cellular enzyme content in the leukocytes and not plasmatic activity. The blood-brain barrier is concerned with plasma proteins. Therefore, theoretical consideration of this barrier as an impediment to correction of CNS disease is presumed to be erroneous.

Turnover of Microglial Cell

"A central question for our understanding of interactions between the mononuclear phagocyte system and the CNS is information about the kinetics of the microglial population." The Oxford research team[24] has begun a systematic assessment of microglia turnover by injecting mice with a single pulse of [^{3}H]thymidine and by identifying the microglia with immunocytochemical techniques with the antibody F4/80+. After a single pulse of [^{3}H]thymidine and a 1-hour survival period, small numbers of cells with typical microglial morphology labeled with [^{3}H]thymidine were seen throughout the brain. Although the labeling index is low after a single pulse, the large numbers of microglia present in the brain mean that within a few months, there will be significant turnover of resident cells. In addition, "preliminary results show that 3 months after reconstitution of female mice irradiated and reconstituted with bone marrow from male mice of the same strain, small numbers of male cells are found in the choroid plexus and in the parenchyma in small groups."[24]

The kinetics of substrate removal must take into account the amount of abnormal material already accumulated. In disease states that occur in infancy, such as metachromatic leukodystrophy or Krabbe disease, or if there is already a major accumulation, as in neurologically disabled patients, the enzyme-substrate relationships cannot be anticipated to be in favor of clearance. Thus, the clinical diseases will not abate unless enzymatic reconstitution is present for many months or longer and there is minimal or no clinical neurologic disease before transplantation.

SPECIFIC DISEASE ENTITIES TREATED

Mucopolysaccharidoses

Abnormalities in the gray matter, as well as macrocephaly, are characteristic of patients with mucopolysaccaridoses (MPS). Patients with Hurler syndrome demonstrate a general delay in development and progressive deterioration of overall cognitive function that varies in rapidity. The range of IQ (Bayley-MDI) scores in children with Hurler syndrome at the University of Minnesota, before transplantation, ranged from 50 to 102 at ages 1.0 to 3.1 years. No specific pattern of neuropsychological deficit was present, although memory deficits and attentional problems have been described. Deafness is characteristic of many of these syndromes and may complicate the assessment of cognitive development. Hurler-Scheie syndrome is characterized by somewhat less retarded intellectual development.

Among several other factors that need to be considered are age and intellectual capability at time of transplant, zygosity of donor, problems related to GVHD, and inability to compare the transplant results with the historical controls appropriately because of the vast difference in medical-social care and attention provided to the former and not to the latter group of patients. Genetic heterogeneity, of all the confounding factors, is most important. Investigators agree that the use of modern molecular biologic techniques will demonstrate a broad spectrum of DNA defects.[25–27] In the future, we should be able to stratify patients on the basis of molecular and clinical data, which will permit selection of those who will most benefit from bone marrow transplantation.

Results of Bone Marrow Transplantation in Mucopolysaccharidoses

The clinical results of bone marrow transplantation in MPS may be summarized as follows.[28–31] First, permanent correction of the metabolic defect in these enzymatic deficiencies of the MPS syndromes results after engraftment of normal donor cells in the recipient. Leukocyte enzymatic activity has been shown to be the same as that of the donor. There is a strong positive correlation between the enzymatic level in the leukocytes and other commonly used measures of engraftment, such as restriction fragment-length polymorphism (RFLP), sex chromosome, red blood cell genotype, and human leukocyte antigen (HLA) typing. No patient that has been completely engrafted for 1 year has lost enzyme status. The longest periods of observation have extended over 10 years.

Second, no further medication is required to maintain this long-lived engraftment status. Although the immune dysfunction engendered by the marrow transplant process is present initially, it is ameliorated considerably by the first year, especially if GVHD was minimal. Only if there is residual and ongoing GVHD will medications be needed.

Third, the visceral disease present in the MPS syndromes abates after long-term engraftment. The accumulation of mucopolysaccharide in the urine, CSF, liver, lymph nodes, tonsils, conjunctiva, and skin decreases. Previous descriptions of heart failure, pneumonias, and hypoxic episodes (sleep apnea) are improved or reversed. The ventriculomegaly and elevated CSF pressure are not major problems after a long period of engraftment. Electroretinography has been shown to improve after engraftment.[32]

Fourth, the severe form of the facial dysmorphology disappears. Patients who are now 8 to 10 years post-transplant have a distinctive facies. The bridge of the nose is flattened. There is an elfin quality about the high cheek bones. Hirsutism is decreased considerably. The thickness of the facial features typical of untreated Hurler syndrome is not present.

Fifth, bone growth continues in the upper and lower extremities and in the vertebral column above and below the gibbus area. However, the gibbus formation remains a major problem. Orthopaedic invasive procedures may be necessary in many of these children. In addition, the London group has noted epiphyseal damage in the femoral head.[29,30]

Intelligence and neurologic outcomes, however, vary considerably. CNS dysfunction and deterioration is a frequent occurrence immediately after bone marrow transplantation. This has been variously attributed to either the preparative or anti-GVHD regimens. However, recuperation from this post-transplant effect occurs within months. In most children, developmental or intelligence quotients stabilize.

Intellectual status post-transplantation has been a source of major concern, since many patients have remained with significant mental impairment.

MPS I: α-L-Iduronidase Deficiency

A number of patients with α-L-iduronidase deficiency (Hurler: MPS IH, IH/S, IS) have been successfully transplanted throughout the world. Despite this large clinical experience, attempts at careful interpretation and prediction of intellectual and neurologic outcome are thwarted because of the number of variables present in these original phase I and II studies.

Foremost among the variables is that the natural history of the disease process is not well characterized. Intrafamilial variation has been documented.[33] Therefore, individual case reports comparing results between transplanted and nontransplanted affected siblings must be viewed with caution. The failure to deteriorate intellectually over a period of many years has been seen by clinical investigators.[34] Similar data are found in the transplantation experience. Yet, in a patient who has lost engraftment and who continues to fail to exhibit enzymatic activity after bone marrow transplantation, intellectual achievement has not deteriorated[34] (Krivit W: personal communication).

Amazingly, there has been no longitudinal study of a group of patients with α-L-iduronidase deficiency in a carefully standardized neuropsychologic manner, on a yearly basis. Bias can naturally prejudice results. In the transplant series, patients and their parents benefit from the enhanced stimulation of recurrent examinations, intrinsic optimism, and potentially improved medical care. This is different from the experience with nontransplanted children.

The intrinsic inability of knowing whether the patients with a better response were Hurler-Scheie patients also clouds interpretation. We need to document prospectively the characteristics of enzymatic activity of the deficient α-L-iduronidase in vitro by uniformly examining the enzyme and molecular lesions for functional differences.[35,36] In addition, the mucopolysacchariduria needs to be prospectively assessed on a qualitative and quantitative basis to classify patients into the appropriate disease category.[37]

MPS II: Iduronate Sulfatase Deficiency

Disease caused by a deficiency of iduronate sulfatase has been divided into a severe form (A) and mild form (B).[38,39] The mild form has a very slow onset during the childhood years, and intelligence is minimally affected. Skeletal and other morphometric observations similarly cause minimal problems until the second and third decades of life.

The severe form of iduronate sulfatase deficiency (MPS II or Hunter syndrome) is characterized by pronounced and devastating psychomotor retardation. The developmental delay leads to severe mental deficiency. This is a result of the accumulation of mucopolysaccharide in the CNS. The hepatosplenomegaly, skeletal deformity, and cerebral ventricular dilatation are quite reminiscent of other severe forms of mucopolysaccharidoses.

Specific skin lesions are typical of iduronate sulfatase deficiency. The viscera are involved, but not to the degree of other mucopolysaccharidoses. The ongoing progression of disease results in significant morbidity and early demise. The deposition of substrate in the connective tissues and bones causes significant dwarfism, restriction in the skeletal and integumentary systems, loss of ambulation, and loss of fine and major motor capability. In the severe Hunter patients, macrocephaly within the first years of life is marked. Enlargement of the head beyond the 97th percentile of head circumference occurs.

Ten patients have received bone marrow transplantation for iduronate sulfatase deficiency. The enzymatic activity in leukocytes has become normal in four patients and low in three patients. The mucopolysacchariduria has been reduced to almost normal levels. Biopsies of the liver and bone marrow have shown a disappearance or marked reduction of the lysosomal abnormalities seen before transplantation. There is considerable discussion and concern relative to the intelligence subsequent to bone marrow transplantation.

Seattle[40]: Biopsies were obtained of hepatic and rectal tissue before bone marrow transplantation, and at 4 months and 1 year after bone marrow transplantation in a 4-year-old boy with MPS II. After transplantation, hepatocytes showed a reduction in the size of lysosomal vacuoles containing amorphous, fibrillogranular material. Fibroblasts within the rectal submucosa showed a marked reduction in the number of lysosomal structures containing fibrillogranular material. Neurons and Schwann cells demonstrated numerous mitochondria-sized vacuoles with polymorphous, filamentous, concentrically stacked plates that were closely packed. The appearance of the neurons (rectal) did not change after transplantation. Sequential histologic evaluation documented improvement in hepatocytes and fibroblasts 1 year after bone marrow transplantation. Evidence for histologic improvement in peripheral nerve and ganglion cells cannot be substantiated. The intellectual prognosis remains guarded (Scott R: personal communication).

Cleveland[41]: A 3½-year follow-up of a patient with Hunter syndrome (MPS II) treated at 7.5 years of age with allogeneic bone marrow transplantation (from his HLA-MLC-identical sister) demonstrated improvement. Sustained engraftment was documented by donor sex chromosomes and ABO group on several occasions. Iduronate sulfatase was undetectable before bone marrow transplantation in leukocytes or serum; normal leukocyte levels and low but definite serum levels were present on multiple determinations 2.5 to 38 months after bone marrow transplantation. Urinary glycosaminoglycan (GAG) excretion markedly decreased toward normal. After bone marrow transplantation, hepatomegaly resolved. Hepatic storage lysosomes disappeared and hepatic GAG content normalized 24 months after bone marrow transplantation from 2200 µg/g liver to a mean of 160 µg/g (nl = 42 to 293 µg/g). CSF GAG was normal 24 months after bone marrow transplantation. Progressive clinical improvement has been apparent since 8 weeks after bone marrow transplantation. Subcutaneous nodules and storage material in skin disappeared, hirsutism and coarse facial features diminished, joint contractures showed a marked decrease in severity, and fine motor skills improved significantly.

His temperament became progressively less aggressive. He grew approximately 14 cm in height after bone marrow transplantation. The tracheostomy was removed 38 months after bone marrow transplantation. His performance IQ of 87 at 23 months after bone marrow transplantation was unchanged from pretransplant status. His verbal skills decreased slightly, resulting in a full-scale IQ of 73. He was doing well in the 4th grade. The patient died suddenly at home 3.5 years after bone marrow transplant. Autopsy confirmed chronic pneumonia and revealed a patent airway, normal liver histology, and presence of storage GAG in heart valves, chondrocytes, and brain. We conclude that bone marrow transplant successfully reversed many of the features of Hunter syndrome in this patient and at least stabilized his CNS.

London[42]: "Two survivors had severe disease and had their bone marrow transplantations at only 5 years of age. In both patients, severe chronic GVHD developed with low levels of leukocyte enzyme. Neither showed any evidence of mental improvement and, although their hepatosplenomegaly has gone, their IQs are steadily falling."

Philadelphia: One patient has just recently been transplanted and maintains IQ status with low enzyme levels (August C: personal communication).

Minnesota: One patient was transplanted at 4.3 years of age. Before bone marrow transplant, at age 4 years 2 months, this child demonstrated a Verbal Reasoning score of 84 on the Stanford Binet Intelligence Scale, an Abstract Visual Reasoning score of 79, and a Short-Term Memory score of 74. No change was noted on retesting 3 months after transplant. Ten months after transplant, his Verbal Reasoning score dropped to 77, his Abstract Visual Reasoning score to 67, and his Short-Term Memory score to 65. This drop was likely the result of no additional learning during the 10-month period. He was tested 3 years after transplant and his Verbal Reasoning score had dropped to 65, his Abstract Verbal Reasoning score had dropped to 55, and his Short-Term Memory score had dropped to 47. It appears that no recovery of function had taken place and, in fact, he was demonstrating no new learning (Chang P-N, Whitley CB, Krivit W: unpublished results).

Japan[43]: The early reports on this 9-year-old child are still in preliminary status.

If bone marrow transplantation has any role, earlier diagnosis and intervention will be needed. The plans for the Consortium Study include bone marrow transplant restricted to before the third birthday, with a minimal developmental quotient of 75 or greater. The future patients need to have genotype testing performed.

MPS III: Sanfilippo Syndromes

The Sanfilippo syndromes (MPS type III) include four separate enzymatic disorders:

Type A: Heparan *N*-sulfatase
Type B: α-*N*-Acetylglucosaminidase
Type C: Acetyl CoA:α-glucosaminidase acetyltransferase
Type D: *N*-Acetylglucosamine 6-sulfatase

Patients with the Sanfilippo types show a wide variation in intelligence with evidence of progressive mental deterioration, usually at 2 to 3 years of life, but in some cases at a later age.[44–46]

As a general observation, the behavioral problems presented in this syndrome are extreme. The hyperactive aggressive personality associated with delayed development make the care and management of these children almost impossible. There are mild radiographic changes, but the skeletal system remains primarily intact. The musculature is intact. The speech is delayed with poor articulation, and there is severe hearing loss. The patients are subject to temper tantrums, destructive behavior, and physical aggression. There is coarse hair, hirsutism, and mild hepatosplenomegaly. Computed tomography (CT) scan of the head demonstrates mild to moderately increased ventricular size with considerable atrophy. In all the patients studied to date, there has been biochemical evidence of metabolic correction. The abnormalities noted in the bone marrow and liver have disappeared.

Fourteen patients have been transplanted: four at the University of Minnesota; two at the St. Petersberg All Children's Hospital; two (twins) at Westminster; two in Nancy, France[30]; three at the University of Kentucky, Lexington; and one at Children's Hospital of Los Angeles. The range of IQ in the four patients before transplant at the University of Minnesota was 50 to 85 at ages 2.6 to 7.0.

After careful review and analysis of all the data presented at the various meetings, and after interviewing the investigators, the plans for the Consortium Study of Sanfillipo patients will include transplant by at least the third birthday and a developmental quotient of 75 or greater.

MPS IV: Morquio Disease

Despite observations that cartilaginous cells will be too distant from blood vessels to acquire normal enzyme, Morquio disease has been treated with bone marrow transplantation.[47–49] Five patients have now had bone marrow transplants. In each case, the visceral disease manifestations have improved, but the skeletal abnormalities have not changed. In the future, earlier marrow transplants may be more successful in remodeling bone. This is based on the assumption that growth cartilage and epiphyses in infancy will be exposed directly to the new bone marrow cells. This might allow for changes in eventual skeletal defects heretofore accepted as untreatable in Morquio disease.

MPS VI: Arylsulfatase B Deficiency

The deficiency of arylsulfatase B (*N*-acetylglucosamine-galactosamine 4-sulfatase) in MPS VI (Maroteaux-Lamy) causes accumulation of dermatan sulfate throughout the body. The disease presents in a multisystem fashion and has considerable variation in the degree of severity. The skeletal system presents with mild to moderate degrees of dysostosis multiplex. Stiffening of the joints and thickening of the skin causes considerable disability. The cardiorespiratory system is impaired by the presence of pulmonary hypertension and resultant cor pulmonale and heart failure.

In most case reports, the CNS impairment is secondary to accumulation of the mucopolysaccharide material in the meninges. The obstruction around the cervical vertebra and base of the brain has frequently resulted in moderate to severe hydrocephalus. Surgical removal of the obstructive mucopolysaccharide has helped during placement of a ventriculoperitoneal shunt. In addition to this obstructive phenomenon, the frequent episodes of hypoxia and heart failure compound the problems of full development of intellectual capacity. Although these patients with Maroteaux-Lamy syndrome have the highest intelligence of all the MPS syndromes, there are reports of mental retardation.[50]

Our original patient was transplanted almost 10 years ago at the age of 13.[51,52] At transplantation, the patient was in severe congestive heart failure and having sleep apneic spells 30 seconds in length, with PaO_2 values in the severely hypoxic range. Subsequently, as noted in our two published reports, these signs and symptoms have disappeared, and she has matriculated in college. We had to do a tracheostomy at the time of bone marrow transplant. Cor pulmonale relapse has occurred once. The tracheostomy has not been amenable to removal. There has been considerable improvement in the facial dysmorphology. Yet, because all the epiphyseal centers were formed and involved with the mucopolysaccharidosis before the transplant, the skeletal defects remain as before.

A second patient at the University of Minnesota has been fully engrafted and now has metabolic correction of the disease process. His marrow transplant was done at the age of 2.3 years and he is now 18 months since transplant (Whitely CB, Krivit W: personal communication).

The third patient with Maroteaux-Lamy syndrome has been successfully engrafted in London (COGENT Report).[53] At 9 years of age, the engraftment went smoothly, and the resolution of the skeletal system defects has provided an increase in the more than the usual range of joint movements. There is decreased corneal clouding and disappearance of hepatosplenomegaly. Echocardiography has shown a change in the lesions in the heart valves. Exercise tolerance has improved. Metabolic correction has been achieved.

A fourth patient has been transplanted in Japan.[43] A fifth patient has been transplanted at Tulane Medical Center and is reported to have had marked resolution of clinical signs and symptoms.[54] A sixth patient was 8 years old but died of cardiac insufficiency and sepsis 11 days post-transplant.[55] A seventh patient has been transplanted in Malaysia (Krivit W: personal communication).

The clinical description of all these patients is varied, and there is ambiguity in discerning mild from moderate and severe involvement. Considerable discussion and review of each of these patients will be necessary before one can discern a pattern to determine which patient would most likely benefit from transplantation.

Leukodystrophies

Metachromatic Leukodystrophy

Metachromatic leukodystrophy has been reviewed and recommendations provided for appropriate timing of bone marrow transplantation.[21] Again, consummate judgment must be used concerning when not to transplant. If there are ongoing and rapidly developing clinical signs and symptoms, the patient should not undergo transplantation. However, if the patient is asymptomatic (i.e., the problem has been detected because other family members have the disease), one must confirm the diagnosis by ensuring absence of pseudodeficiency allele by enzyme analysis or by performing careful sulfatide loading tests before attempting transplantation.

The decision to transplant a patient with ongoing signs and symptoms of neurologic disease must be carefully reviewed. Juvenile and adult forms of metachromatic leukodystrophy offer the opportunity of CNS enzymatic replenishment before the disease process has proceeded too far. Conversely, in most instances of the late-infantile form of metachromatic leukodystrophy, the disease process has proceeded to the point where transplant is contraindicated. Somewhere between the juvenile and late-infantile form, there are patients whose disease process is so slow that they may benefit from bone marrow transplantation.

Globoid Cell Leukodystrophy

Similar caveats must be observed for globoid cell leukodystrophy as for metachromatic leukodystrophy, concerning the appropriate and correct choices of whom to transplant. Infantile Krabbe disease should be excluded, as the course is too rapid and the damage too great to anticipate any significant improvement. Juvenile and adult forms of globoid cell leukodystrophy should be considered. Globoid cell leukodystrophy is reviewed in detail in Chapter 13.

X-Linked Adrenoleukodystrophy in Children

Adrenoleukodystrophy (ALD) is a family of disorders with severe neonatal to milder adult-onset forms. These disorders show increased levels of very long chain fatty acid (VLCFA) in plasma (see below). Those persons not under consideration for bone marrow transplantation are adult patients with adrenomyeloneuropathy or symptomatic heterozygotes. Those with neonatal adrenoleukodystrophy are also excluded, as are patients who have Addison disease alone.

No symptoms whatsoever are noted during the first 3 years of life in a boy who will later succumb to a severe and rapid neurologic incapacitation. The first signs and symptoms are those of an attention-deficit disorder. These symptoms progress despite medication and become more prominent. Thereafter, psychomotor retardation becomes severe. The interval between onset of the first neurologic symptoms to deterioration of the neurologic system to a vegetative state is rapid. The mean time from the beginning of changes to severe neurologic symptoms is approximately 2 years. There is considerable variation from patient to patient, relative to the rate of neurologic degeneration. Death results from the decerebrate status and its complications. The mean time of death is 7.2 years.

The presence of increased levels of VLCFA in the plasma has become the standard method of identifying affected persons. Evidence of an increase in VLCFA in the brain of those affected persons has been reported. This disease of the white matter is associated with colorblindness, and the gene for this disorder has been localized to chromosome Xq28. A defect of C26:0 lignoceryl-CoA ligase has also been identified. A DNA probe, DXS52, has been useful for carrier detection.

VLCFA are derived from both dietary and endogenous sources. In order to reduce the levels of VLCFA, an artificial diet has been constructed and is being clinically evaluated. In boys with rapidly progressive childhood ALD, biochemical normalization has been achieved. Unfortunately, in those already ill with neurologic disease, clinical progression was not arrested. Consequently, to prevent neurologic deterioration, dietary treatment was begun in asymptomatic affected males.

Identification of asymptomatic affected males was undertaken by screening families at risk and by running public awareness campaigns. The diet was then started and normalization of levels of VLCFA obtained. Nevertheless, clinical progression was noted despite presymptomatic dietary therapy. The first signs were identified by magnetic resonance imaging (MRI) abnormalities. MRI scans of the brain of patients who continued to deteriorate neurologically were shown to be superior to CT scans in detecting minimal lesions. MRI scans demonstrated small unilateral lesions in the lateral lemniscus or the inferior colliculus. Later, CT scans were able to demonstrate periventricular white matter lesions in the posterior parietal and occipital lobes.

A recent report by Aubourg et al.[56] demonstrated that bone marrow transplantation led to reversal of early neurologic and neuroradiologic manifestations of X-linked adrenoleukodystrophy. Previous reports of bone marrow transplantation for adrenoleukodystrophy[57,58] had demonstrated reversal to normalization of the VLCFA abnormalities. The first patient with ALD who was treated with bone marrow transplantation was already incapacitated neurologically and so severely involved that no clinical change was noted.

Similarly, the initial patient treated by Parkman and Weinberg[59] had progressed beyond potential for recovery and died some time after bone marrow transplantation. However, the brother of the latter patient was then transplanted when still presymptomatic. According to recent personal communications, the clinical neurologic and MRI scans now 2 years post-transplantation are different than those of his brother. The final outcome is not yet evident.

Transplantation should be contraindicated for all patients with clinically existing severe neurologic disability or rapidly advancing neurologic disability. Patients to be included in the Consortium Study will be derived from the population of identified affected males from previous screening. All patients will have been on dietary therapy for adrenoleukodystrophy. All these patients will have undergone careful neurologic examination and neurophysiologic studies and, most importantly, an MRI scan will have been performed every 6 months or at the first sign of clinical disease.

Appropriate planning for the identification of a donor is already in progress. HLA-identical siblings who are not affected with ALD will be sought in the families followed in the dietary study. Since HLA typing has already been done in the dietary study, the identification of appropriate donor-recipient matches can be established more readily than in other circumstances.

Those to be considered for transplant will be asymptomatic, having minimal MRI abnormalities. As soon as the MRI scan has been confirmed to be abnormal and neuropsychological and neurophysiologic tests have been completed, transplantation will be undertaken as soon as possible. Follow-up measurements would obviously include the neurophysiologic and neuropsychological testing and MRI examinations.

FORMATION OF A CONSORTIUM

The decision concerning bone marrow transplantation is one of the more complex and difficult in the medical field, both for the family and for all the physicians concerned. This becomes of even greater difficulty in the discussions concerning the use of bone marrow transplantation for storage diseases. This is because we have multiple equations of disease type, donor match, age, intelligence quotient, and genetic heterogeneity. The desperation of the parents who are suddenly faced with these diagnoses is of major concern, as they are so willing to "try anything" when faced with the course of disease outlined for their children by their physicians.

Because of all these complex questions, we have formed a consortium to provide more satisfactory answers than we have at present. Twenty centers at which such transplants have been done in the United States and Canada have come together to participate in a study project. Increasingly, the need for a transplant/nontransplant study has been recognized. The method of comparative analysis cannot be done on a truly randomized basis, from an ethical point of view, since most investigators and parents have seen evidence of regression of disease. The natural distribution of compatible donors has been discussed in these Consortium meetings. Fifty percent of the patients can be anticipated to have a donor that would be generally considered acceptable. The risk includes mortality and long-term serious and very consequential morbidity. Thus, the intrinsic availability of appropriate donors within the HLA-histocompatibility system will provide an unbiased method for doing a comparative study. In both groups, identical protocol studies for neuropsychological, neuroradiologic, and neurophysiologic measurements will be conducted and followed for several years.

Furthermore, because of the intrinsic nature of the genetic heterogeneity in these diseases, we must be able to use the probes being developed and that are now available. This is now being done for the Gaucher, metachromatic leukodystrophy, and Hunter patients and will be followed within a very short period for the other diseases. While waiting for these to be ascertained and confirmed, our intention is to obtain and preserve fibroblasts prospectively from each of the patients entered into the study for later analysis, when specific probes become available.

We have formulated rules of eligibility. These include specific age limits (2 years in Hurler and 3 years in Hunter and Sanfilippo syndromes) and an IQ or DQ of 75 or greater. End points for determining neuropsychological failure have been set. These include three criteria: (1) IQ/DQ falling below 50 or a 33 percent loss, (2) deterioration in adaptive functioning below 50 or 25 percent loss, and (3) deterioration in language functioning below 50 or 25 percent loss.

The rationale for these end points was developed after considerable discussion. The end points represent regression from the beginning of the transplant of a significant degree. To be considered worthwhile, the transplant process should have an intellectual outcome that is robust enough to sustain scrutiny. The choice of end point considered for statistical analysis and failure also stems from a projected viewpoint of what an IQ of 50 actually means in functional terms, many years later.

A person with an IQ of 50 is functioning at a level of 7 to 8 years of age. This means that such persons will never be self-sufficient, economically or otherwise. They will not be responsible as a parent, nor will they be capable of having a responsible heterosexual relationship. They may be able to make simple monetary change but could not be trusted to manage a checkbook or pay bills. They could work in a supervised situation but could not be independent and responsible on a job. Although they could manage simple food preparation and cleaning, they would be unable to take charge of meals, cleaning, and care of clothing. They could dress themselves and keep clean and neat, but could not monitor their own health needs and might need reminders about hygiene. They could read and write simply but would be unlikely to understand the world around them well. Moreover, coping in an emergency is unlikely, nor could they give directions or describe the means to achieve a goal. They would be unlikely to weigh the consequences of their actions independently before making a decision, nor would they be likely to control their anger or hurt feelings when denied their own way.

CONCLUSION

This retrospective review has provided the theoretical considerations and some of the positive and negative aspects of bone marrow transplantation for storage diseases. There is reason to be optimistic, but at the same time cautious. Treatment of lysosomal storage diseases with bone marrow transplantation has proved efficacious. The decision process has become more complex. Questions arise as to which disease is amenable to transplantation, the appropriate age, the limitations imposed by donor availability, and the long-term results. We have learned what we may and may not anticipate from the bone marrow transplantation process in several of these disorders. The time has come to do a prospective and comparative analysis, so that we can provide more specific recommendations for the future. The Consortium Study should provide enough data to help us stratify patients for the maximum responses for both the short term and long term.

REFERENCES

1. DeDuve C: Lysosomes revisited. Eur J Biochem 123:391, 1983
2. Hers HG: Inborn lysosomal diseases. Gastroenterology 48:625, 1965
3. Neufeld E: Replacement of genotype-specific proteins in mucopolysaccharidosis. In Bergsma D, Desnick RJ, Krivit W (eds): Enzyme Therapy in Genetic Diseases. Birth Defects 9(2):27, 1973; Williams & Wilkins, Baltimore, 1973
4. Kornfeld S: Trafficking of lysosomal enzymes in normal and disease states. J Clin Invest 77:1, 1986
5. Kihara H, Porter M, Fluharty A: Enzyme replacement in cultured fibroblasts from metachromatic leukodystophy. In Bergsma D, Desnick RJ, Krivit W (eds): Enzyme Therapy in Genetic Diseases. Birth Defects 9(2):19, 1973; Williams & Wilkins, Baltimore, 1973
6. Olsen I, Muir H, Smith R, et al: Direct enzyme transfer from lymphocytes is specific. Nature 306:75, 1983
7. Gruber HE, Koenker R, Luchtman V, et al: Glial cells metabolically cooperate. Potential requirements for gene replacement therapy. Proc Natl Acad Sci USA 82:6662, 1985
8. Gale RP, Sparkes RS, Golde DW: Bone marrow origin of hepatic macrophages (Kupffer cells) in humans. Science 201:937, 1978
9. Paradis K, Sharp HL, Vallera DA, Blazar BR: Kupffer cell engraftment across the major histocompatibility barrier in mice: Bone marrow origin, class II antigen expression and antigen-presenting capacity. J Pediatr Gastroenterol Nutr 11:525, 1990
10. Hoogerbrugge PM, Poorthius BJHM, Wagemaker G, Van Bekkum DW: Bone marrow correction of lysosomal enzyme deficiency in various organs of β-glucuronidase deficient mice by allogeneic bone marrow transplantation. Transplantation 43(5):609, 1987
11. Whitley CB, Ramsay NKC, Kersey JH, Krivit W: Bone morrow transplantation for Hurler syndrome: Assessment of metabolic correction. p. 7. In Krivit W, Paul NW (eds): Bone Marrow Transplantation for Treatment of Lysosomal Storage Diseases. Alan R Liss, New York, 1986
12. Lawson LJ, Perry VH, Dri P, Gordon S: Heterogeneity in the distribution and morphology of microglia in the normal and adult mouse brain. Neuroscience 39:151, 1990
13. Perry VH, Gordon S: Macrophages and microglia in the nervous system. Trends Neurosci 11:273, 1988
14. Ting JP-Y, Nixon DF, Weiner LP, Frelinger JA: Brain Ia antigens have a bone marrow origin. Immunogenetics 17:295, 1983
15. Hickey WF, Kimura H: Perivascular microglial cells of the CNS are bone marrow derived and present antigen in vivo. Science 239:20, 1988

16. Hoogebrugge PM, Suzuki K, Suzuki K, et al: Donor-derived cells in the central nervous system of twitcher mouse after bone marrow transplantation. Science 239:1035, 1988
17. Shull RM, Hastings NE, Selcer RR, et al: Bone marrow transplantation in canine mucopolysaccharidosis I: Effects within the central nervous system. J Clin Invest 79:435, 1987
18. Taylor RM, Stewart GJ, Farrow BRH: Correction of enzyme deficiency by allogeneic bone marrow transplantation following total lymphoid irradiation in dogs with lysosomal storage diseases (fucosidosis). Transplant Proc 18:326, 1986
19. Krivit W, Whitley CB, Chang PN, et al: Lysosomal storage diseases treated by bone marrow transplantation: Review of 21 patients. p. 261. In Johnson E, Pochedly C (eds): Bone Marrow Transplantation in Children. Raven Press, New York, 1990
20. Krivit W, Whitley CB, Lund G, et al: Improvement of clinical expression of central nervous system manifestation in lysosomal storage diseases treated by bone marrow transplantation. p. 189. In Baum SJ, Santos GW, Takaku F (eds): Experimental Hematology Today—1987: Recent Advances and Future Directions in Bone Marrow Transplantation. Proceedings of a symposium held in conjuction with the Sixteenth Annual Meeting of the International Society for Experimental Hematology, August 23–28, Tokyo, Japan. Springer-Verlag, New York, 1987
21. Krivit W: Recommendations for treatment of metachromatic leukodystrophy by bone marrow transplantation based on a review of seven patients who have been engrafted for at least 1 year. Am J Med Genet (in press)
22. Krivit W, Shapiro E, Kennedy W, et al: Effective treatment of late infantile metachromatic leukodystrophy by bone marrow transplantation. N Engl J Med 322:28, 1990
23. Krivit W, Whitley CB, Chang PN, et al: Lysosomal storage diseases treated by bone marrow transplantation. p. 367. In Gale RP, Champlin RE, (eds): Bone Marrow Transplantation: Current Controversies. UCLA Symposia, 1988. Alan R Liss, New York, 1989
24. Perry VH, Gordon S: Macrophages and the nervous system. International Rev Cytol 125:203, 1991
25. Hopwood J, Ashton L, Brooks P, et al: Biochemistry and molecular genetics of α-L-iduronidase. p. 16. Presented at the Second International Symposium on the Mucopolysaccharidoses and Related Diseases, Manchester, England, August 31 to September 3, 1990 (abst)
26. Hopwood J, Anson D, Bielicki J, et al: Hunter syndrome and the iduronate-2-sulphatase gene. p. 18. Presented at the Second International Symposium on the Mucopolysaccharidoses and Related Diseases, Manchester, England, August 31 to September 3, 1990 (abst)
27. Stolzfus LJ, Uhrhammer N, Soda-Pineda B, et al: Mucopolysaccharidosis I; cloning and characterization of cDNA encoding canine α-L-iduronidase. J Hum Genet 47:167, 1990 (abst)
28. Whitley CB, Krivit W, Kersey JH, et al: Bone marrow transplantation for mucopolysaccharidosis disease: The Minnesota experience. p. 27. Presented at the Second International Symposium on the Mucopolysaccharidosis and Related Disorders, Manchester, England, August 31 to September 3, 1990 (abst)
29. Hobbs JR: Bone marrow transplantation for mucopolysaccharide disease: The Westminster experience. p. 27. Presented at the Second International Symposium on the Mucopolysaccharidosis and Related Disorders, Manchester, England, August 31 to September 3, 1990 (abst)

30. Hugh-Jones K, Hobbs JR, Vellodi A, et al: Long-term follow-up of children with Hurler's disease treated with bone marrow transplantation. p. 103. In Hobbs JR (ed): Correction of Certain Genetic Diseases by Transplantation. Westminster Medical School Research Trust, London, 1989
31. Bordigoni P, Vidailhet M, Lena M, et al: Bone marrow transplantation for Sanfilippo syndrome. p. 114. In Hobbs JR (ed): Correction of Certain Genetic Diseases by Transplantation. Westminster Medical School Research Trust, London, 1989
32. Summers CG, Purple RL, Whitley CB, et al: Ocular changes in mucopolysaccharidosis following bone marrow transplantation. Ophthalmology 96:977, 1990
33. McDowell GA, Cowan TM, Greene CL, Blitzer MG: Intrafamilial clinical variability in lysosomal storage disease: Implications for studies of treatment. Am J Hum Genet 47:A163, 1990
34. Henslee-Downey PJ, Pettigrew AL, Ciocci G, et al: Bone marrow transplantation in Hurler syndrome: Am J Hum Genet 47:A158, 1990
35. Matalon R, Deanching M, Omura K: Hurler, Hurler-Scheie and Scheie compound: Residual activity of α-L-iduronidase toward natural substrate suggesting allelic mutations. J Inher Metab Dis 6(suppl 2):133, 1983
36. Muller VJ, Hopwood JJ: α-L-Iduronidase deficiency in MPS type I. Clin Genet 26:414, 1984
37. Matalon R: Characteristics of urinary glycosaminoglycan excretion in Hurler, Hurler-Scheie, and Scheie patients. p. 253. In Wapnir R (ed): Congenital Metabolic Diseases. Vol. 15. Dekker, New York, 1985
38. Young ID, Harper PS: Mild form of Hunter's syndrome: Clinical delineation based on 31 cases. Arch Dis Child 57:828, 1987
39. Young ID, Harper PS: The natural history of the severe form of Hunter's syndrome: A study based on 52 cases. Dev Med Child Neurol 25:481, 1983
40. Persyk A, Rutledge JC, Saunders JE, Scott CR: Bone marrow transplantation for Hunter syndrome (MPS II): Evidence for selective histologic improvement one-year post transplant. p. 80. Presented at the Fifth International Congress of Inborn Errors of Metabolism, Asilomar, June 1–5, 1990 (abst)
41. Warkentin PI, Strandjord SE, Whitley CB, Coccia PF: Correction of Hunter syndrome with bone marrow transplantation: A three year follow-up report. p. 71. Presented at the First International Congress on Mucopolysaccharidoses and Related Diseases, Minneapolis, Minnesota, May 20–22, 1988 (abst)
42. Hobbs JR: Hunter's disease p. 13. In Hobbs JR (ed): Correction of Certain Genetic Diseases by Transplantation. Westminster Medical School Research Trust, London, 1989
43. Imaizumi M, Gushi K, Suzuki J, et al: Bone marrow transplantation for inborn errors of metabolism. Jpn Soc Inher Metab Dis 6(suppl 41):140, 1990
44. van Schrojenstein-de Valk HM, van de Kamp JJ: Followup on seven adult patients with mild Sanfilippo B-disease. Am J Med Genet 28:125, 1987
45. Wraith JE, Danks DM, Rogers JG: Mild Sanfilippo syndrome: A further cause of hyperactivity and behavioural disturbance. Med J Aust 147:450, 1987
46. Nidiffer FD, Kelly TE: Developmental and degenerative patterns associated with cognitive, behavioural, and motor difficulties in the Sanfilippo syndrome: An epidemiological study. J Ment Defic Res 27:185, 1983
47. Russell S, Vowels M, McDonald J, et al: Bone marrow transplantation for Morquio's syndrome. p. 62. Presented at the Proceedings of the Second

International Symposium on the Mucopolysaccharidoses, Manchester, England, August 31 to September 3, 1990
48. Kato S, Kubota C, Yabe H, et al: Bone marrow transplantation in Morquio's disease. In Hobbs JR (ed): Correction of Certain Genetic Diseases by Transplantation. Westminster Medical School Research Trust, London, 1989
49. Desai S, Hobbs JR, Hugh-Jones, et al: Morquio's disease (MPS type VI) treated by bone marrow transplant. Exp Hematol 2(suppl 13):98, 1983
50. Vestermark S, Tonnesen T, Anderson MS, Guttler F: Mental retardation in a patient with Maroteaux-Lamy. Clin Genet 31:114, 1987
51. Krivit W, Pierpont ME, Ayaz K, et al: Bone marrow transplantation in Maroteaux-Lamy disease (mucopolysaccharidosis VI): Correction of the enzymatic defect. N Engl J Med 31:1606, 1984
52. McGovern MM, Ludman MD, Short MP, et al: Bone marrow transplantation in Maroteaux-Lamy syndrome (MPS type VI): Status 40 months after BMT. p. 41. In Krivit W, Paul NW (eds): Bone Marrow Transplantation for Treatment of Lysosomal Storage Diseases. Alan R Liss, New York, 1985
53. Jurges E, Jones S, Hancock M, et al: Bone marrow transplantation for Maroteaux-Lamy mucopolysaccharidoses. p. 127. In Hobbs JR (ed): Correction of Certain Genetic Diseases by Transplantation. Westminster Medical School Research Trust, London, 1989
54. Kirkpatrick DV, Barrios NJ, Shapira E, Humbert JR: Treatment of enzyme storage diseases with matched and partially matched bone marrow transplantation: Tulane experience. Blood 76(suppl l):548, 1990
55. Ringden O, Groth CG, Aschan J, et al: Bone marrow transplantation for metabolic disorders at Huddinge Hospital. Transplant Proc 22:198, 1990
56. Aubourg P, Blanche S, Jambaque I, et al: Reversal of early neurologic and neuroradiologic manifestations of X-linked adrenoleukodystrophy by bone marrow transplantation. N Engl J Med 322:1860, 1990
57. Moser HW, Tutschka PJ, Brown FR III, et al: Bone marrow transplant in adrenoleukodystrophy. Neurology 34:1410, 1984
58. Yeager AM, Moser HW, Tustchka PJ, et al: Allogeneic bone marrow transplantation in adrenoleukodystrophy: Clinical, pathologic, and biochemical studies. In Krivit W, Paul NW (eds): Bone Marrow Transplant for Treatment of Lysosomal Storage Diseases. Birth Defects 22:79, 1986; March of Dimes Foundation—Alan R Liss, New York, 1986
59. Weinberg K, Moser A, Watkins PA, et al: Bone marrow transplantation for adrenoleukodystrophy. Pediatr Res 23(suppl):334A, 1988

# 13

BONE MARROW TRANSPLANTATION AS TREATMENT FOR GLOBOID CELL LEUKODYSTROPHY

Elsa G. Shapiro · Lawrence Lockman ·
William Kennedy · Donald Zimmerman ·
Edwin H. Kolodny · Srinivasa Raghavan ·
German Wiederschain · David A. Wenger ·
Joo-Ho Sung · C. Gail Summers ·
William Krivit

INTRODUCTION

Bone marrow transplantation as a method for the treatment of the central nervous system (CNS) pathology in several storage diseases remains controversial. This divergence of opinions is attributable to the varying interpretations of reports of individual patients in small series of cases. The marked genetic heterogeneity, even in siblings in the same families, provides reason for critical review of published material. A prospectively planned comparative study of patients in transplant and nontransplant groups with appropriate statistical analysis has been initiated. The importance of this latter study is underlined because of the increasing potential to be realized from autologous gene therapy in the future if allogeneic transplantation is successful.

In Chapter 12 the requisites for considering bone marrow transplantation depended on the salutary effects of improvement in visceral organ function. These changes can have a positive effect on the CNS in an indirect manner, for example, by decreasing hypoxic episodes. Conversely, the pattern of deterioration in the leukodystrophies, which is asserted primarily in processing and conceptual skills, is independent of visceral disease. Yet, bone marrow transplantation has provided positive effects as noted in late infantile metachromatic leukodystrophy and adrenoleukodystrophy patients.

Because the time sequence of transplanted hematopoietic cells traversing the brain is measured in many months, a positive effect can only be expected when the disease process is minimal and the rate of decline is relatively slow. Several leukodystrophies meet these criteria. Among these, the juvenile form of globoid cell leukodystrophy was considered potentially reversible after bone marrow transplantation. Globoid cell leukodystrophy is due to a deficiency of galactocerebrosidase (galactocerebroside β-galactosidase). This deficiency results in the accumulation of psychosine and white matter degeneration associated with the formation of large macrophages in the brain. These large (globoid) cells provide a distinctive morphologic marker of the disease process. The white matter undergoes progressive demyelination. Gliosis and continued destruction of the neuronal system occur, leading to decerebrate posturing and death.[1]

Clinical expression of the globoid cell leukodystrophy varies considerably, depending on the patient's age at onset. The classic disease presents in infancy with rapid development of incapacitation and seizures leading to death by 18 months of age. Juvenile and adolescent globoid cell leukodystrophy are descriptive terms denoting onset at later ages and a protracted clinical course.

Clinically, bone marrow transplantation has been shown to halt disease progression and to provide a treatment for metachromatic leukodystrophy and adrenoleukodystrophy.[2-4] Successful treatment of these diseases depends on engraftment from the normal donor, which then provides a permanent source of normal enzymatic activity.

Globoid cell leukodystrophy has an experimental animal counterpart, the *twitcher mouse*. This mouse model is characterized by globoid cells in the brain and the deficiency of galactocerebrosidase.[5,6] After bone marrow transplantation in the twitcher mouse, normal galactocerebrosidase is present in peripheral leukocytes. Disappearance of the globoid cells in the brain has been noted. The life span is changed from 100 percent demise at 28 days to 50 percent alive at 1 year. The neurologic behavior is improved, and the hindlimb paralysis is diminished considerably after bone marrow transplantation.[7-9] Biochemical and pathologic abnormalities in the CNS are corrected or markedly different after bone marrow transplantation.[10-12] There is one report to the contrary.[13]

Furthermore, donor cells do enter the CNS from the new hematopoietic system after bone marrow transplantation.[14,15] Mac-1 antigen, a marker of hematopoietic cell origin, has been noted on globoid cells in these experimental animals.[16] There has also been corroboration of this effect in the human experience by use of in situ hybridization techniques for the Y chromosome. Female recipients who receive a male marrow donor in transplantation for leukemia and malignancies have been studied. In these appropriate male donor-female recipient pairs, the male hematopoietic cells have been found in the CNS subsequent to engraftment over 100 days.[17] Because of these findings, we believe that bone marrow transplantation in human globoid cell leukodystrophy is appropriate for clinical trial. We report the results in one patient.

BONE MARROW TRANSPLANTATION IN GLOBOID CELL LEUKODYSTROPHY

Case History

The patient was 12 years of age (DOB 3/22/77) at the time she underwent bone marrow transplantation on April 14, 1989. Impaired vision caused by increasing optic atrophy had first been noted at 5 years of age. Colorblindness was noted early. School performance had been gradually deteriorating since 8 years of age. Tremors of the hands and ataxia began at 9 years of age and became more noticeable during the past 2 years. Motor incoordination and stumbling were increasing. Magnetic resonance imaging (MRI) of the brain at the Mayo Clinic showed an increase in the T_2 signal from October 1986 to December 1988. A sural nerve biopsy in November 1986 showed slight diffuse reduction in density of myelinated fibers.

Muscle biopsy revealed prominent renaut bodies with occasional onion bulbing. Occasional fibers were noted within folded loops of myelin and reduplicated myelin. Teased fibers revealed 32 percent type A, 7 percent type C, 12 percent type D, 12 percent type E, and 38 percent type F. The pathologic process was considered most compatible with a demyelinating disorder with remyelination.

Several abnormalities were first noted on neurologic examination at 9 years of age. These included intention tremor, more prominent on the left and a postural tremor on the left. Ataxia, a positive Babinski sign, positive Romberg test, and diminished vibratory and position sensation in the lower extremities were noted. These were obvious immediately before transplantation.

There was significant wasting of the thenar, hypothenar, quadriceps, hamstrings, anterior and posterior tibialis muscles, extensors of the toes, and peronei muscles. The brachioradialis tendon reflexes were slightly diminished, but the Achilles reflex was increased and accompanied by two to three beats of clonus. The patient displayed a diminished ability to walk on her heels, as well as to skip and hop. There was incoordination on the toe-to-finger test. Rapid alternating movements were decreased. Two-point discrimination was 4 mm in the fingertips. Neurophysiologic data from the Mayo Clinic obtained before transplantation are incorporated below.

Visual acuity was 20/60 OD, and 20/70 OS before transplantation. Eye examination was significant for a small left exotropia, abnormal color vision tested with the Ishihara isochromatic plates, and bilateral moderate diffuse optic atrophy.

A verbal IQ of 102 and a performance IQ of 72 were noted before transplantation on the Wechsler Intelligence Scale for Children, Revised (WISC-R) (administered at the Mayo Clinic). Kaufman Assessment Battery for Children scores were Sequential, 106 and Simultaneous Processing, 86. Reading ability was at the 56th percentile and mathematics at the 62nd percentile. Further neuropsychological testing at the University of Minnesota Pediatric Neurology Clinic disclosed significant deficits on tests of visual-spatial ability, namely, face recognition and spatial orientation. She had difficulties in object recognition, which could be described as a partial visual agnosia. Below-average performance was also noted in verbal fluency and receptive vocabulary. Intact performance on measures of complex language and problem-solving ability was noted. Behaviorally, she was noted to be quite serious and mildly restricted in her range of affect. She had excellent motivation to perform well and was exceedingly cautious and careful in her approach to tasks.

Family History

At the same time that juvenile globoid cell leukodystrophy was diagnosed in the patient, her two sisters (M and G) were also noted to have deficiency of galactocerebrosidase. Confirmation was obtained by study of fibroblast cultures from all three girls. Three brothers were unaffected, and a fourth brother and her parents were found to be carriers. The levels of galactocerebrosidase noted in the leukocytes and fibroblasts in the family are noted in Table 13-1.

The major difficulty suffered by the affected sisters has been pes cavus, for which they have had corrective surgery. Muscle strength in the thenar and hypothenar muscle groups, tibialis anterior, extensors of the toes, and peronei groups was diminished. Deep tendon reflexes were diminished for the biceps, brachioradialis, triceps, quadriceps, and gastrocnemius-soleus muscles. Diminished vibratory sensation was noted in the toes. MRI showed increased signal along the pathways of the optic radiations in both posterior temporal and occipital regions. Beginning atrophy of the posterior third of the corpus callosum was minimal. Motor nerve conduction studies revealed a demyelinating neuropathy (median nerve, 27 m/s; peroneal nerve, 24 m/s on December 9, 1988 for affected sibling G). The intelligence testing for affected sibling G

Table 13.1 Galactocerebrosidase Levels in Leukocytes and Fibroblasts From Family Members[a]

Source	Leukocytes	Fibroblasts
	(nmol/mg protein/h)	
Patient, recipient	0	0.3, 0.04, 0.2
Mother, carrier	0.3	1.0
Father, carrier	1.5	1.6
Sister [M], affected	0	0.1
Sister [G], affected	0	0.04
Brother [K], donor, normal		2.7
Brother [A], normal	3.5	
Brother [E], carrier	0.8	
Brother [N], normal	3.3	
Normal control subjects	2.6 ± 1.6 (2.4, 2.7, 1.8, 2.3)	1.5 ± 0.4
Prior case of globoid cell leukodystrophy		0.05
Prior carrier		2.0, 2.3

[a] Determined by Dr. Kolodny and Dr. Natowicz.

resulted in a verbal IQ of 90 and a performance IQ of 93 on December 7, 1988. The same tests revealed a verbal IQ of 96 and a performance IQ of 91 for the other affected sibling, M.

Bone Marrow Transplant

The preparative regimen consisted of busulfan 1.25 mg/kg qid PO for 4 days, followed by cyclophosphamide 60 mg/kg IV over 1 hour for each of 4 consecutive days. MESNA was also given during the latter 4 days to prevent hemorrhagic cystitis.

The bone marrow, taken from patient's HLA-identical, MLC-nonreactive, homozygous, normal brother (K), amounted to nucleated cell count of 5.0×10^8/kg of recipient weight. Engraftment was noted to be occurring by day 9. A white blood cell (WBC) count of 3800, with an absolute neutrophil count of 2500, was present on day 13. On day 16 post-transplant, donor engraftment was documented by presence of normal galactocerebrosidase levels (Table 13-2), male sex chromosome, DNA markers (restriction fragment length polymorphism), and red blood cell genotypic analysis.

Prevention of graft-versus-host disease (GVHD) was provided by prednisone and cyclosporine, given intravenously during the first 2 weeks, and then orally thereafter. GVHD was documented by skin rash with a positive skin biopsy along with a mild watery diarrhea on day 18. Prednisone at 60 mg/m^2 was therefore given, per protocol. She was discharged from hospital on day 26 after the bone marrow transplantation.

On day 28 post-transplant, a generalized seizure was noted. This had been preceded by several days of increasing listlessness and headache. No evidence of infection was detected. No vascular events were observed in a computed tomography (CT) scan of the head. Prednisone dosage was also reduced rapidly.

Table 13-2. Galactocerebrosidase Leukocyte Activity Levels Before/After Transplantation[a]

Date	Days Post-transplant	Enzymatic Activity (nmol/mg protein/h)
Pretransplant		
March 1989		0.06
Post-BMT		
May 1989	16	4.2
June 1989	55	4.8
June 1989	70	4.4
October 1989	180	7.1
June 1990	310	3.9
Brother (donor) normal		5.0
Normal control subjects		4.64 ± 1.92 SD
Prior cases of GLD		0–0.3 (range)
Prior cases of carrier		0.8–5.3 (range)

Abbreviations: BMT, bone marrow transplantation; GLD, globoid cell leukodystrophy.
[a] Determined by Dr. D. Wenger.

Phenytoin treatment was begun. Cyclosporine level was 228. Cyclosporine was discontinued until levels were in the 75 to 125 range. Subsequently there has been no recurrence of seizure. Later phenytoin was discontinued and carbamazepine (Tegretol) was given for two dosages only.

On day 100 post-transplant, a lip biopsy was positive for GVHD. Schirmer's test was normal even though dryness of eyes were noted clinically. Teardrops were provided as therapy. These ocular difficulties have subsequently completely abated.

RESULTS OF BONE MARROW TRANSPLANTATION

Endocrinologic Data

Autoimmune thyroiditis had been diagnosed in October 1986 at the Mayo Clinic. The initial visit had shown antithyroid microsomal antibodies, anti-insulin antibodies, and a normal glucose tolerance test. The large palpable thyroid regressed to normal size after the patient received thyroxine 0.075 mg/d for 2 years. Thyroid studies before and after transplant were normal. She continues on thyroid hormone replacement.

The patient's height at transplant was between the 10th and 25th percentiles and weight was at the 5th percentile. Subsequent to bone marrow transplantation, the height percentile has declined to between the 5th and 10th percentiles. Weight for height has remained between the 10th and 25th percentiles. After bone marrow transplantation, primary ovarian failure has supervened: estradiol less than 10 pg/ml; follicle-stimulating hormone (FSH) 234 IU/L; luteinizing hormone (LH) 115 IU/L.

Neurologic Examination

On neurologic examination, 14 months after bone marrow transplantation, cranial nerve function was normal. Although muscle bulk was reduced, strength was normal. Deep tendon reflexes were slightly reduced in the biceps, triceps, and brachioradialis but brisk at the quadriceps and gastrocnemius-soleus, where there was unsustained clonus. The Babinski sign was equivocal. There was no postural or intention tremor, but mild dysmetria was noted on finger-nose-finger testing. Gait was narrow-base and shuffling, in part attributable to tight heel cords. The Romberg sign was negative. Tandem gait was still difficult but improved as compared with her previous examination.

Ophthalmologic Examination

Fourteen months after transplant, visual acuity was measured at 20/50 OD, and 20/60 OS, and an intermittent exotropia persisted. The diffuse optic atrophy was unchanged, and Goldman perimetry revealed stable paracentral shallow scotomas.

Motor Nerve Conduction

Motor nerve conduction studies have shown a slight but clear decline in velocity. Ulnar nerve motor conduction velocity was 31 m/s in October 1986 and December 1988, 32 m/s in June 1989, and after transplant, 31 m/s in October 1989 and 26 m/s in June 1990. Conduction velocity studies in the median, peroneal, and tibial nerves have followed a similar pattern. Sensory nerve conduction as well as F and H waves have also shown slight decline.

Visual Evoked Potentials

Visual evoked potentials, using a black-and-white checkboard pattern as stimulus, reversing at a rate of 2.1 Hz to stimulate each eye independently, showed a P-100 wave latency on the left of 177.5 ms and on the right of 160 ms on December 9, 1988 (Mayo Clinic). Six months after transplant on October 12, 1989, the P-100 was 145.2 ms on right and 130.8 ms on the left. One year after transplant, the P-100 latency was 141 ms on the right and 119 ms on the left. Although these are improved, the latencies are still considerably prolonged. The amplitude and morphology of P-100 was considered better the year after transplantation.

Brain Stem Auditory Evoked Potentials

The brain stem auditory evoked potential (BAEP) test was considered within normal limits bilaterally before and after transplant. Wave latencies subsequent to bone marrow transplant have not increased, as would be expected if there were an ongoing leukodystrophic process. Instead, the wave latencies have decreased toward the median of normal limits.

Somatosensory Evoked Potentials in Lower Extremities

The somatosensory evoked potentials in the lower extremities were poorly resolved, with no cervical waves identified after stimulation of either side before transplant. On December 8, 1988, the N/P37 was 55.2 ms on the left and 43.2 ms on the right. Both values were abnormally prolonged. One year after transplant, there were no latency changes. However, the responses were better formed and reproducible as compared with the original findings.

Somatosensory Evoked Responses in Upper Extremities

The somatosensory evoked potentials (SSER) in the upper extremities were well configured and reproducible, as compared with the lower extremities noted above. The left side, but not the right, suggested a mild peripheral dysfunction. There also was a suggestion of involvement of the somatosensory pathway on the left side above the craniocervical junction. The SSER in the upper extremities was within the same range, both at 6 months and 1 year after transplantation.

Magnetic Resonance Imaging

Before transplantation, MRI showed changes in the intracranial neural tissues, including severe atrophy of the posterior third of the corpus callosum, with demyelinating changes extending laterally from that structure into the white matter of the cerebral hemispheres. The changes in the white matter extended forward to the posterior frontal region and posteriorly to the occipital poles. Demyelination was also seen in the descending tracks in the brain stem, anteriorly in the pons, and posterolaterally in the cerebral peduncles. These changes were moderately more advanced in December 1988, as compared with the study in October 1986, and were considered evidence of progression of disease. The MRI taken in June 1990 showed no progression or change in any of the above T_2 signal characteristics.

Electroencephalogram

An electroencephalogram (EEG) taken on March 21, 1989 showed significant abnormalities with diffuse slowing and frequent sharp waves from the temporal areas. One year after transplantation, despite the one episode of seizures on day 28 after transplantation caused by cyclosporine and prednisone toxicity, there was considerable improvement in the electrical activity as compared with the original observations.

Neuropsychological Data

Six months after transplantation, moderate improvement was seen on intellectual testing with a verbal IQ of 108 and a performance score of 85 on the WISC-R. This performance was maintained with a verbal scale of 107 and a performance scale of 81 one year after transplantation (Fig. 13-1). (The four-point decrement in performance IQ from 6 months to 1 year after transplant was within the error of measurement on this scale and was insignificant.)

On achievement testing, the patient's performance in reading showed major improvement with a change from the 56th to the 99th percentiles for her age. Math performance remained at the average level for her age (Fig. 13-2).

On verbal testing, she showed significant improvement in fluency and in receptive vocabulary (Fig. 13-3). Nonverbal and perceptual tasks were stable (Fig. 13-4). However, she continued to demonstrate a mild visual agnosia. Fine motor coordination was stable, although coordinated movement with both hands showed some improvement. Behaviorally, her range of affect was slightly better and, according to her parents' subjective judgment, improvement was seen in her overall adjustment.

Cerebrospinal fluid

Cerebrospinal fluid protein measurements were 76 mg percent on July 7, 1988, 123 mg percent immediately before transplant, 95 mg percent 28 days

Bone Marrow Transplantation as Treatment for Globoid Cell Leukodystrophy

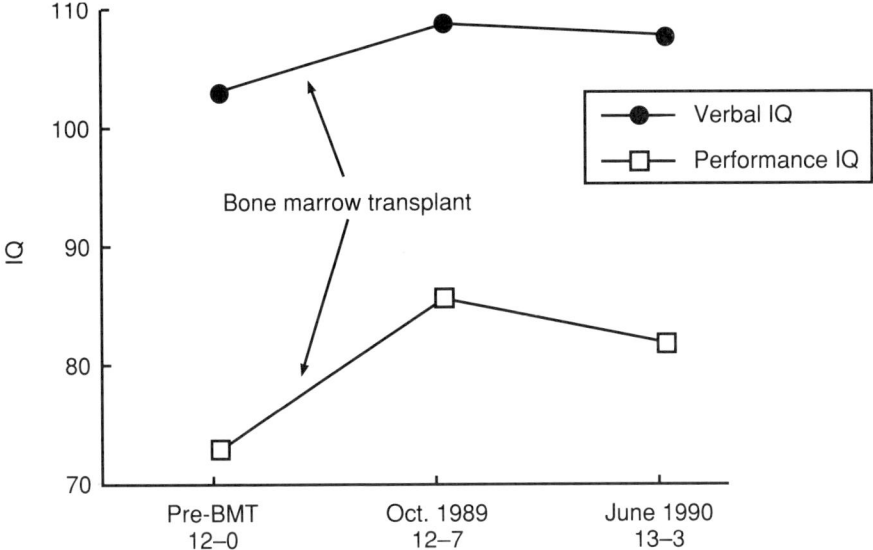

Fig. 13-1. WISC-R verbal and performance.

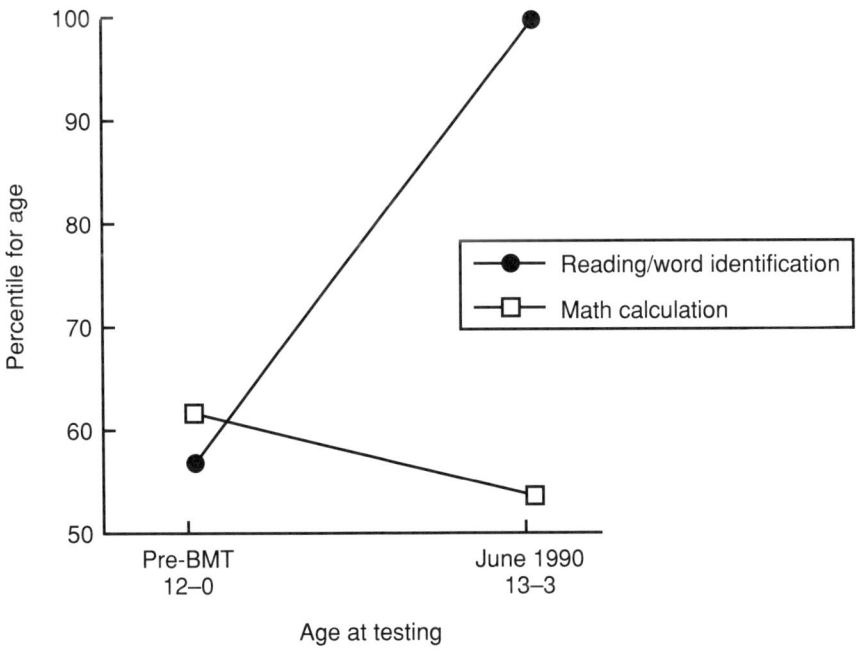

Fig. 13-2. Achievement test percentiles.

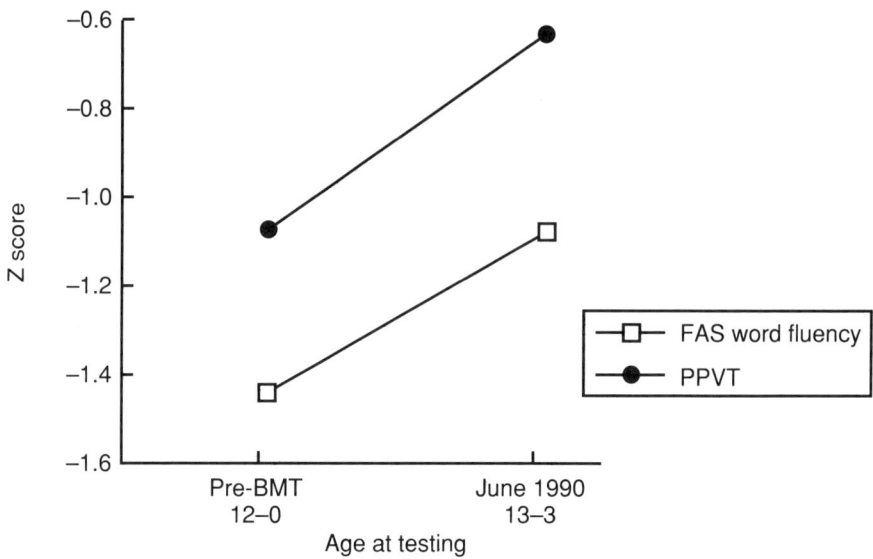

Fig. 13-3. Scores on language tests.

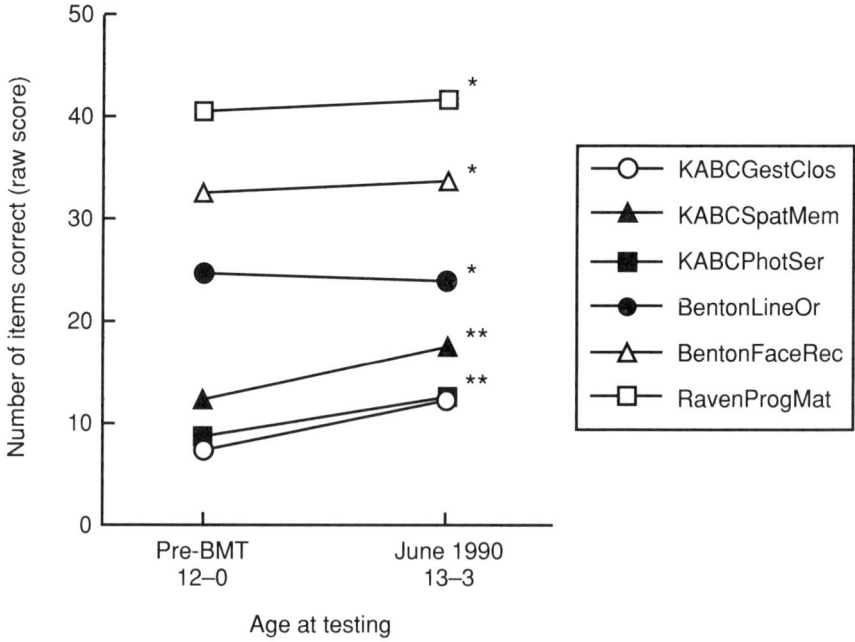

Fig. 13-4. Nonverbal neuropsychological test results. One asterisk, within the normal range; two asterisks, below the normal range.

after transplant, and 60 mg percent 6 months after transplant. Myelin basic protein was normal throughout.

DISCUSSION

The patient has maintained engraftment for more than 1 year after transplantation. A normal level of galactocerebrosidase is now present, as noted in Table 13-2. The increase to above donor levels of enzymatic activity has been noted in all successfully engrafted patients in each of the metabolic diseases. This observation is interpreted as evidence of enrichment of peripheral blood by younger cells from the regenerating marrow stem cell population. Other evidence of full engraftment from the donor in this patient includes continued 100 percent DNA restriction fragment-length polymorphism (RFLP) of the donor as well as appropriate changes in leukocyte sex chromosome and red blood cell genotype markers. Increased CSF protein, typical for globoid cell leukodystrophy, was noted before transplantation and has subsequently become normal.

Improvement in school performance has been documented. Excellent marks have been obtained since the transplant, as compared with previous marks, when she was almost failing her grade. Indeed, this positive effect is noted, even though the school system provided a challenge by allowing the patient to advance a grade level in order to stay with her age peers subsequent to transplantation.

Supporting this observation in school are the changes in scores on reading achievement from a normal to a superior level. Improvement in verbal fluency and receptive vocabulary were seen. The patient appears to be making progress in all areas that involve language and academic performance. Gains have been noted in visual memory, perceptual acuity, and organization. Her mild visual agnosia and problems in perceptual recognition remain stable. Although below average for her age, stability in fine motor coordination is seen. Overall, her intellectual performance has kept up with her age. Previous deficiencies in verbal, achievement measures, and some specific visual-processing abilities have diminished. Loss of the ataxia and tremors occurred during the year after transplantation. The positive changes in the visual evoked response as well as stability in the results of the EEG and BAEP tests were noted.

A previous report of bone marrow transplantation in the treatment of globoid cell leukodystrophy from Minnesota included the effect of transplantation on two patients with infantile Krabbe disease.[18] Our first patient was transplanted at 6 months of age from a histocompatible identical sibling donor. Complications of CNS viral and bacterial infections occurred during the peritransplant period.

Stability at the arrested neurologic status during the 3 years of life was marked by failure to gain any developmental milestones. Galactocerebrosidase leukocyte levels remained normal throughout. Death was caused by an overwhelming respiratory infection. Postmortem biopsy of the cerebrum and cerebellum revealed diffuse degeneration of the white matter and gliosis without globoid cells. Enzymatic activity of galactocerebrosidase was detected in the cerebral cortex and cerebellum, but not in the brain stem. The methods used for assay of galactocerebrosidase included labeled 6-[^{3}H]galactocerebroside ([^{3}H]-Gal-Cer) and a novel flurogeneic substrate, 6-hexadecanoyl-amino-4-methylumbelliferyl-β-D-galactopyranoside (HMGal) (Table 13-3). The amount of activity was approximately 10 percent of the control value, as detected by the latter method (patient UPN #0603).

The second infant was transplanted at 3 months of age with a T-cell-depleted marrow from a sibling and remained engrafted for 2 years. Seizures and decerebrate status, which had occurred in an affected older sibling, were not present during the time our patient was engrafted. He subsequently lost engraftment. The failure to retain permanent engraftment is to be anticipated with an increased frequency when the donor marrow is T-cell depleted.

The third and fourth patients were twin girls, aged 18, with rapidly developing neurologic signs. The diagnosis of globoid cell leukodystrophy was established by the demonstration of deficiency of galactocerebrosidase activity in both leukocytes and fibroblast culture. Unfortunately, marrow transplantation from unrelated donors in both patients resulted in their demise as a result of overwhelming GVHD. Galactocerebrosidase activity in the peripheral leukocytes was noted to be that of donor normal levels indicative of full engraftment. At postmortem examination, classic globoid cells were present in the CNS. Enzymatic activity in the CNS was detectable at an extremely low level (Table

Table 13-3. Galactocerebrosidase Activity in Brain of Control and Patients With Krabbe Disease[a]

Source and Brain Region	Enzymatic Activity (nmol/mg protein/h)	
	[^{3}H]-Gal-Cer	HMGal
Control		
Cerebral cortex	2.04	3.3
Brain stem	1.97	2.9
Cerebellum	2.60	4.35
Patient UPN #0603 (3 yr post-transplantation)		
Cerebral cortex	0.05	0.28
Brain stem	0.00	0.00
Cerebellum	0.03	0.33
Patient UPN #1159 (45 d post-transplantation)		
Cerebral cortex	0.0	0.05
White matter	0.0	0.00
Cerebellum	0.0	0.15
Patient UPN #1179 (70 d post-transplantation)		
Frontal cortex	0.03	0.05
Cerebellum	0.05	0.0

[a] Determined by Dr. Raghaven, Dr. Wiederschain, and Dr. Kolodny.
Correlation coefficient (r) = 0.995 for comparison between substrates.

13-3). This amount could be considered contamination from WBCs in the blood vessels of the brain, and not as absolute evidence of CNS enzyme.

The latter observation is compatible with the concept that enzyme replacement in the brain requires several months after full engraftment. The physiologic capability for migration of monocyte-macrophage cells from blood vessels into the brain has been demonstrated to occur slowly in experimental animals. By using a Y-chromosome DNA probe for in situ hybridization, transplanted donor cells in humans have been shown to enter the CNS parenchyma at a gradual rate.

Also, motor nerve conduction studies in both animal and human subjects have demonstrated that remyelination will take place slowly over years. Stabilization of motor nerve conduction in the clinical setting is compatible with that found in the experimental literature.[19-23]

The concept of months of time required for recipient-donor cellular CNS exchange must be underscored. This complex pathophysiologic process of the "time-needed for migration" of cellular entry into the brain provides an understanding of the experimental and clinical data obtained to date.

The rapidity of the disease process in infantile Krabbe disease will not permit sufficient time to prevent catastrophic clinical events. Similarly, in the experimental twitcher mouse model, the events following transplantation cannot be expected to be fully changed to normal because of the rapidity of disease occurrence as compared with the life span of mice. True clinical improvement can be anticipated only when the balance is in favor of the time needed for migration of normal donor cells into brain, as opposed to the natural rate of disease progression. This has been fully discussed and reviewed in relationship to other animal models concerning the effect of bone marrow transplantation on the CNS.[18] Failure to take this time sequencing into appropriate consideration will lead to illogical conclusions as applied to experimental animal studies and to human clinical conditions.

Our recommendation is to exclude infantile Krabbe disease patients from consideration for bone marrow transplantation. This is because there is not enough time for new hematopoietic cells to enter the CNS and correct the rapidly advancing disease in these infants.

By contrast, the data obtained in our patient with juvenile globoid cell leukodystrophy appear to indicate that, in selected instances, bone marrow transplantation should be considered. The strongest recommendation would be to do such procedures in juvenile and adult forms if the rate of disease progression is slow and if there is an HLA-identical homozygous donor sibling available for transplantation. These criteria were met in our patient after an excellent pretransplant review conducted at the Mayo Clinic.[24]

On the other hand, individualization relative to recommendations would be needed if an HLA-identical sibling is not available and if there is presence of significant clinical disease. For instance, several centers are aggressively pursuing and accomplishing transplantation for leukemias and other diseases in less than HLA-identical sibling matches by appropriate T-cell depletion of marrow and by using monoclonal therapy against GVHD.

Patients with the adult form of globoid cell leukodystrophy should definitely be considered bone marrow transplant candidates. The demise of our two latter patients was not caused by the ongoing neurologic disease but rather to the GVHD engendered from the use of fully matched, but reactive, unrelated donors. We personally await the demonstration of improved results with the bone marrow transplant methods, using T-cell depletion or other approaches, before we would consider similar patients for transplantation in the future.

ACKNOWLEDGMENTS

This work was supported by grants from The Eagles (Kroc Foundation), the Bone Marrow Transplant Research Fund and Program Project National Institutes of Health (PO1CA21737), Children's Cancer Research Fund, the Minnesota Medical Foundation, Deans BRSG Fund, and the Graduate School of the University of Minnesota.

We wish to thank the parents and siblings of our patients for their intense support of these endeavors. The bone marrow transplantation team has included major roles in patient care by the nursing and pediatric residents staffs. In particular, the attending pediatric staff, which has provided important clinical input for all these patients, has included Dr. Kersey, Dr. Ramsay, Dr. Woods, Dr. Whitley, Dr. Neglia, Dr. Blazar, and Dr. Nesbit. We particularly wish to thank Dr. Timothy Braverman of St. Paul Children's Hospital for skills as a clinical pathologist in helping obtain autopsy brain biopsies in one of our patients. Acknowledgments are specifically due to Mrs. E.M. Petersen of the Department of Human Genetics of the University of Capetown and to Dr. Marvin Natowicz of the Shriver Center for the fibroblast cultures and enzymatic diagnosis.

REFERENCES

1. Suzuki K, Suzuki Y: Galactosylceramide lipidosis: Globoid-cell leukodystrophy (Krabbe disease). p. 1699. In Scriver CR, Beaudet AL, Sly WS, Valle D (eds): The Metabolic Basis of Inherited Disease. 6th Ed. McGraw-Hill, New York, 1989
2. Krivit W, Shapiro E, Kennedy W, et al: Treatment of late infantile metachromatic leukodystrophy by bone marrow transplantation. N Engl J Med 322:28, 1990
3. Krivit W, Whitley CB, Lund G, et al: Improvement of clinical expression of central nervous system manifestation in lysosomal storage diseases treated by bone marrow transplantation. p. 189. In Baum SJ, Santos GW, Takaku F (eds): Experimental Hematology Today—1987: Recent Advances and Future Directions in Bone Marrow Transplantation. Proceedings of a symposium held in conjunction with the Sixteenth Annual Meeting of the International Society for Experimental Hematology, August 23–28, 1976, Tokyo, Japan. Springer-Verlag, New York, 1987
4. Aubourg P, Blanche S, Jambaque I, et al: Reversal of early neurologic and neuroradiologic manifestations of X-linked adrenoleukodystrophy by bone marrow transplantation. N Engl J Med 322:1860, 1990

5. Duchen LW, Eicher EM, Jacobs JM, et al: A globoid cell type of leukodystrophy in the mouse: The mutant twitcher. p. 107. In Baumann N (ed): Neurological Mutants Affecting Myelination. Elsevier, New York, 1980
6. Bourque BA, Bornstein MB, Peterson ER, Suzuki K: The twitcher mouse: Myelinogenesis in organotypic culture. Brain Res 261:295, 1983
7. Hoogerbrugge PM, Poorthuis BJHM, Wagemaker G, et al: Alleviation of neurologic symptoms after bone marrow transplantation in twitcher mice. Transplant Proc 21:2980, 1989
8. Hoogerbrugge PM, Poorthuis BJHM, van Bekkum DW: Improved neuro-behavior in twitcher mice after bone marrow transplantation. p. 331. In Gale RP, Champlin RE (eds): Bone Marrow Transplantation: Current Controversies. Alan R Liss, New York, 1989
9. Yeager AM, Brennan S, Tiffany C, et al: Prolonged survival and remyelination after hematopoietic cell transplantation in the twitcher mouse. Science 225:1052, 1984
10. Hoogebrugge PM, Poorthuis BJHM, Romme Ad E, et al: Effect of bone marrow transplantation on enzyme levels and clinical course in the neurologically affected twitcher mouse. J Clin Invest 81:1790, 1988
11. Suzuki K, Hoogebrugge PM, Poorthuis BJHM, et al: The twitcher mouse. Central nervous system pathology after bone marrow transplantation. Lab Invest 58:302, 1988
12. Ichioka T, Kishimoto Y, Brennan S, et al: Hematopoietic cell transplantation in murine globoid cell leukodystrophy (the twitcher mouse): Effects on levels of galactosylceramidase, psychosine, and galactocerebrosides. Proc Natl Acad Sci USA 84:4259, 1987
13. Seller MJ, Perkins KJ, Fensom AH: Galactosylcerebrosidase activity in tissues of twitcher mice with and without bone marrow transplantation. J Inher Metab Dis 9:234, 1986
14. Hickey WF, Kimura H: Perivascular microglial cells of the CNS are bone marrow-derived and present antigen in vivo. Science 239:290, 1988
15. Hoogerbrugge PM, Suzuki K, Suzuki K, et al: Donor-derived cells in the central nervous system of twitcher mice after bone marrow transplantation. Science 239:1035, 1988
16. Kobayashi S, Katayama M, Bourque E, et al: The twitcher mouse: Positive immunohistochemical staining of globoid cells with monoclonal antibody against Mac-1 antigen. Dev Brain Res 20:49, 1985
17. Unger E, Sung JH, Krivit W: The use of human Y chromosome in situ hybridization technique to demonstrate that the central nervous system has donor male hematopoietic cells in female recipient after successful engraftment in bone marrow transplantation. (submitted for publication)
18. Krivit W, Whitley CB, Chang PN, et al: Lysosomal storage diseases treated by bone marrow transplantation: Review of 21 patients. p. 261. In Johnson FL, Pochedly C (eds): Bone Marrow Transplantation in Children. Raven Press, New York, 1990
19. Dhuna A, Toro C, Kennedy W, Krivit W: Longitudinal neurophysiological studies in a patient with metachromatic leukodystrophy following bone marrow transplantation. (submitted for publication)
20. Krivit W: Recommendations for treatment of metachromatic leukodystrophy with bone marrow transplantation: A review of 7 patients engrafted for over a year. Symposium on Batten's Disease. Am J Med Genet (in press)

21. Kondo A, Hoogerbrugge PM, Suzuki K, et al: Pathology of the peripheral nerve in the twitcher mouse following bone marrow transplantation. Brain Res 460:178, 1988
22. Toyoshima E, Yeager AM, Brennan S, et al: Nerve conduction studies in the twitcher mouse (murine globoid cell leukodystrophy). J Neuro Sci 74:307, 1986
23. Scaravilli F, Jacobs JM: Improved myelination in nerve grafts from the leucodystrophic twitcher into trembler mice: Evidence for enzyme replacement. Brain Res 237:163, 1982
24. Baker H, Trautmann JC, Younge BR, et al: Late juvenile onset Krabbe's disease. Ophthalmology 97:1176, 1990

14

HUMAN GENE THERAPY: STRATEGIES AND PROSPECTS FOR INBORN ERRORS OF METABOLISM

Robert J. Desnick · Edward H. Schuchman

INTRODUCTION

During the past two decades, major advances have been made in elucidating the molecular pathologies of many inherited metabolic diseases.[1] To date, the catalytic, receptor, transport, or structural protein defects in more than 200 inborn errors of metabolism have been identified, and the specific molecular lesions in the genes encoding these proteins are being characterized at an ever-increasing rate.[2] Implementation of biochemical and molecular techniques has made the diagnosis and prenatal diagnosis of these disorders a reality. A variety of therapeutic approaches have been undertaken, including dietary restriction, the use of chelators, cofactor supplementation, and, most recently, gene product replacement and allotransplantation (Table 14-1). The limitations, as well as the encouraging successes, of these strategies have been the subject of recent reviews. However, in spite of these major achievements, most patients and families with these debilitating disorders have become increasingly disappointed by the absence of specific therapies.

The only means to cure, rather than treat, these devastating disorders would be to replace or correct the defective genes. If a normal gene could be properly targeted and expressed in the cellular sites of pathology, the genetic defect would be corrected and the disease course either slowed, stabilized, or reversed. Such is the objective of current efforts to develop gene therapy for the treatment of inherited metabolic diseases. This review summarizes the recent advances and current limitations of gene therapy and highlights the most likely avenues this field will follow during the next decade.

Table 14-1. Strategies for the Treatment of Inherited Metabolic Diseases

Metabolic manipulation
    Dietary restriction
    Substrate depletion techniques
    Chelation-enhanced excretion
    Plasmapheresis/affinity binding
    Surgical bypass procedures
    Metabolic inhibition
    Product replacement

Gene product therapy
    Cofactor supplementation
    Enzyme induction
    Allotransplantation
    Enzyme replacement therapy

Gene therapy
    Production of human gene products
    Gene replacement

Preventive therapy
    Heterozygote screening
    Genetic counseling
    Prenatal diagnosis

APPROACHES FOR GENE THERAPY

Genes can be transferred into the germline or into somatic cells. Germline gene therapy involves the insertion of a normal gene into the germ cell or one-cell embryo such that it will correct the genetic defect and be transmitted, in a mendelian fashion, from generation to generation. By contrast, the goal of somatic gene therapy is to insert genes into somatic cells, particularly those with primary disease pathology. The foreign gene must gain access to the cell nucleus for stable integration into the host chromosome and for the regulated expression of the desired gene product. In this way, an exogenous gene can be inserted only into the somatic cells of the patient, without the possibility of being transmitted through the germline. Thus, therapy is limited to the diseased patient.

Germline Gene Therapy in Mice

Experimental germline gene therapy in humans has been prohibited in concept, philosophy, and fact by the Guidelines of the NIH Recombinant DNA Advisory Committee on Human Gene Therapy.[3] Such experiments have been permitted in mice, however, and demonstrate the feasibility of inserting functional genes into the germline, with subsequent cure of the murine disease.

The technique for inserting a foreign gene or DNA segment into the mouse germline is shown in Figure 14-1.[4] One-cell mouse embryos are removed immediately after fertilization. The male pronucleus is visible at this stage, and for-

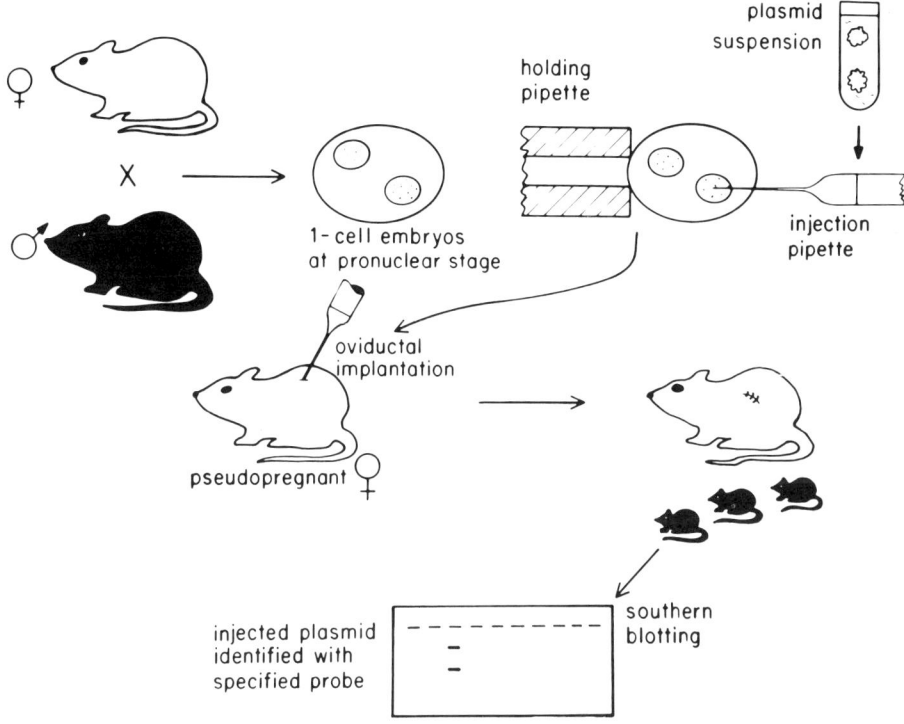

Fig. 14-1. Procedure for introducing a foreign gene into the germline of mice (i.e., production of transgenic mice). Immature albino females are superovulated and mated to males homozygous for the wild-type allele at the albino locus. One-cell embryos at the pronuclear stage of development (day 1 of gestation) are removed from the oviducts, after which one of the pronuclei is microinjected with the recombinant-plasmid DNA. Embryos are then reimplanted into the oviducts of the albino females rendered pseudopregnant by mating to albino vasectomized males. The pigmentation markers are employed to eliminate the possibility of the embryos giving rise to live young from oocytes fertilized by spermatozoa from inadequately vasectomized males. In this scheme, such embryos would be of the albino pigmentation phenotype. After normal delivery, pigmented mice are evaluated by Southern hybridization for retention of microinjected DNA. (From Gordon et al.,[4] with permission.)

eign DNA can be inserted into the male pronuclei by microinjection. After being injected, about 100 embryos are placed into the uterus of a pseudopregnant female mouse. Only a few injected embryos will successfully undergo fetal development. After birth of the pups, the presence of the foreign gene can be determined by Southern hybridization analysis, using DNA isolated from tail biopsies. These "founder mice" may be chimeric, the foreign gene having only gained access to certain cell types. Those that have the foreign gene (one to several hundred copies) integrated into their sperm- or egg-producing cells will transmit the gene in a mendelian fashion. Moreover, the integrated genes tend to be expressed with the appropriate tissue specificity. For example, the human β-globin gene, normally expressed only in erythrocytes, was expressed only in this cell lineage after transfer into the mouse germline.[5] Because these mice can express the foreign genes, they are called transgenic mice.

The ability to insert foreign genes into the germline of mice has provided the opportunity to test the feasibility of gene therapy. Using mice with specific genetic defects as the test systems, normal genes from other species have been injected. Their incorporation into the mouse genome has effectively and dramatically cured the murine disease (Table 14-2). For example, transfer of the rat growth hormone gene into growth hormone-deficient dwarf little mice resulted in the cure of the growth hormone deficiency and the generation of larger-than-normal mice.[6] Similarly, transfer of the human β-globin gene into mice with β-thalassemia (caused by a gene deletion in the murine β-globin gene) resulted in the regulated expression of β-globin molecules in these mice, association of the murine α-globin chains with the human β-globin chains, and correction of the hematologic disease.[5] These and other examples[7–9] (Table 14-2) illustrate the feasibility of gene therapy and that, at least in the mouse, it can effectively cure a genetic disease.

It should also be noted that insertion of the foreign gene is random and that it is possible to incur damage by inserting the foreign gene into an essential gene, rendering it inactive. Although such events occur infrequently, "insertional mutagenesis" has provided the basis for studies in developmental biology.[11] Thus, the random integration of injected exogenous genes can be harmful as well as helpful.

Table 14-2. Correction of Genetic Defects in Transgenic Mice

Murine Disease	Defective Gene	Microinjected Gene	Investigators
Growth hormone deficiency Dwarf little mouse (*lit*)	Growth hormone (deletion)	Rat growth hormone	Hammer et al.[6]
β-Globin deficiency β-Thalassemia mouse (hbb^{th-1})	β-Globin (deletion)	Human β-globin	Costantini et al.[5]
Hypogonadism Hypogonadal mouse (*hpg*)	18GnRH-GAP (3' deletion)	Murine GnRH-GAP	Mason et al.[7]
MBP deficiency Shiverer mouse (*shi*)	MBP (3' deletion)	Murine mBP	Readhead et al.[8]
OTC deficiency Sparce mouse (*spf*)	OTC (point mutation, N117H)	Human OTC	Jones et al.[9]
β-Glucuronidase deficiency Mucopolysaccharidosis VII mouse (gus^{MPS})	β-Glucuronidase	Human β-glucuronidase	Kyle et al.[10]

Abbreviations: GnRH-GAP, gonadotropin-releasing hormone and GnRH-associated peptide; MBP, myelin basic protein; OTC, ornithine carbamyltransferase.

Somatic Cell Gene Therapy

Somatic gene transfer involves the insertion of the normal gene into the target cells of pathology to produce sufficient quantities of the gene product to correct the metabolic defect. During the past 5 years, efforts have focused on the use of retroviral vectors to transfer cDNA into a variety of cell types, including hematopoietic cells,[12–18] fibroblasts,[19–24] hepatocytes,[25–29] endothelial cells,[30] and keratinocytes.[31,32] Hematopoietic cells have been the main target of gene transfer studies, since bone marrow transplantation (BMT) in patients with certain genetic disorders has proved curative.

The Trojan horse of somatic cell gene therapy is the retrovirus. Retroviruses are uniquely suited for somatic gene transfer, since integration of the retroviral genome into the host cell chromosome is an obligatory part of their life cycle and is a highly efficient process.[33–37] In addition, with few exceptions, integration of the virus into the host chromosome, expression of its genes from the integrated provirus, and the generation of progeny have no deleterious effects on the viability and function of the infected host cell. Figure 14-2 shows the life cycle of the retrovirus. The virus binds to a cell membrane receptor and is then internalized and transported to the lysosome, where it is uncoated. The viral RNA genome is then released into the cytoplasm. Reverse transcriptase then converts the single-stranded viral RNA genome into a double-stranded DNA

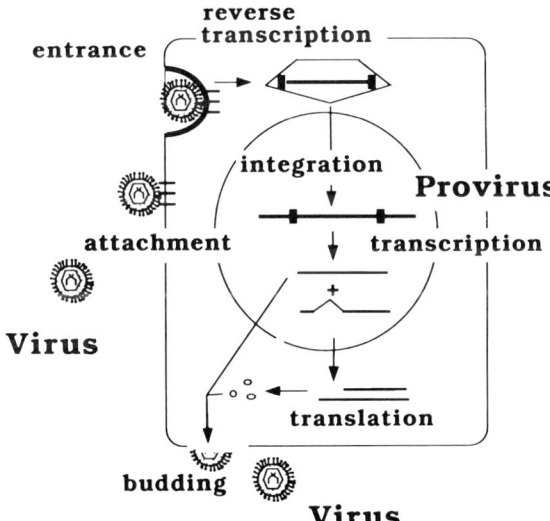

Fig. 14-2. Retroviral life cycle. After adsorption and uptake (i.e., binding and infection) of the viral particle (virion), the single-stranded retroviral RNA genome is converted to double-stranded DNA by reverse transcriptase. After uptake into the nucleus, the viral DNA integrates randomly into a host chromosome, forming a provirus. The provirus serves as a template for the host RNA polymerase II to transcribe mRNAs for the viral proteins and the viral genomic RNA. Two copies of the genomic RNA and the translated viral proteins assemble and, by budding, release a new infectious virus to repeat the cycle. (From Hu and Temin,[37] with permission.)

copy that can be integrated (by recombination using a virally encoded integrase) into the host genome as a *provirus*. The provirus is able to use the host cell machinery to transcribe and translate the viral genome and produce copies of the RNA genome, as well as the essential viral proteins. The viral particle can then be assembled within the cytoplasm, and infectious viral particles can cross the host cell membrane by budding and infect other cells.

The structure of the retroviral genome is shown in Figure 14-3A. At each end is a sequence called the long terminal repeat (LTR). This sequence contains the promoter and elements needed for translation of the three major viral genes, *gag, pol,* and *env,* which encode the viral core, polymerase (reverse transcriptase), and envelope glycoproteins, respectively. In addition, there is a sequence near the 5' LTR known as the ψ (psi) site, which is essential if the viral RNA transcript is to be packaged into a viral particle. If the ψ sequence is altered or deleted (i.e., ψ-minus), that RNA genome cannot be encapsulated in the viral particle.

Investigators have taken clever advantage of the ψ sequence to develop a method to package a foreign gene into a defective retroviral particle for gene therapy[38,39] (Fig. 14-3). For example, the retroviral *gag, pol,* and *env* genes can be replaced by a cDNA that encodes a therapeutic protein (Fig. 14-3). After integration of this recombinant construct, it would be transcribed as a viral RNA genome with an intact ψ region, permitting it to be packaged. A second retroviral construct required for the production of a defective retrovirus is made by removing the ψ sequence (Fig. 14-3C). This ψ-minus construct has the remainder of the retroviral genome including the *gag, pol,* and *env* genes. Cells

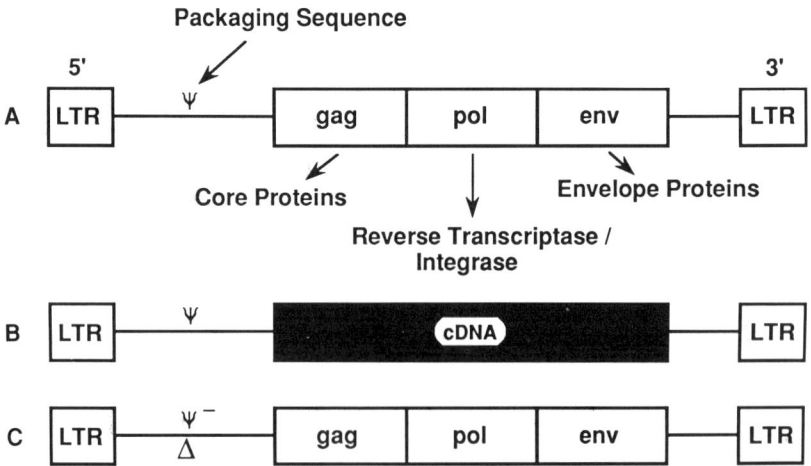

Fig. 14-3. Modification of the retroviral genome for gene transfer experiments. Schematic illustration of the retroviral genome (**A**). Retroviral gene construct with a human cDNA replacing the *gag, pol,* and *env* genes (**B**), and retroviral genome with a deletion of the ψ sequence, which is normally required for packaging the retroviral genome (**C**).

transfected with the ψ-minus construct are called packaging cells (or "helper" cells), since they produce the proteins needed for assembly of an infectious viral particle. Because the viral genome in these packaging cells does not have the ψ sequence, the ψ-minus viral genome cannot be packaged. However, when the packaging cells are transfected with a ψ-plus recombinant construct containing the foreign cDNA sequence, a defective viral particle containing only the recombinant RNA sequence will be assembled (Fig. 14-4). This viral particle is defective, since it does not contain the viral *gag*, *pol*, and *env* genes necessary to produce the viral proteins required to assemble an infectious viral particle. However, it does package enough viral proteins (reverse transcriptase, integrase) to permit insertion of a cDNA copy of this genome into the host cell chromosome. Thus, this recombinant virus can be used to infect somatic cells

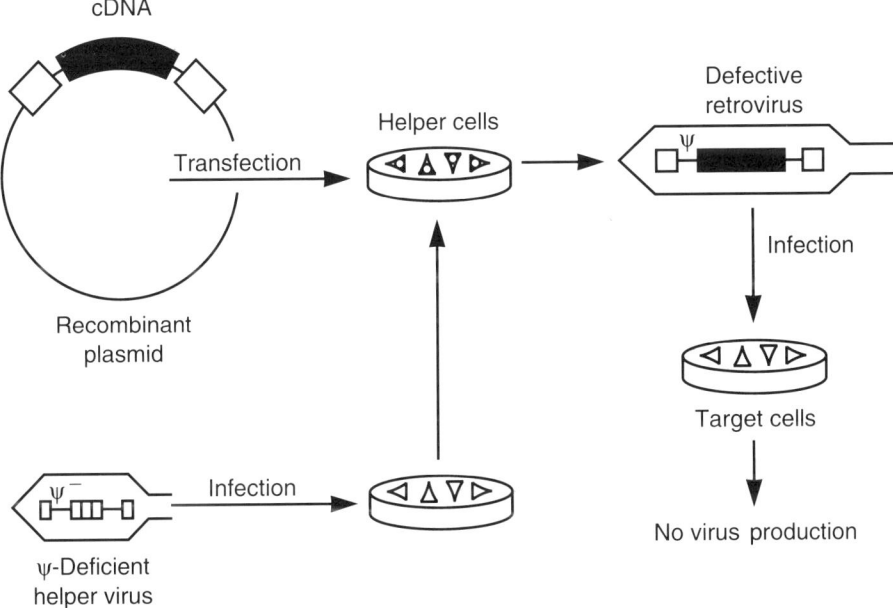

Fig. 14-4. Strategy for retrovirally mediated somatic gene transfer. The ψ-deficient (ψ⁻) helper virus produces all the normal viral proteins, but cannot package its own RNA because it lacks the appropriate packaging signal. Helper cells are made by inserting the helper provirus into the genome of the helper cell. This is accomplished by infection of the cells with helper virus. Recombinant vector DNA, including a human cDNA sequence, is inserted into helper cells by transfection with a plasmid containing the human cDNA flanked by retroviral LTR sequences. Because the recombinant vector contains the ψ sequence, the vector RNA genome is automatically encapsulated by viral proteins produced by the helper provirus DNA in the helper (ψ_2) packaging cells. The resulting viral particles are released by budding from the helper cell membrane. The vector virus is capable of only one infection because it lacks the information needed to make viral proteins. Target cells (e.g., human bone marrow cells) are then infected with the virus. Once the vector construct is integrated into the target cell DNA as a provirus, the foreign gene sequence can be expressed. However, no viral particles will be produced by the transfected target cells, since the genes for the viral proteins have been deleted from the recombinant viral construct.

efficiently and introduce a foreign gene without subsequent production of infectious viral particles.

One of the major problems encountered with the earlier generation of retroviral packaging cell lines was that recombination between the introduced retroviral vectors with the previously integrated retroviral helper sequences occasionally led to the occurrence of wild-type virus. Therefore, a new generation of packaging cell lines has been constructed in which the helper viral genome contains numerous rearrangements (e.g., insertions and deletions) that reduce the likelihood of a recombinational event.[40,41] Current efforts are focused on the development of retroviral packaging cell lines that produce higher titers of recombinant virus. Exposure of cells to higher viral titers may result in superinfection, hence, higher levels of gene expression. For example, Kozak and Kabat[42] developed a "ping-pong amplification" method that permits amplification of retroviral vectors and higher viral titers.

The design of improved retroviral vectors for use in gene transfer experiments also is an active area of investigation. For example, Lim et al.[43] reported the construction of a simplified vector that contained only one internal transcriptional unit for the expression of human adenosine deaminase (ADA) and no dominant selectable marker such as *neo*. Using this simplified vector, human ADA was efficiently expressed in murine bone marrow both in vitro and in vivo. Such vector constructs containing a single gene have wide application, since various investigators have reported that incorporation of the *neo* gene for preselection of hematopoietic stem cells with G418, followed by autologous BMT may be disadvantageous and may result in incomplete hematopoietic reconstitution in vivo. In addition, the presence of a second transcriptional unit in retroviral vectors may have deleterious effects on the expression of the nonselected gene. More detailed discussions of the various vector constructs and packaging cell lines designed for gene transfer are available.[34–36,44]

GENE TRANSFER INTO HEMATOPOIETIC CELLS

Most investigators have concentrated on retroviral gene transfer into pluripotent hematopoietic stem cells (PHSCs) as the prototype cell for somatic gene therapy. These cells are readily accessible in sufficient numbers for infection, can be grown in culture, and provide the opportunity to generate a variety of blood elements and bone marrow-derived cells (e.g., Kupffer cells, osteoclasts, pulmonary macrophages) expressing the transferred gene. Retrovirally infected PHSCs may be readily reintroduced into affected individuals by autologous BMT. Transplanted stem cells will then migrate from the bone marrow and establish a lineage of differentiated cells throughout the body (e.g., Kupffer cells in the liver), all of which express the foreign gene product. The disease candidates for such therapy include those in which successful allogenic BMT has already proved therapeutic or curative, such as severe combined immunodeficiency disease (SCID) caused by ADA deficiency.

In addition, it has been suggested that other metabolic diseases in which bone marrow-derived cells are not the primary sites of pathology may also be effectively treated by the transplantation of stem cells that contain the appropriate gene, particularly if the gene product can be efficiently secreted and taken up by the primary cellular sites of pathology.

Since the defective retroviruses used for gene transfer are produced in relatively low titers by the packaging cells (10^6 to 10^7 cfu/ml), efforts have been directed to develop optimal conditions for stem cell growth. For example, Bodine et al.[45] demonstrated that when the hematopoietic growth factors interleukin-3 (IL-3) and interleukin-6 (IL-6) were added to the media of cultured murine bone marrow cells, the number of colony forming units-spleen (CFU-S) stem cells increased more than 10-fold over that obtained using standard cell culturing procedures. IL-3 and IL-6 appear to act synergistically to stimulate the division of PHSCs, a prerequisite for the stable integration of the retroviral vector. In the same study, the purified CFU-S stem cells were infected with a retroviral vector containing the human β-globin gene and were reintroduced into mice. Six of the eleven mice analyzed showed expression of the human β-globin gene 12 months post-transplantation, demonstrating effective gene transfer into murine PHSCs. Most recently, the isolation of cDNA clones encoding a novel growth factor, known as kit ligand, mast cell growth factor, or stem cell factor, has been reported.[46,47] This growth factor appears to act synergistically with IL-3 and IL-6 to promote proliferation of very primitive bone marrow stem cells.

Various protocols are being developed to enrich for PHSCs, which constitute less than 1 percent of murine bone marrow cells. For example, Szilvassy et al.[18] reported a procedure for the single-step isolation of PHSCs from mouse bone marrow cells treated with 5-fluorouracil (5-FU). About 25 percent of the cells obtained in this manner were day 12 CFU-S, and at least 1 percent were capable of long-term marrow repopulation after transplantation into lethally irradiated mice. These investigators also showed that transplantation of 5-FU-treated stem cells containing the *neo* gene into irradiated recipients resulted in the expression of the *neo* gene in lymphoid and myeloid tissues 6 months post-transplantation. Another approach to isolate, purify, and concentrate murine PHSCs for gene transfer is the use of monoclonal antibodies that recognize murine PHSCs.[48,49] However, it remains to be seen whether these antibodies also recognize similar stem cell populations in other mammals.

In summary, recent advances in the understanding of mammalian hematopoiesis have stimulated new experimental protocols that permit the proliferation and enrichment of murine stem cell populations for the targeting and long-term expression of retrovirally transferred genes in mice. However, it should be appreciated that the feasibility of bone marrow-mediated gene therapy has not been fully demonstrated in monkeys[50] or humans. Further knowledge of human stem cell biology is needed to facilitate the successful application of this approach for the treatment of human disorders by stem cell gene transfer.

GENE TRANSFER INTO HEPATIC CELLS

Since many inherited diseases result from the deficiency of hepatic-specific enzymes or proteins, the introduction and expression of human genes in hepatic cells have become active targets of investigation. Somatic gene therapy may be accomplished by the isolation and culture of hepatocytes obtained by partial hepatectomy, modification of these cells by introduction of the normal gene using retroviral vectors, and transfer of the corrected cells into the liver by transplantation or by direct injection. Retroviral vectors have been used successfully for the in vitro transfer of recombinant genes into primary cultures of mouse, rat, and rabbit hepatocytes.[25,27,51]

Several strategies have been employed to introduce modified hepatocytes into the liver of animals. Recently, Ponder et al.[52] demonstrated that a large fraction of hepatocytes that were introduced into mice by intrasplenic injection migrated to the liver parenchyma and continued to express liver-specific gene products for up to 1 year, the life span of a normal mouse. Demetriou et al.[53,54] showed that hepatocytes attached to collagen-coated microcarrier beads injected into the peritoneal cavity could survive for up to 6 months. Analogously, Anderson et al.[55] developed a three-dimensional support matrix (coated with purified collagen type IV) that permitted engineered primary hepatocytes to be grown and selected in vitro and then subcutaneously and/or intraperitoneally implanted into adult animals. The viability of the cells contained within these support matrices, termed *organoids*, relies primarily on the ability of these structures to become vascularized. Therefore, Thompson et al.[56,57] created organoid neovascular structures using polytetrafluoroethylene fibers coated with collagen and heparin-binding growth factor 1 (HBGF-1), an initiator of angiogenesis during development. The organoids were seeded with normal hepatocytes and implanted into the peritoneal cavity of Gunn rats, which are deficient in glucuronyltransferase activity. These implants developed visible vascular lumina and nonvascular structures resembling nerve tissue. In addition, the serum bilirubin levels in the transplanted rats were reduced more than 60 percent for up to 180 days post-transplantation. Although the long-term viability of the cells contained within the implants is unknown, organoids hold promise for the future delivery of retrovirally infected cells other than bone marrow into affected individuals.

Another strategy for hepatic gene transfer has employed the asialoglycoprotein receptor for hepatocyte-specific gene uptake and internalization in vivo[58] (see Ch. 16). A high-affinity ligand for the asialoglycoprotein receptor, asialoorosomucoid, was covalently attached to polylysine. The protein conjugate was then noncovalently bound to a DNA expression vector, forming a soluble complex. The asialoorosomucoid-coated vector was efficiently taken up by the receptor-mediated process, and β-galactosidase activity was stably expressed in vivo (see Ch. 16). This technique would be an alternative gene delivery system to hepatocyte transplantation for the treatment of diseases resulting from the deficiency of liver-specific proteins.

RETROVIRAL GENE TRANSFER INTO OTHER CELL TYPES

Retroviral vectors have been used to introduce genes, by stable means, into a variety of other cell types, including skin fibroblasts, keratinocytes, endothelial cells, and muscle cells. Selden et al.[20] introduced a retroviral vector containing the human growth hormone (hGH) gene into cultured mouse fibroblasts; these cells were than implanted into various locations in mice. The implants synthesized and secreted hGH; however, their function depended on their location and size. With appropriate immunosuppression, the implants survived for more than 3 months. This approach to gene therapy was termed *transkaryotic implantation*. Subsequently, other researchers turned to cultured skin fibroblasts as a target for retroviral gene transfer and various heterologous genes, including α_1-antitrypsin,[60] ADA,[19,24] β-glucosidase,[21] β-glucuronidase,[60] and factor IX,[23] have been expressed in vitro. For example, St. Louis and Verma[23] showed that a retroviral vector expressing human factor IX could be introduced into mouse primary skin fibroblasts; the infected cells could be embedded in collagen and then implanted under the epidermis. Sera from the transplanted mice contained human factor IX protein for at least 10 to 12 days, at which time mouse anti-human factor IX antibodies were produced. Similar results were obtained by Palmer et al.[19] These workers inserted the human ADA gene into rat fibroblasts and then reintroduced the genetically modified cells on a collagen matrix into a full-thickness skin wound. The retroviral sequence was detected for 8 months, indicating that the transplanted cells survived. However, gene expression was markedly decreased after 1 month, suggesting that the foreign sequence was specifically inactivated. These studies signal a general concern. Are transferred genes subject to species-specific inactivation, or can the use of alternative regulatory elements in the vector (e.g., enhancers) support continuous expression?

Other investigators have targeted keratinocytes, the principal cell type of the epidermis, for gene transfer studies.[31,32] Fenjves et al.[32] infected human cultured keratinocytes with a retroviral vector containing the human apolipoprotein E (apoE) gene and then engrafted these cells onto athymic mice and rats. The expressed apoE crossed the epidermal-dermal barrier, entered the systemic circulation, and produced effects on distal tissues, providing further evidence that skin may be an important target organ for somatic gene therapy. However, in view of the recent results reported by Palmer et al.,[24] further studies are required to determine whether gene expression can be maintained in fibroblasts and keratinocytes modified by retrovirally mediated gene transfer.

Vascular and endothelial smooth muscle cells and skeletal muscle cells also have been evaluated for retrovirally mediated gene transfer. For example, the β-galactosidase gene was expressed in endothelial and vascular smooth muscle cells and then introduced into specific arterial segments by means of a catheter delivery system.[30] β-Galactosidase was specifically expressed in the arteries for at least 5 months post-transplantation, demonstrating the ability to target genes directly to the vascular endothelial system. Wolff et al.[61] showed that mouse skeletal muscle cells could be directly injected in vivo with DNA encoding chloramphenicol aminotransferase (CAT), luciferase, or β-galactosidase. The expression of these foreign gene products was demonstrated for at least 2 months.[61] In addition, Nabel et al.[62] demonstrated the direct infection of arterial endothelial cells, using liposomes containing the β-galactosidase gene, which were administered intravenously with a double balloon catheter. These and related findings[63] indicated that genes could be directly injected or lipofected into certain cell types, with subsequent expression of the gene products. Clearly, further investigation of this attractive strategy is required.

ALTERNATIVE VECTORS FOR SOMATIC GENE THERAPY

The past 5 years have witnessed remarkable progress in the design and use of retroviral vectors for gene transfer. However, retroviral vectors are limited by the following factors: (1) they cannot efficiently infect nondividing cells (e.g., neurons); (2) they can only accommodate up to about 10 kb of foreign DNA; and (3) the viral LTRs may be shut down after infection, interfering with long-term expression. Moreover, sufficient viral titers required to increase the efficiency of infection have not been obtained. This latter point is of concern particularly when attempting to infect stem cells that may represent less than 1 percent of the cell population. Thus, investigators are developing other viral-based vector systems for gene transfer. Two alternative systems are the herpes simplex virus type 1 (HSV-1) (see Ch. 17) and the adeno-associated virus (AAV). The HSV-1 vectors are attractive, since they permit the delivery of genes directly into nondividing cells. Furthermore, the natural neurotropism of HSV-1 may permit efficient targeting of foreign genes to the neurons in the peripheral and central nervous systems (CNS), thus overcoming a major obstacle to somatic gene therapy for neurogenetic disorders. Prototype HSV-1 vectors (e.g., pHSV1ac) have been developed in which the *lacZ* gene, encoding β-galactosidase, was expressed under the regulatory control of an intermediate-early gene promoter (1E415).[64] After stereotactic injection, pHSV1ac stably expressed β-galactosidase in rat peripheral and CNS neurons, as well as in adult rat brains. To facilitate this work, HSV-1 packaging cell systems recently have been developed that contain temperature sensitive mutations[65] or deletions[66] in the IE3 gene of HSV-1. These mutations suppress lytic growth (which would ordinarily kill the cells) and permit early gene synthesis, to create a form of persistent infection. Thus, foreign gene expression can continue for long periods without the need for integration into host cells and without causing damage to the infected cells. Analogous to the new-generation retroviral packaging cell lines, these latter HSV-1 packaging cells should prevent reversion to a wild-type phenotype by recombination.

The human DNA virus, AAV, offers another promising alternative vector for somatic gene transfer.[67–72] AAV can be propagated as a lytic virus or maintained as a provirus integrated into the host cell DNA. In the lytic cycle, replication requires co-infection with either adenovirus or HSV. Without a helper virus, AAV can persist in the genome as an integrated provirus for up to 100 passages in cultured cells. Because it requires a helper virus for replication, AAV is classified as a defective member of the parvovirus family. The *cis*-acting elements, required for packaging, integration/rescue, and replication of the viral DNA, are located within a 284-bp sequence of the approximately 4.6-kb genome. The major advantages of AAV for somatic gene transfer are its wide host range (i.e., virtually all mammalian cells can be infected with AAV), its lack of pathologic effects, and the fact that its *cis*-acting control elements are well defined. Furthermore, AAV is preferentially integrated into a specific region of the host genome. In humans, this site is located on chromosome 19.[72] AAV has been used as a vector system to express a variety of eukaryotic genes. For example, Hermonat et al.[67] produced a recombinant AAV in which the *neo* gene was substituted for the AAV capsid gene. After infection of cultured cells, G418 resistance was demonstrated in both murine and human cell lines. The stably integrated vector could be rescued to produce replicating sequences after superinfection with adenovirus. Similarly, AAV vectors have been used to express other gene products (e.g., CAT, β-globin) in a variety of human cells, and the generation of high-titer, helper-free viral stock has been reported.[69–71]

In addition to HSV-1 and AAV, other viral vectors have been used to express heterologous genes in cultured cells. Among these are the influenza virus,[73] vaccinia virus,[74] and cytomegalovirus (CMV).[75] Although the development of these viral vectors has not progressed as rapidly as HSV-1 and AAV, many hold significant promise for somatic gene therapy. For example, the genome of the influenza virus, a negative-strand RNA virus, has been modified to insert foreign genes.[73] Recombinant RNA is expressed from plasmid DNA in which the coding sequence of the influenza A virus nonstructural (NS) gene is replaced with the CAT gene. When transfected with purified influenza A virus polymerase proteins, in the presence of helper virus, the recombinant RNA is amplified, expressed, and packaged into infectious viral particles. This viral vector system may be a particularly efficient way to deliver genes to the respiratory tract. However, each of these vectors replicates through a lytic cycle. Although vaccinia virus and influenza virus may be especially well suited for use as vaccine vectors,[76] further developments are needed to allow for their use as a long-term gene expression vehicle.

Investigators also have begun to determine the value of episomal vectors for somatic gene therapy. For example, gene transfer vectors have been described on the basis of bovine papillomavirus,[77] simian virus 40 (SV40),[78] and Epstein-Barr virus (EBV).[79] Although these episomal vectors are potentially useful for somatic gene therapy, a number of obstacles remain to be overcome, including (1) inactivation of their transforming properties, (2) long-term maintenance and expression, both in vitro and in vivo, and (3) latent integration properties. Although retrovirally based vectors are currently used by most investigators for somatic gene transfer studies, it is likely that alternative vectors will be developed and evaluated. These alternative vector systems may overcome some of the inherent problems associated with retrovirally mediated gene transfer and, in fact, offer certain advantages, such as gene targeting to specific cell types.

DISEASE CANDIDATES FOR SOMATIC GENE THERAPY

Among the genetic diseases considered as potential candidates for somatic gene therapy are those for which BMT has proved therapeutic. In these disorders (e.g., hemoglobinopathies, immunodeficiencies, other blood cell defects), the symptoms result from deficient function or lack of a bone marrow-derived gene product. Successful BMT results in the cellular replacement of normal blood elements that can produce the normal gene product and correct the metabolic disease. Among such disorders, the prime candidate for gene therapy trials is SCID attributable to deficient ADA activity. Although successful BMT from an enzymatically normal, HLA-identical sibling would cure this lethal disease, most patients do not have such a sibling and are therefore candidates for other therapeutic endeavors, including enzyme replacement (see Ch. 10), BMT from a parent, or autologous transplantation of bone marrow cells that have been infected with a retroviral construct containing the human ADA gene. Studies have demonstrated that retroviral-ADA gene constructs can express the enzyme in cultured T and B cells from SCID patients and reconstitute their immunologic function,[17] indicating the likelihood that autologous transplantation with retrovirally modified stem cells will correct this disease.

PROSPECTS FOR HUMAN GENE THERAPY

Initial Trials Undertaken

The first human gene therapy protocols have recently been approved by the National Institutes of Health (NIH) and the Food and Drug Administration (FDA), and clinical trials have begun. The approval process required for these studies is outlined in Figure 14-5. The first approved trial for human gene transfer was not for the treatment of an inherited metabolic disease but involved the use of a retrovirally transferred marker gene to determine the in vivo fate of tumor infiltrating lymphocytes (TILs) administered to patients with malignant melanoma.[80] In these studies, the *neo* gene was retrovirally transferred into the TIL cells, which were then intravenously administered to the patient. The presence of the marked cells in the tumor was determined at autopsy. Not only did the TIL cells target to the tumors, but these studies demonstrated the short-term safety of this approach for gene transfer.[81] In October 1990, the Recombinant Advisory Committee approved the next step, the use of TIL cells to deliver tumor necrosis factor (TNF) directly to the tumor site in vivo. In January 1991, the first TIL-TNF expressing cells were

Human Gene Therapy: Strategies and Prospects for Inborn Errors of Metabolism   253

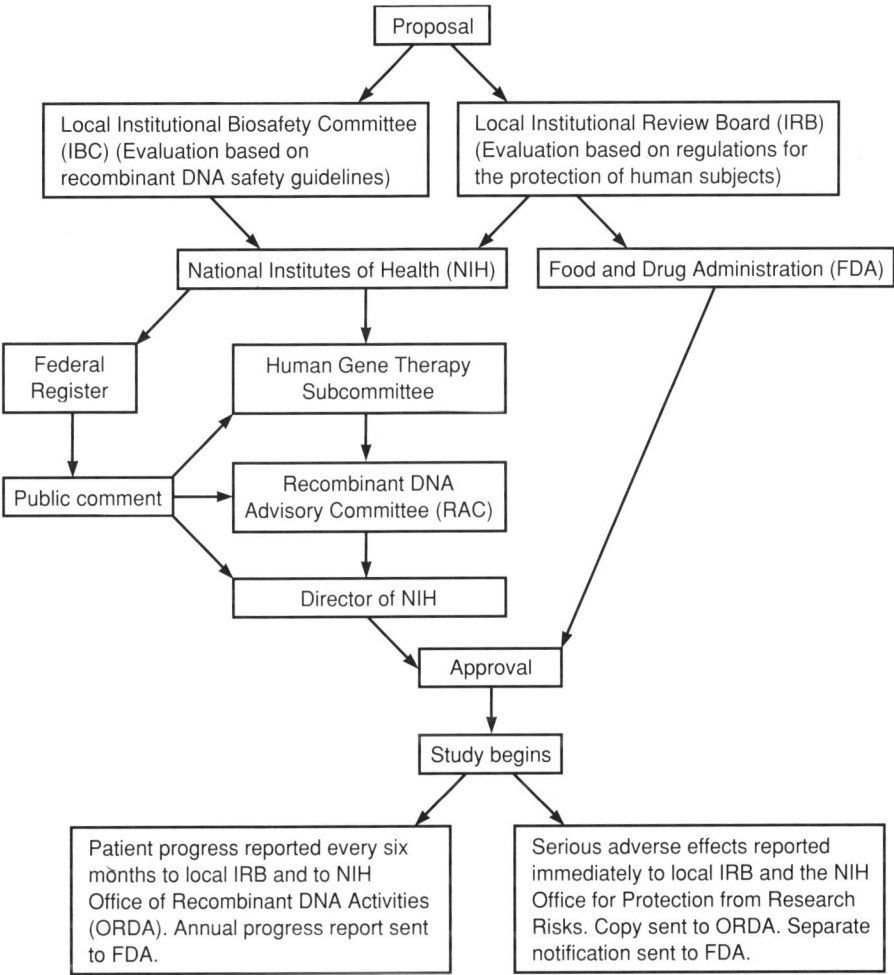

Fig. 14-5. Review process for clinical studies involving human somatic cell gene therapy. (From Nichols,[91] with permission.)

administered, initiating the first use of gene therapy for the treatment of this rapidly progressive cancer.

The first human trial of somatic gene replacement for an inherited metabolic disease also was approved in October 1990 and was recently undertaken at the NIH. An ADA-deficient 8-year-old girl with SCID is currently receiving infusions of T lymphocytes that have been modified by infection with a retroviral construct expressing human ADA. If these studies prove safe and effective, the next step will be the transplantation of this patient's genetically modified bone marrow stem cells. Suffice it to say, human gene therapy has begun, albeit by small first steps. It is likely that the feasibility of this approach to cure certain genetic disorders will be demonstrated in the near future.

Gene Replacement Versus Gene Repair

The strategies for somatic gene therapy discussed above involve the addition of the normal gene into genetically defective cells. In some cases in which the defective alleles are expressed, the addition of a normal gene may not markedly alter the disease pathology. This latter complication may be important for polymeric proteins that require subunit associations for biologic function. Examples of such disorders might include α_1-antitrypsin deficiency, osteogenesis imperfecta, hereditary spherocytosis, and other disorders in which the continued production of a mutant gene product causes disease pathology.[82,83] In addition, the use of retrovirally mediated gene transfer may cause deleterious effects in recipient cells by random gene disruption and/or aberrant gene regulation (e.g., expression of temporally controlled genes or oncogenes). Thus, future efforts must be directed toward developing methods for in situ gene repair (i.e., the ability to target genes to recombine specifically with the defective host gene), thereby replacing the molecular lesion with the normal sequence.

About 10 years ago, several groups independently demonstrated that mammalian cells grown in culture contained the enzymatic machinery to carry out homologous (i.e., site-specific) recombination.[84] Thomas and colleagues[85] demonstrated that mammalian tissue culture cells could use this machinery to recombine genes introduced into them with their cellular homologues, although at a low frequency. These initial results stimulated various laboratories to investigate further the mechanisms underlying mammalian homologous recombination, with the goal of improving the frequency of these site-specific events relative to random integration. Today, it is generally accepted that homologous recombination events occur with a frequency of about 1 in 1000, using most nonviral gene transfer technologies (e.g., calcium phosphate precipitation, electroporation). The most important technologic breakthroughs have come from the development of selection schemes designed to enrich for cells containing homologously recombined DNA. In general, this is a two-step process involving initial selection with a dominant selectable marker, such as *neo*, to identify cells that have stably integrated exogenous DNA, followed by screening for the homologous recombination events. A most sensitive screening method uses the polymerase chain reaction to identify G418-resistant colonies in which two marker sequences have been brought into flanking positions by homologous recombination.[86] Alternatively, the expression of the dominant selectable marker may be made conditional on the homologous recombination event. Another particularly efficient screening method introduced by Capecchi and co-workers is termed *positive-negative selection*.[87,88] Using the above techniques, more than 20 mammalian genes have been "knocked-out" in vitro by homologous recombination. Homologous recombination has already been used to create animal models of specific human genetic diseases by knocking out the normal gene in embryonic stem cells and then inserting the modified stem cells into blastomeres that can produce mice with the deranged gene in the germline.[88,89] Thus, as this technology continues to evolve, the prospect of someday performing gene "repair" therapy seems more and more promising.

ACKNOWLEDGMENTS

This review was supported in part by grants AM34045, DK26824, and DK25759 from the National Institutes of Health and by grants 1-578 and 1-1224 from the March of Dimes Birth Defects Foundation. We express appreciation to Ms. Barbara Rogers for preparation of the manuscript.

REFERENCES

1. Scriver CR, Beaudet AL, Sly WS, Valle D (eds): The Metabolic Basis of Inherited Disease. 6th Ed. McGraw-Hill, New York, 1989
2. McKusick VA: Mendelian Inheritance in Man. 9th Ed. Johns Hopkins University Press, Baltimore, 1990
3. Recombinant Advisory Committee: Points to consider in the design and submission of human somatic-cell gene therapy protocols. Recomb DNA Tech Bull 9:221, 1986
4. Gordon JW, Scangos GA, Plotkin DJ, et al: Genetic transformation of mouse embryos by microinjection of purified DNA. Proc Natl Acad Sci USA 77:7380, 1980
5. Costantini F, Chad K, Magram J: Correction of murine β-thalassemia by gene transfer into the germ line. Science 233:1192, 1986
6. Hammer RE, Palmiter RD, Brinster RL: Partial correction of a murine hereditary growth disorder by germ-line incorporation of a new gene. Nature 311:65, 1984
7. Mason AJ, Pitts SL, Nikolics K, et al: The hypogonadal mouse: Reproductive functions restored by gene therapy. Science 234:1372, 1986
8. Readhead C, Popko B, Takahashi N, et al: Expression of a myelin basic protein gene in transgenic Shiverer mice: Correction of the dysmyelinating phenotype. Cell 48:703, 1987
9. Jones SN, Grompe M, Munir MI, et al: Ectopic correction of ornithine carbamyltransferase deficiency in sparse fur mice. J Biol Chem 265:14684, 1990
10. Kyle JW, Birkenmier ET, Gwynn B, et al: Correction of murine mucopolysaccharidosis VII by a human β-glucuronidase transgene. Proc Natl Acad Sci USA 87:3914, 1990
11. Harbers K, Kuehn M, Delius H, Jaenisch R: Insertion of retrovirus into the first intron of al(I) collagen gene leads to embryonic lethal mutation in mice. Proc Natl Acad Sci USA 81:1504, 1984
12. Eglitis MA, Kantoff PW, Gilboa E, Anderson WF: Gene expression in mice after high efficiency retroviral-mediated gene transfer. Science 230:1395, 1985
13. Dick JE, Magli MC, Phillips RA, Bernstein A: Genetic manipulation of hematopoietic stem cells with retrovirus vectors. Trends Genet 2:165, 1986
14. Hock RA, Miller AD: Retrovirus-mediated transfer and expression of drug resistance genes in human hematopoietic progenitor cells. Nature 320:275, 1986
15. Kwok WW, Schuening F, Stead RB, Miller AD: Retroviral transfer of genes into canine hematopoietic progenitor cells in culture. Proc Natl Acad Sci USA 83:4552, 1986
16. Williams DA, Orkin SH, Mulligan RC: Retrovirus-mediated transfer of human adenosine deaminase gene sequences into cells in culture and into murine hematopoietic cells in vivo. Proc Natl Acad Sci USA 83:2566, 1986
17. Kantoff PW, Kohn DB, Mitsuya H, et al: Correction of adenosine deaminase deficiency in cultured human T and B cells by retrovirus-mediated gene transfer. Proc Natl Acad Sci USA 83:6563, 1986

18. Szilvassy SJ, Fraser CC, Eaves CJ, et al: Retrovirus-mediated gene transfer to purified hematopoietic stem cells with long-term lympho-myelopoietic repopulating ability. Proc Natl Acad Sci USA 86:8798, 1989
19. Palmer TD, Hock RA, Osborne WRA, Miller AD: Efficient retrovirus-mediated transfer and expression of a human adenosine deaminase gene in diploid skin fibroblasts from an adenosine deaminase-deficient human. Proc Natl Acad Sci USA 84:1055, 1985
20. Selden RF, Skoskiewics MJ, Howie KB, et al: Implantation of genetically engineered fibroblasts into mice: Implications for gene therapy. Science 236:714, 1987
21. Sorge J, Kuhl W, West C, Beutler E: Complete correction of the enzymatic defect of type I Gaucher disease fibroblasts by retroviral-mediated gene transfer. Proc Natl Acad Sci USA 84:905, 1987
22. Miyanohara A, Sharkey MF, Witztum JL, et al: Efficient expression of retroviral vector-transduced human low density lipoprotein (LDL) receptor in LDL receptor-deficient rabbit fibroblasts in vitro. Proc Natl Acad Sci USA 85:6538, 1988
23. St. Louis D, Verma IM: An alternative approach to somatic cell gene therapy. Proc Natl Acad Sci USA 85:3150, 1988
24. Palmer TD, Rasman GJ, Osborne WRA, Miller AD: Genetically modified skin fibroblasts persist long after transplantation but gradually inactivate introduced genes. Proc Natl Acad Sci USA 88:1330, 1991
25. Ledley FD, Grenett HE, McGinnis-Shelnutt M, Woo SLC: Retroviral-mediated gene transfer of human phenylalanine hydroxylase into NIH 3T3 and hepatoma cells. Proc Natl Acad Sci USA 83:409, 1986
26. Ledley FD, Grenett HE, Bartos DP, Woo SL: Retroviral-mediated transfer and expression of human α-1-antitrypsin in cultured cells. Gene 61:113, 1987
27. Wolffe JA, Yee J-K, Skelly HF, et al: Expression of retrovirally transduced genes in primary cultures of adult rat hepatocytes. Proc Natl Acad Sci USA 84:3344, 1987
28. Wilson JM, Jefferson DM, Chowdhury JR, et al: Retrovirus-mediated transduction of adult hepatocytes. Proc Natl Acad Sci USA 85:3014, 1988
29. Anderson KD, Thompson JA, DiPietro JM, Montgomery KT: Gene expression in implanted rat hepatocytes following retroviral-mediated gene transfer. Somat Cell Mol Genet 15:215, 1989
30. Zweibel JA, Freeman SM, Kantoff PW, et al: High-level recombinant gene expression in rabbit endothelial cells transduced by retroviral vectors. Science 243:220, 1988
31. Morgan JR, Barrandon Y, Green H, Mulligan RC: Expression of an exogenous growth hormone gene by transplantable human epidermal cells. Science 237:1476, 1987
32. Fenjves ES, Gordon DA, Pershing LK, Williams DL: Systemic distribution of apolipoprotein E secreted by grafts of epidermal keratinocytes: Implications for epidermal function and gene therapy. Proc Natl Acad Sci USA 85:8803, 1989
33. Weiss R, Teich N, Varmus H, Coffin J (eds): RNA Tumor Viruses. Cold Spring Harbor Laboratory, Cold Spring Harbor, NY, 1984
34. Bernstein A, Berger S, Huszar D, Dick J: Gene transfer with retroviral vectors. p. 235. In Setlow JK, Hollaender A (eds): Genetic Engineering: Principles and Methods. Vol. 7. Plenum Press, New York, 1985
35. Gilboa E, Eglitis MA, Kantoff PW, Anderson WF: Transfer and expression of cloned genes using retroviral vectors. BioTechniques 4:504, 1986
36. Gilboa E: Retrovirus vectors and their uses in molecular biology. Bioessays 5:252, 1987

37. Hu WS, Temin HM: Retroviral recombination and reverse transcription. Science 250:1227, 1990
38. Mann R, Mulligan RC, Baltimore D: Construction of a retrovirus packaging mutant and its use to produce helper-free defective retrovirus. Cell 33:153, 1983
39. Cone RR, Mulligan RC: High efficiency gene transfer into mammalian cell: Generation of helper-free retroviruses with broad host range. Proc Natl Acad Sci USA 81:6549, 1984
40. Markowitz D, Goff S, Bank A: A safe packaging line for gene transfer: Separating viral genes on the different plasmids. J Virol 62:1120, 1988
41. Markowitz D, Goff S, Bank A: Construction and use of a safe and efficient amphotropic packaging cell line. Virology 167:400, 1988
42. Kozak SL, Kabat D: Ping-pong amplification of a retroviral vector achieves high-level gene expression: Human growth hormone production. J Virol 64:3500, 1990
43. Lim B, Apperley JF, Orkin SH, Williams DA: Long-term expression of human adenosine deaminase in mice transplanted with retrovirus-infected hematopoietic stem cells. Proc Natl Acad Sci USA 85:8892, 1989
44. Ledley FD: Human gene therapy. Biotechnology 89:401, 1989
45. Bodine DM, Karlsson S, Nienhuis AW: Combination of interleukins 3 and 6 preserves stem cell function in culture and enhances retrovirus-mediated gene transfer into hematopoietic stem cells. Proc Natl Acad Sci USA 7:8901, 1989
46. Williams DE, Eisenman J, Baird A, et al: Identification of a ligand for the C-kit proto-oncogene. Cell 63:167, 1990
47. Zsebo KM, Wypych J, McNiece IK, et al: Identification, purification and biological characterization of hematopoietic stem cell factor from buffalo rat liver conditioned medium. Cell 63:195, 1990
48. Spangrude GJ, Heimfeld S, Weissman IL: Purification and characterization of mouse hematopoietic stem cells. Science 241:58, 1988
49. van De Rign, Heimfeld S, Spangrude GJ, Weissman IL: Mouse hematopoietic stem-cell antigen Sca-1 is a member of the Ly-6 antigen family. Proc Natl Acad Sci USA 86:4634, 1989
50. Kantoff PW, Gillio A, McLachlin JR, et al: Expression of human adenosine deaminase in non-human primates after retroviral-mediated gene transfer. J Exp Med 166:219, 1987
51. Peng H, Armentano D, Graham L, et al: Tissue specific expression of human phenylalanine hydroxylase in mouse hepatocytes following retroviral mediated gene transfer. Proc Natl Acad Sci USA 85:8146, 1988
52. Ponder K, Gupta S, Leland F, et al: Mouse hepatocytes migrate to liver parenchyma and function indefinitely after intrasplenic transplantation. Proc Natl Acad Sci USA 88:1217, 1991
53. Demetriou AA, Whiting JF, Feldman D, Levenson S: Replacement of liver function in rats by transplantation of microcarrier-attached hepatocytes. Science 233:1190, 1986
54. Demetriou AA, Levenson SM, Novikoff PM, Novikoff AB: Survival, organization, and function of microcarrier-attached hepatocytes transplanted in rats. Proc Natl Acad Sci USA 83:7475, 1986
55. Anderson KD, Thompson JA, DePietro JM, Montgomery KT: Gene expression in implanted rat hepatocytes following retroviral-mediated gene transfer. Somat Cell Mol Genet 15:215, 1989
56. Thompson JA, Anderson KD, DePetro JM, et al: Site-directed neovessel formation in vivo. Science 241:1349, 1988

57. Thompson JA, Haudenschild CC, Anderson KD, et al: Heparin-binding growth factor 1 induces the formation of organoid neovascular structures in vivo. Proc Natl Acad Sci USA 86:7928, 1989
58. Wu GY, Wu CH: Receptor-mediated gene delivery and expression in vivo. J Biol Chem 263:14621, 1988
59. Garver RI, Chytil A, Karlsson S, Fells GA: Production of glycosylated physiologically "normal" human α_1-antitrypsin by mouse fibroblasts modified by insertion of a human α_1-antitrypsin cDNA using a retroviral vector. Science 239:752, 1987
60. Wolfe JH, Schuchman EH, Stramm LE, et al: Retroviral transfer of the β-glucuronidase cDNA corrects the metabolic defect in cultured mucopolysaccharidosis VII cells. Proc Natl Acad Sci USA 87:2877, 1990
61. Wolff JA, Malone RW, Williams P, et al: Direct gene transfer into mouse muscle in vivo. Science 247:1465, 1990
62. Nabel EG, Plautz G, Nabel G: Site-specific gene expression in vivo by direct gene transfer into the arterial wall. Science 249:1285, 1990
63. Felgner PL, Rhodes G: Gene therapeutics. Nature 349:351, 1991
64. Geller AI, Freese A: Infection of cultured central nervous system neurons with a defective herpes simplex virus 1 vector results in stable expression of *Escherichia coli* β-galactosidase. Proc Natl Acad Sci USA 87:1149, 1990
65. Davison MJ, Preston VG, McGeoch DJ: Determination of the sequence alteration in the DNA of the herpes simplex virus type 1 temperature-sensitive mutant ts K. J Gen Virol 65:859, 1984
66. Dobson AT, Margolis TP, Sedarati F, et al: A latent, nonpathogenic HSV-1-derived vector stably expresses β-galactosidase in mouse neurons. Neuron 5:353, 1990
67. Hermonat PL, Labow MA, Wright R, Berns KI: Genetics of adeno-associated virus: Isolation and preliminary characterization of adeno-associated virus type 2 mutants. J Virol 51:329, 1984
68. McLoughlin SK, Labow MA, Wright R, et al: Adeno-associated virus general transduction vectors: Analysis of proviral structures. J Virol 62:1963, 1984
69. Sieg G, Bates RC, Berns KI, et al: Characteristics and taxonomy of Parvoviridae. Intervirology 23:51, 1985
70. Laface D, Hermonat P, Wakeland E, Peck A: Gene transfer into hematopoietic progenitor cells mediated by an adeno-associated virus vector. J Virol 162:483, 1988
71. Ohi S, Dixit M, Tivery MK: Construction and characterization of recombinant adeno-associated virus genome containing human β-globin cDNA. J Cell Biol 107:304A, 1988
72. Kohn RM, Siniscalco M, Samulski RJ, et al: Site-specific integration by adeno-associated virus. Proc Natl Acad Sci USA 87:2211, 1990
73. Luytjes W, Krystal M, Enami M, et al: Amplification, expression, and packaging of a foreign gene by influenza virus. Cell 59:1107, 1990
74. Mackett M, Smith GL, Moss B: General method for production and selection of infectious vaccinia recombinants: Expression of foreign genes. J Virol 49:857, 1984
75. Spaete RR, Mocarski ES: Insertion and deletion mutagenesis of the human cytomegalovirus genome. Proc Natl Acad Sci USA 84:7213, 1987
76. Paoletti E: Poxvirus recombinant vaccines. Ann NY Acad Sci 590:309, 1990
77. Mecsas J, Sugden B: Replication of plasmids derived from bovine papilloma virus type 1 and Epstein-Barr virus in cells in culture. Annu Rev Cell Biol 3:87, 1987
78. Karlsson S, Humphries RK, Gluzman Y, Nienhuis AW: Transfer of genes into hematopoietic cells using recombinant DNA viruses. Proc Natl Acad Sci USA 82:158, 1985

79. Sugden B, Marsh K, Yates J: A vector that replicates as a plasmid and can be efficiently selected in B-lymphoblasts transformed by Epstein-Barr virus. Mol Cell Biol 5:410, 1985
80. Kasid A, Morecki S, Aebersold P, et al: Human gene transfer: Characterization of human tumor-infiltrating lymphocytes as vehicles for retroviral-mediated gene transfer in man. N Engl J Med 87:473, 1990
81. Rosenberg SA, Aebersold P, Cornetta K, et al: Gene transfer into humans—Immunotherapy of patients with advanced melanoma, using tumor-infiltrating lymphocytes modified by retroviral gene transduction. N Engl J Med 323:570, 1990
82. Dycaico MJ, Grant SGN, Felts K, Nichols WS: Neonatal hepatitis induced by α_1-antitrypsin: A transgenic mouse model. Science 242:1409, 1988
83. Bryan TM, Townes TM, Reilly MP, Asakura T: Human sickle hemoglobin in transgenic mice. Science 247:566, 1990
84. Bollag AJ, Waldman AS, Liskay RM: Homologous recombination in mammalian cells. Annu Rev Genet 23:199, 1989
85. Thomas KR, Folger KR, Capecchi MR: High frequency targeting of genes to specific sites in the mammalian genome. Cell 44:419, 1986
86. Kim HS, Smithies O: Recombinant fragment assay for gene targetting based on the polymerase chain reaction. Nucleic Acids Res 15:8887, 1988
87. Mansaur SL, Thomas KR, Capecchi MR: Disruption of the proto-oncogene *int*-2 in mouse embryo-derived stem cells: A general strategy for targeting mutations to non-selectable genes. Nature 336:345, 1988
88. Thomas KR, Capecchi MR: Introduction of homologous DNA sequences into mammalian cells induces mutations in the cognate gene. Nature 324:34, 1988
89. Kuehn MR, Bradley A, Robertson EJ, Evans MJ: A potential animal model for Lesch-Nyhan syndrome through introduction of HPRT mutations into mice. Nature 326:295, 1987
90. Koller BH, Hagemann LJ, Doetschman T, et al: Germ-line transmission of a planned alteration made in hypoxanthine phosphoribosyl transferase gene by homologous recombination in embryonic stem cell. Proc Natl Acad Sci USA 86:8927, 1989
91. Nichols EK: Human Gene Therapy. Institute of Medicine, National Academy of Sciences, Harvard University Press, Cambridge, MA, 1988

15

GENE TRANSFER INTO HEMATOPOIETIC AND SKIN CELLS

A. Dusty Miller · Michael Kaleko ·
J. Victor Garcia · Arthur R. Thompson ·
William R.A. Osborne · Theo D. Palmer

INTRODUCTION

Gene transfer into somatic cells of humans, or gene therapy, has great potential for the treatment of inherited as well as acquired disease. Attempts to apply this technology to humans are just beginning, and many problems have yet to be resolved. These problems fall into two categories. One involves the control of gene expression and the design of gene transfer vectors that provide appropriate expression of the transferred gene. The other involves the delivery of genes to the appropriate somatic cells, or the transplantation of cells that have already been modified in culture. This chapter addresses these issues by discussing the status of gene transfer into two somatic cell targets for gene therapy, hematopoietic cells and skin fibroblasts.

GENE TRANSFER INTO HEMATOPOIETIC CELLS

An obvious target for gene therapy is the hematopoietic system because of the availability of methods for bone marrow transplantation, the many cell types derived from primitive hematopoietic cells, and the wide distribution of these cells. Ideally, one would like to transfer genes into hematopoietic stem cells that persist for the life of the individual and that give rise to all the differentiated hematopoietic cells. Alternatively, it is possible to transfer genes into more differentiated hematopoietic cells for the treatment of certain diseases.

We have used the human adenosine deaminase (ADA) gene as a model for use in gene therapy. ADA deficiency in humans leads to a severe immunodeficiency involving both B and T lymphocytes that is lethal unless treated.[1,2] Available treatments include bone marrow transplantation, which is curative, and enzyme replacement therapy.[3] However, not all patients have suitable bone marrow donors; also, enzyme replacement therapy requires repeated enzyme injections for life and is not entirely effective. Gene therapy offers the hope that a properly functioning ADA gene would be delivered to the patient's own cells to correct the defect permanently.

To transfer ADA efficiently into hematopoietic cells, many research groups have used retroviral vectors. We have tested a variety of vector designs for expression of ADA in hematopoietic cell lines.[4] These vectors (Fig. 15-1) include a human ADA cDNA driven by a variety of enhancer-promoter combinations, including the simian virus 40 (SV40) early promoter (LNSA), a lymphotropic papovavirus promoter (LNLA), a human β-globin promoter (LNBBA), a human cytomegalovirus (CMV) immediate early promoter (LNCA), and the retroviral long terminal repeat (LTR) (LASN). Of particular relevance is the high level of ADA expression observed in ADA-deficient lymphoblasts infected with the LASN virus (Table 15-1), since lymphoid cells are

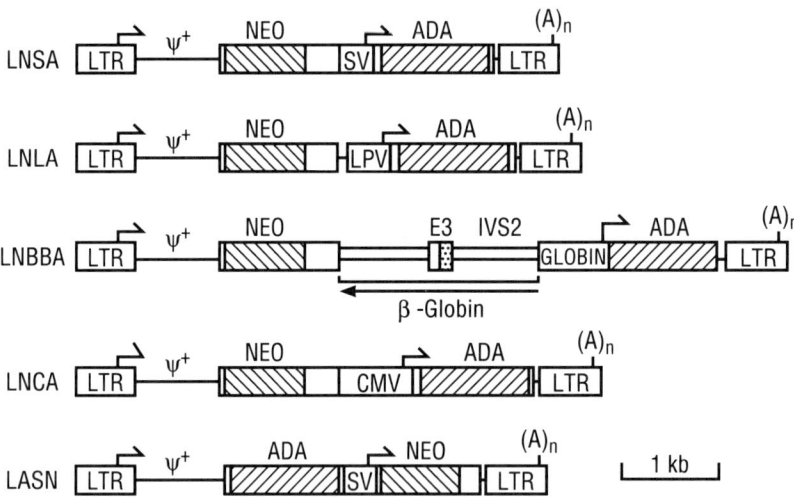

Fig. 15-1. Vectors for transfer and expression of human adenosine deaminase (ADA). Promoters are indicated as large open boxes with arrows denoting the site of RNA transcription initiation; SV indicates the SV40 early region promoter and enhancers, CMV indicates CMV immediate early promoter and enhancers, LPV indicates the lymphotropic papovavirus enhancer linked to an enhancerless SV40 early promoter, GLOBIN indicates the human β-globin promoter, and LTR indicates the retroviral long terminal repeat that contains the retroviral promoter and enhancers. Large boxes with cross-hatched or stippled regions indicate fragments of the *neo* or β-globin genes or an ADA cDNA, and the cross-hatched areas represent the coding regions. Small boxes indicate the second intervening sequence (IVS2) or sequences downstream of the β-globin gene. Lines indicate retroviral sequences. E3, exon 3 of the human β-globin gene; ψ^+, extended retroviral packaging signal; $(A)_n$, polyadenylation site.

Table 15-1. ADA Activity in Infected DHL-9 Lymphoblasts[a]

Virus	ADA Promoter	ADA	PNP	ADA/PNP
None	None	<0.01	2.66	<0.004
LNSA	SV40	0.15	3.08	0.05
LNLA	LPV	0.33	3.94	0.08
LNBBA	β-Globin	0.03	3.24	0.009
LASN	MoMLV	3.86	3.01	1.3
LNCA	CMV	0.69	3.18	0.22

Abbreviations: ADA, adenosine deaminase; CMV, cytomegalovirus; PNP, purine nucleoside phosphorylase.

[a] Enzymatic activity units are μmol/h/mg protein.

primarily affected in human ADA deficiency. LASN also directed the synthesis of high levels of ADA in the K562 myeloid cell line.[4]

We have used the LASN vector to infect hematopoietic cells from mice[5] according to the protocol depicted in Figure 15-2. Bone marrow donors were treated with 5-fluorouracil (5-FU) 4 days before bone marrow harvest to ablate more differentiated hematopoietic cells and to stimulate the replication of early hematopoietic stem cells, a requirement for retrovirus infection.[6] To promote efficient infection, the bone marrow cells were cocultivated with vector-producing cells for 4 days. This procedure results in constant exposure of the bone marrow cells to virus during the 4-day period. In some experiments, the bone marrow cells were then exposed to selection in the antibiotic G418 to enrich for cells that had been infected with the vector, since the vector also carries the *neo* gene, conferring resistance to G418. The genetically modified cells were then injected into W/W^v mice. These mice carry a genetic defect that affects the hematopoietic stem cells such that marrow from normal mice will engraft in these animals without prior irradiation of the recipient W/W^v animals.

After bone marrow transplantation, blood from the animals was examined for human ADA expression after starch gel electrophoresis, which separates

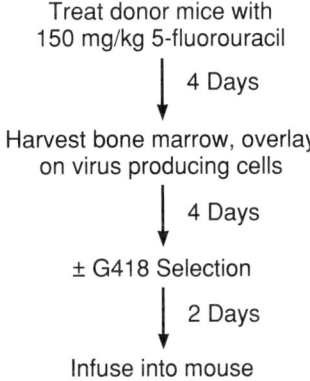

Fig. 15-2. Protocol for gene transfer into mouse bone marrow.

mouse and human ADA[5] (Fig. 15-3). All mice expressed human ADA in whole blood at levels comparable to the endogenous mouse ADA for up to 1 month after transplantation. Some animals continued to express human ADA at the same level for 5.5 months, when the animals were sacrificed for analysis of other tissues. The mice that received G418-selected marrow generally showed higher levels of human ADA expression than did those receiving nonselected marrow. While human ADA could be detected in blood and bone marrow of animals sacrificed at 5.5 months after transplantation, we were unable to detect human ADA in the spleen, thymus, or lymph nodes of the animals. DNA analysis confirmed that vector DNA was present in all tissues analyzed, so this result was not because of incomplete reconstitution, and might reflect tissue-specific repression of vector expression.

To summarize, we have been able to obtain long-term human ADA expression in hematopoietic cells of mice at levels that should be therapeutically useful, since they are similar to endogenous levels of mouse ADA in blood. However, we see clear evidence for suppression of vector expression in some animals, suggesting that improvements in vector design to provide higher levels of transduced gene expression may be possible.

Several groups have documented long-term expression of human ADA in mice after bone marrow transplantation.[5,7–11] In addition, similar techniques have been used to show long-term expression of human β-globin[12–14] and a mutant dihydrofolate reductase gene (DHFR*).[15] In the case of β-globin, the

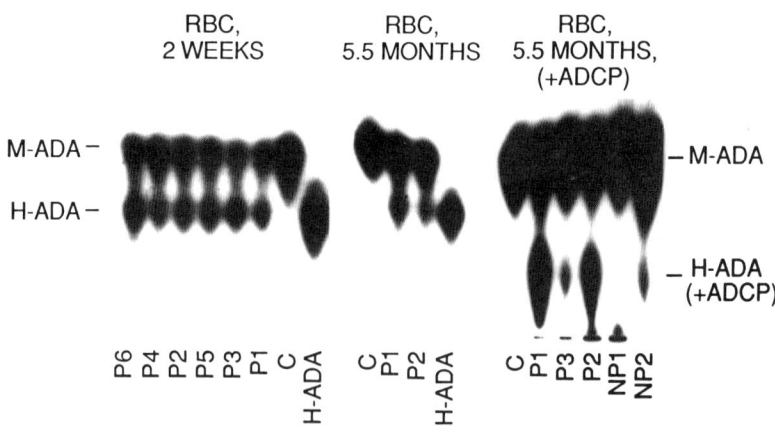

Fig. 15-3. Detection of human adenosine deaminase (ADA) in blood of mice transplanted with genetically modified bone marrow. Blood was obtained from tail veins at 2 weeks and 5.5 months after transplantation. Mouse ADA (M-ADA) and human ADA (H-ADA) were resolved by starch gel electrophoresis. For more sensitive detection of H-ADA, human ADA combining protein (ADCP) was added to the lysates before analysis (right panel). ADCP retards the migration of H-ADA and provides better separation of H-ADA from M-ADA. Animals P1–P6 received marrow preselected in G418, while animals NP1 and NP2 received marrow that was not preselected. Lanes labeled C indicate analyses of blood from a control animal not infected with the LASN vector, and lanes labeled H-ADA indicate analyses of extracts from human T cells as a human ADA standard.

best levels obtained were about 1 to 2 percent of the endogenous mouse β-globin levels, which would not be therapeutically beneficial. More recently, elements required for high-level β-globin expression have been identified[16] that were not included in the original vectors. Efforts are under way to include these sequences and to increase β-globin expression to useful levels.[17] In the case of DHFR*, the mice became resistant to toxic concentrations of methotrexate, suggesting that the DHFR* gene might be useful in protecting transplanted marrow from methotrexate toxicity during chemotherapy post-transplantation.

While bone marrow-directed gene transfer studies in mice have been encouraging, successful infection of stem cells from large animals has not yet been demonstrated.[18,19] While a recent report indicates that this situation may be improving,[20] the evidence from large animal studies does not permit extension of these studies to humans with a reasonable certainty of success. Several differences between the way that experiments are conducted in large animals compared with the way they are done in mice are worth noting. First, the large animals models studied are outbred, necessitating autologous marrow transplantation. As a result, the amount of marrow available for transplantation is limited. It is known, from the mouse studies, that stem cells are lost during the infection process and during selection in G418. These losses probably cannot be tolerated in an autologous transplantation model; therefore, selection of marrow in G418 before transplantation has not been performed. Another difference in the models is that 5-FU has not yet been used in the large animal models because of the additional toxicity that the test animal would endure. It is hoped that these problems can be solved by improved knowledge of stem cell biology, in particular the factors that cause these cells to replicate and thus become susceptible to retroviral vector infection.

Given the difficulty experienced in infection of hematopoietic stem cells, attention has turned toward infection of more differentiated hematopoietic cells, in particular, T cells. T cells provide a important target for treatment of ADA deficiency because this immune dysfunction primarily affects lymphoid cells. In addition, recent experiments that involved marking tumor infiltrating lymphocytes (TIL) in human patients undergoing immunotherapy for malignant cancer[21] have shown that retrovirally mediated gene transfer is suitable and safe for gene transfer into T cells. In these experiments, T cells that infiltrate tumor masses are removed and cultured in the presence of interleukin-2 (IL-2) and can be expanded to large numbers. The cells are marked with a retroviral vector carrying the *neo* gene to track the distribution of the cells after reinfusion. The results have shown that TIL home to tumor masses and persist for long periods of time in patients, albeit at low frequency. The cells used to produce the vector, vector stocks, and all the patients that have received TIL have been free of helper virus. The patients have shown no ill effects from the vector-infected cells; thus, the procedure appears safe.

In the case of the TIL experiments, the transferred gene did not have to be expressed. The vector DNA was used only as a tag to follow the cells. For ADA deficiency, the ADA gene must be expressed in the infected T cells, and we have shown that peripheral T cells from ADA-deficient patients can be infected with the LASN vector and converted to ADA-positive. These genetically modified cells then become resistant to the toxic effects of 2'-deoxyadenosine, a substrate of ADA that is greatly elevated in ADA-deficient patients and that is responsible for the toxic manifestations of the disease. An attempt to treat ADA-deficient human patients by in vitro infection of their peripheral T cells, followed by reinfusion of these cells, has recently been initiated. In addition, the Human Gene Therapy Subcommittee of the Recombinant DNA Advisory Committee has approved a protocol in which the tissue necrosis factor gene will be introduced into TIL for targeted delivery of this cytotoxic protein to malignant cells.

GENE TRANSFER INTO SKIN FIBROBLASTS

Owing in part to the difficulties of gene transfer into hematopoietic stem cells, a variety of other somatic tissues have been considered as targets for gene transfer. Skin fibroblasts provide an attractive target because they are easily obtained and transplanted and can be grown in culture. One might also predict that since these cells appear not to undergo further differentiation, unlike the extensive differentiation of hematopoietic stem cells, suppression of gene expression might be less of a problem in skin fibroblasts.

Skin fibroblasts slowly lose their ability to divide during cultivation; thus, it is important to transfer genes into these cells with high efficiency, to avoid prolonged cultivation and potential loss of transplantability. Many gene transfer techniques, such as DNA coprecipitation with calcium phosphate, result in very low frequencies of gene transfer. By contrast, retroviral vectors promote gene transfer efficiencies of more than 50 percent into fetal, newborn, or adult skin fibroblasts.[22] Cultured human skin fibroblasts have been shown to make a variety of therapeutically beneficial proteins after transfer of the respective genes, including ADA,[22,23] glucocerebrosidase,[24] purine nucleoside phosphorylase,[25] low-density lipoprotein (LDL) receptors,[26] and factor IX.[27]

Genetically modified fibroblasts continue to synthesize transferred genes after transplantation back into animals. Fibroblasts that secrete clotting factor IX have been implanted under the skin in collagen matrices or injected intraperitoneally on collagen beads.[27,28] Up to 190 ng/ml factor IX was detected in the plasma of some animals for short periods after transplantation.[27] Normal levels of factor IX in humans are about 5 µg/ml, but 500 ng/ml would alleviate most of the symptoms of severe hemophilia B. However, expression was observed for no longer than 1 month in these animals.

We have examined this problem of decreased gene expression after transplantation of rat skin fibroblasts, using a human ADA gene to follow gene expression.[23] Rats were used as a model because rat fibroblasts, like human fibroblasts, tend not to spontaneously immortalize in culture as easily as mouse skin fibroblasts, and therefore provide a better model for ultimate use of these techniques in humans. ADA was chosen as a marker for gene expression because it is an intracellular enzyme, which could therefore be analyzed in tissue samples. By contrast, clotting factor IX is secreted and is distributed throughout the body.

Rat skin fibroblasts were infected with the LASN vector (Fig. 15-1), selected in G418, and transplanted back to syngeneic rats. The patches containing the transplanted cells were removed at various times and cut into pieces; the pieces were analyzed for human and rat ADA expression, the presence of vector sequences, and the presence and activity of vector sequences in cells grown out of the grafts. The decrease in ADA/purine nucleoside phosphorylase (PNP) level with time (Table 15-2) was also reflected in starch gel analysis of ADA in tissue samples (Fig. 15-4) and showed that human ADA dropped from very high levels to background levels within about 1 month. We calculate that this decrease represents a greater than 1500-fold reduction in human ADA expression. By contrast, vector DNA persists in the tissue at undiminished levels for at least 8 months (Table 15-2). Therefore, the genetically modified fibroblasts persist after transplantation, but gene expression is dramatically suppressed.

Cells cultured from the patches of grafted tissue also contained vector sequences, although at somewhat lower levels than the tissue itself (Table 15-2). Cultivation of the cells did not restore expression of human ADA or the cotransferred *neo* gene as measured by the ability of the cells to form colonies in G418 (Table 15-2). Thus, the suppression of gene expression was not simply the result of the altered status of the cells in animals, since gene suppression persisted when the cells were recultured. Nor did suppression of gene expression in animals appear to be the result of immune reaction against ADA.[23] In

Table 15-2. ADA Vector Persistence and Expression in Transplanted Rat Fibroblasts[a]

Time	Transplanted Tissue		Cells Cultured from Tissue		% G418^r Colonies[b]
	% Vector Positive	ADA/PNP	% Vector Positive	ADA/PNP	
0	—	—	100	3.6	60
2 d	40	1.9	30	3.2	46
2 wk	30	0.34	13	0.87	8
1 mo	30	0.23	9	0.13	0.03
8.5 mo	60	0.13	6	0.14	<0.001

[a] Vector DNA was analyzed by using PCR. ADA activities were normalized to PNP activities to correct for variability in protein extraction from tissue. ADA/PNP in normal rat skin is 0.17 ± 0.04 and in cultured rat fibroblasts is 0.17 ± 0.03. ADA/PNP in uninfected control transplants was indistinguishable from those of normal skin or cultured cells.

[b] The percentage of G418^r colonies was calculated as follows:

$$\frac{\text{Colonies that grew in G418}}{\text{colonies that grew in the absence of G418}} \times 100$$

The plating efficiency of G418^r cells in G418 is 20–50% lower than that in the absence of G418; thus, even a population of cells that has been grown in G418 does not give 100% G418^r colonies (e.g., day 0). Values are the means of two or three separate explants for each time point.

Abbreviations: ADA, adenosine deaminase; PNP, purine nucleoside phosphorylase; PCR, polymerase chain reaction.

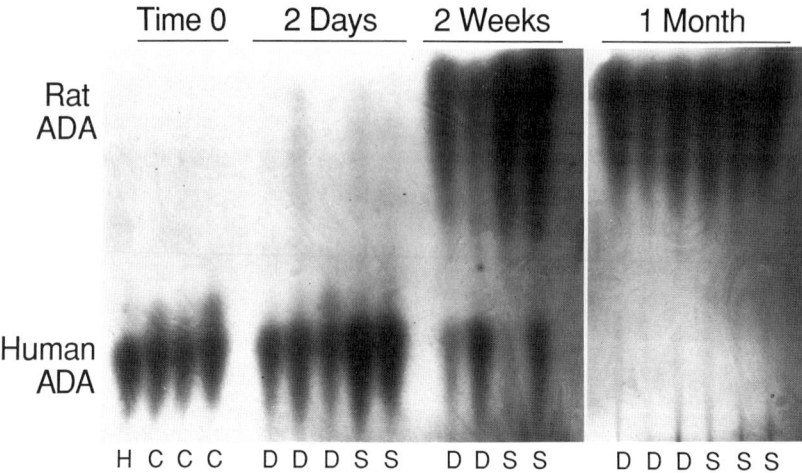

Fig. 15-4. Human and rat adenosine deaminase (ADA) in transplanted tissues. Equal amounts of total ADA were analyzed by starch gel electrophoresis. Although not shown here, loading of equal amounts of total protein showed that rat ADA remained relatively constant in the rat cell samples. Lane descriptions: H, human T-cell sample; C, cultured cells before transplantation; D, dermal equivalent transplants; S, subcutaneous transplants of fibroblast-containing collagen matrices.

addition, growth of the genetically modified fibroblasts in culture for prolonged periods did not result in a significant decrease in expression of human ADA, showing that gene suppression only occurred in animals.

SUMMARY

We have shown that genes can be efficiently transferred into skin fibroblasts and that these cells persist long term after transplantation. While the cells can produce therapeutically useful proteins, gene expression did not persist for more than 1 month. We expect that the use of alternative gene regulatory elements will permit prolonged gene expression, as endogenous genes are expressed in these fibroblasts (e.g., rat ADA). It should be possible to incorporate elements from these endogenous genes into retroviral vectors to promote long-term expression of the transferred genes. Thus fibroblasts may still prove a suitable target for gene therapy.

It is surprising that expression of transferred genes was suppressed so efficiently in skin fibroblasts, while the use of the same vector (LASN) to infect hematopoietic stem cells resulted in long-term gene expression. We expected to encounter more difficulties with gene expression in hematopoietic cells because of their extensive differentiation program and corresponding changes in patterns of gene expression, which might result in vector inactivation. That suppression of gene expression in fibroblasts occurs only after transplantation of the cells into animals indicates the complexity of the control of gene expression in cells in animals, an issue that is crucial for successful application of gene therapy.

CONCLUSION

While the application of gene therapy to the treatment of human disease has taken longer than expected, given that many diseases are understood in great detail and that the normal genes have been isolated, we have made significant progress in this field. We can now achieve levels of human proteins in the somatic tissues of animals that are likely to be therapeutic in humans. In addition, the first genes have been transferred into humans without apparent ill effects. While many investigators have focused on retroviral vectors for gene transfer, many new techniques appear useful as well, such as direct DNA injection[29] or targeted DNA delivery (see Ch. 16). A number of different somatic cell types are now being assessed as potential targets for gene therapy. In the near future, gene therapy will justifiably take its place beside other treatments for inborn errors of metabolism, next to transplantation, nutrition, and enzyme replacement therapies.

ACKNOWLEDGMENTS

This work was supported by grants HL07808, HL36444, AI19565, AI27291, and DK38531 from the National Institutes of Health.

REFERENCES

1. Giblett ER, Anderson JE, Cohen F, et al: Adenosine deaminase deficiency in two patients with severely impaired cellular immunity. Lancet 2:1067, 1972
2. Kredich NM, Hershfield MS: Immunodeficiency diseases caused by adenosine deaminase deficiency and purine nucleoside phosphorylase deficiency. p. 1157. In Stanbury JB, Wyngaarden JB, Fredrickson DS, et al (eds): The Metabolic Basis of Inherited Disease. 5th Ed. McGraw-Hill, New York, 1983
3. Hershfield MS, Buckley RH, Greenberg ML, et al: Treatment of adenosine deaminase deficiency with polyethylene glycol-modified adenosine deaminase. N Engl J Med 316:589, 1987
4. Hock RA, Miller AD, Osborne WRA: Expression of human adenosine deaminase from various strong promoters after gene transfer into human hematopoietic cell lines. Blood 74:876, 1989
5. Kaleko M, Garcia JV, Osborne WRA, Miller AD: Expression of human adenosine deaminase in mice after transplantation of genetically-modified bone marrow. Blood 75:1733, 1990
6. Miller DG, Adam MA, Miller AD: Gene transfer by retrovirus vectors occurs only in cells that are actively replicating at the time of infection. Mol Cell Biol 10:4239, 1990
7. Belmont JW, MacGregor GR, Wagner-Smith K, et al: Expression of human adenosine deaminase in murine hematopoietic cells. Mol Cell Biol 8:5116, 1988
8. Lim B, Apperley JF, Orkin SH, Williams DA: Long-term expression of human adenosine deaminase in mice transplanted with retrovirus-infected hematopoietic stem cells. Proc Natl Acad Sci USA 86:8892, 1989
9. Wilson JM, Danos O, Grossman M, et al: Expression of human adenosine deaminase in mice reconstituted with retrovirus-transduced hematopoietic stem cells. Proc Natl Acad Sci USA 87:439, 1990

10. Osborne WRA, Hock RA, Kaleko M, Miller AD: Long term expression of human adenosine deaminase in mice after transplantation of bone marrow infected with amphotropic retroviral vectors. Hum Gene Ther 1:31, 1990
11. Moore KA, Fletcher FA, Villalon DK, et al: Human adenosine deaminase expression in mice. Blood 75:2085, 1990
12. Dzierzak EA, Papayannopoulou T, Mulligan RC: Lineage-specific expression of a human β-globin gene in murine bone marrow transplant recipients reconstituted with retrovirus-transduced stem cells. Nature 331:35, 1988
13. Bender MA, Gelinas RE, Miller AD: A majority of mice show long-term expression of a human β-globin gene after retroviral transfer into hematopoietic stem cells. Mol Cell Biol 9:1426, 1989
14. Bodine DM, Karlsson S, Nienhuis AW: Combination of interleukins 3 and 6 preserves stem cell function in culture and enhances retrovirus-mediated gene transfer into hematopoietic stem cells. Proc Natl Acad Sci USA 86:8897, 1989
15. Corey CA, DeSilva AD, Holland CA, Williams DA: Serial transplantation of methotrexate-resistant bone marrow: Protection of murine recipients from drug toxicity by progeny of transduced stem cells Blood 75:337, 1990
16. Grosveld F, van Assendelft GB, Greaves DR, Kollias G: Position-independent, high-level expression of the human β-globin gene in transgenic mice. Cell 51:975, 1987
17. Novak U, Harris EAS, Forrester W, et al: High level β-globin expression after retroviral transfer of locus activation region-containing human β-globin gene derivatives into murine erythroleukemia cells. Proc Natl Acad Sci USA 87:3386, 1990
18. Kantoff PW, Gillio AP, McLachlin JR, et al: Expression of human adenosine deaminase in nonhuman primates after retrovirus-mediated gene transfer. J Exp Med 166:219, 1987
19. Stead RB, Kwok WW, Storb R, Miller AD: Canine model for gene therapy: Inefficient gene expression in dogs reconstituted with autologous marrow infected with retroviral vectors. Blood 71:742, 1988
20. Bodine DM, McDonagh KT, Brandt SJ, et al: Development of a high titer retrovirus producer cell line capable of gene transfer into rhesus monkey hematopoietic stem cells. Proc Natl Acad Sci USA 87:3738, 1990
21. Rosenberg SA, Aebersold P, Cornetta K, et al: Gene transfer into humans: Immunotherapy of patients with advanced melanoma using tumor infiltrating lymphocytes modified by retroviral gene transduction. N Engl J Med 323:570, 1990
22. Palmer TD, Hock RA, Osborne WR, Miller AD: Efficient retrovirus-mediated transfer and expression of a human adenosine deaminase gene in diploid skin fibroblasts from an adenosine deaminase-deficient human. Proc Natl Acad Sci USA 84:1055, 1987
23. Palmer TD, Rosman GJ, Osborne WRA, Miller AD: Genetically-modified skin fibroblasts persist long-term after transplantation but gradually inactivate introduced genes. Proc Natl Acad Sci USA 88:1330, 1991
24. Sorge J, Kuhl W, West C, Beutler E: Complete correction of the enzymatic defect of type I Gaucher disease fibroblasts by retroviral-mediated gene transfer. Proc Natl Acad Sci USA 84:906, 1987
25. Osborne WR, Miller AD: Design of vectors for efficient expression of human purine nucleoside phosphorylase in skin fibroblasts from enzyme-deficient humans. Proc Natl Acad Sci USA 85:6851, 1988
26. Miyanohara A, Sharkey MF, Witztum JL, et al: Efficient expression of retroviral vector-transduced human low density lipoprotein (LDL) receptor in LDL receptor-deficient rabbit fibroblasts in vitro. Proc Natl Acad Sci USA 85:6538, 1988

27. Palmer TD, Thompson AR, Miller AD: Production of human factor IX in animals by genetically modified skin fibroblasts: potential therapy for hemophilia B. Blood 73:438, 1989
28. St Louis D, Verma IM: An alternative approach to somatic cell gene therapy. Proc Natl Acad Sci USA 85:3150, 1988
29. Wolff JA, Malone RW, Williams P, et al: Direct gene transfer into mouse muscle in vivo. Science 247:1465, 1990

16

DELIVERY AND EXPRESSION OF GENES IN HEPATOCYTES

George Y. Wu · Catherine H. Wu

INTRODUCTION

The Concept

One of the most popular methods for gene transformation in vitro is the $CaPO_4$ precipitation technique. In this procedure, $CaPO_4$ is allowed to precipitate in the presence of DNA, forming insoluble particles containing molecules of foreign DNA trapped within.[1] Exposure of cultured cells to these precipitates results in their internalization by a nonspecific endocytotic process.[1] Although the efficiency is low, clearly some of the DNA in $CaPO_4$ precipitates is able to avoid enzymatic digestion, and eventually reaches the nucleus, resulting in new gene expression by the host cells.

Hepatocytes possess unique cell-surface receptors that can internalize galactose-terminal (asialo-)glycoproteins.[2] In this process, binding of the ligand results in invagination of the plasma membrane and formation of membrane-limited endosomal vesicles containing the ligand.[3] This receptor-mediated endocytotic pathway usually results in fusion of endosomal vesicles with lysosomes and enzymatic degradation of the ligand.[4] However, it has been shown that degradation of asialoglycoprotein conjugates can be incomplete, leading to the release of polypeptides into the cytoplasm.[5] We hypothesized that, under the proper conditions, DNA could similarly survive a receptor-mediated endocytotic event. Based on the use of asialoglycoprotein receptors as natural internalization sites, we demonstrated that a variety of biologic agents bound to carrier molecules can be targeted specifically to cells bearing the appropriate receptors for the carrier.[6–8] In these studies, agents were coupled covalently to asialoglycoproteins. In the case of DNA, however, covalent linkage to an asialoglycoprotein could result in the alteration of bases that could cause impaired transcriptional fidelity in target cells.

Thus, we have developed a soluble DNA carrier system[9,10] consisting of two components: (1) an asialoglycoprotein carrier (ligand) capable of being internalized by unique cell-surface receptors present exclusively on hepatocytes and (2) a polycation (e.g., polylysine) that can bind DNA in a strong nondamaging electrostatic interaction,[11] forming a soluble complex.

We proposed the following hypothesis. If a polycation were covalently bound to an asialoglycoprotein, subsequent addition of DNA to form a soluble complex could permit targeted delivery of foreign DNA specifically to hepatocytes based on recognition of the asialoglycoprotein component of the conjugate. Because of its targetability, the system could result in the concentrated delivery of genetic material to hepatocytes and, because of its solubility, delivery of foreign genes to hepatocytes could be achieved by intravenous injection. In addition, because hepatocytes have a low turnover rate,[12] are active metabolically, and possess an excellent blood supply, these cells may be well suited as recipients for delivery and expression of replacement genes.

PREPARATION OF A TARGETABLE DNA CARRIER SYSTEM

To test this system, DNA in the form of the plasmid, pSV2 CAT, was used. This plasmid contains the gene for the bacterial enzyme chloramphenicol acetyltransferase (CAT) driven by an SV40 viral promoter. Chloramphenicol acetyltransferase catalyzes the acetylation of chloramphenicol. Mammalian cells lack the CAT gene. Therefore, the appearance of acetylated products of chloramphenicol can provide a simple system for detection of the CAT enzymatic activity in target cells and would serve as a convenient marker of successful foreign gene transfection.[13]

The human serum glycoprotein, orosomucoid, was prepared from pooled human plasma[14] as starting material for the preparation of asialoorosomucoid (AsOR), a well-studied protein with high affinity for asialoglycoprotein receptors. Orosomucoid was desialylated[15] and labeled with iodine-125 (^{125}I). To prepare a targetable DNA carrier system, [^{125}I]-AsOR was conjugated to poly-L-lysine using N-succinimidyl 3-(2-pyridyldithio)propionate.[16] The conjugate was dialyzed and subsequently purified by molecular sieve chromatography. Based on the specific activity of [^{125}I]-AsOR and the lysine content of the conjugate, the molar ratio of AsOR to poly-L-lysine in the conjugate was calculated to be 5:1.

TARGETABLE CONJUGATE-DNA COMPLEXES

The first objective was to determine the quantity of DNA that should be mixed with conjugate to form a soluble complex. To accomplish this, a gel-retardation assay system was developed in which samples containing equal amounts of DNA were mixed with increasing amounts of $[^{125}I]$-AsOR-PL conjugate. Each sample was filtered through 0.45-μm membranes to ensure that the complexes did not contain precipitates. The $[^{125}I]$-AsOR-PL-DNA samples were loaded onto an agarose gel for electrophoresis.[17] A decrease in staining intensity of the DNA bands that penetrated the gel was found as the proportion of AsOR-PL conjugates in the samples increased. A corresponding increase was found in staining of DNA that remained at the top of the gel with the conjugate.[17] This suggested that progressive amounts of DNA were bound by the AsOR-PL conjugate in the wells.

To confirm this, the plasmid, pSV2 CAT, was labeled with ^{32}P by nick-translation. Constant quantities of labeled DNA were again mixed with increasing concentrations of unlabeled conjugate.[18] Samples were prepared as described for the previous experiment. Filtrates of the soluble complexes were again electrophoresed on an agarose gel, and an autoradiogram was obtained. This confirmed that with increasing proportions of conjugate in the samples, increasing amounts of $[^{32}P]$-DNA were retained in complexes at the top of the gel. For this particular plasmid and this preparation of conjugate, the DNA content of the complexes was maximized at a conjugate/DNA ratio of approximately 1.87:1. This represented the lowest conjugate/DNA ratio in which DNA entry into the gel was completely abolished.[18]

TARGETED GENE DELIVERY AND EXPRESSION IN VITRO

In order to determine whether the DNA carrier system could, in fact, deliver functional DNA to receptor-bearing cells, an in vitro system was used, consisting of two hepatoma cell lines: Hep G2, which are asialoglycoprotein receptor (+) cells, and SK-Hep 1, which are receptor (−). Each cell line was incubated in medium containing unlabeled AsOR-PL-DNA complex containing the CAT gene in the form of the pSV2 CAT plasmid or controls. Plates containing equal numbers of cells were assayed for CAT activity by incubation with $[^{14}C]$chloramphenicol.[13] As shown in Figure 16-1, incubation of the SK-Hep 1, receptor (−) cells, with either AsOR-PL-DNA complex, or controls containing various components of the complex, did not produce detectable CAT activity under any conditions.[17]

By contrast, Hep G2, receptor (+) cells incubated with the AsOR-PL-DNA complex did result in CAT gene transfection, as demonstrated by the formation of acetylchloramphenicol derivatives. Hep G2 cells, incubated under identical conditions with controls: DNA alone, DNA plus PL, or DNA plus AsOR, all present in the same concentrations as provided by the complete complex did not result in gene transfection[17] (Fig. 16-2).

When an excess of AsOR was added to the complex to compete with the complex for asialoglycoprotein receptors, expression of CAT in Hep G2 cells was not detected, supporting the notion that recognition of the complex was

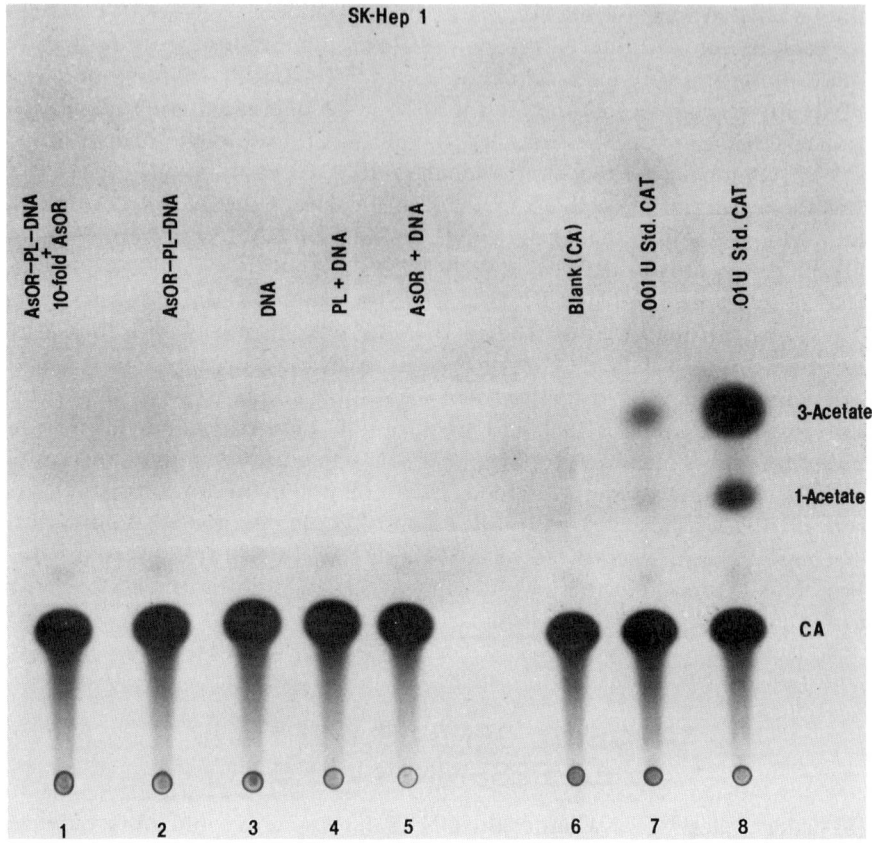

Fig. 16-1. Assay for gene transformation in SK-Hep 1 cells. Cells were grown to one-quarter confluence and then incubated for 48 hours at 37°C under 5 percent CO_2 in the presence of filtered AsOR-poly-L-lysine complex or components of the complex. After harvesting and sonication, cells were assayed for CAT enzymatic activity by thin-layer chromatography.[13] The effects of incubation with AsOR-poly-L-lysine complex alone is shown in lane 2; complex plus AsOR, lane 1; DNA, lane 3; poly-L-lysine plus DNA, lane 4; and AsOR plus DNA, lane 5. [^{14}C]Chloramphenicol alone is shown in lane 6; standard CAT enzyme, 0.001 units, lane 7; and CAT 0.01 units, lane 8. PL, poly-L-lysine. (From Wu and Wu,[17] with permission.)

directed by the asialoglycoprotein component of the conjugate.[17] Potential toxic effects of targetable complex on cells were determined using Trypan blue exclusion as a measure of cell viability. Cells treated with complex proliferated at the same rate as untreated control cells.[9]

These data indicate that (1) a DNA carrier system based on internalization of asialoglycoproteins by hepatocytes can be used to target a foreign gene in a nontoxic soluble form, specifically to receptor-bearing cells[17,18] and (2) the gene can be expressed by cultured hepatic cells.

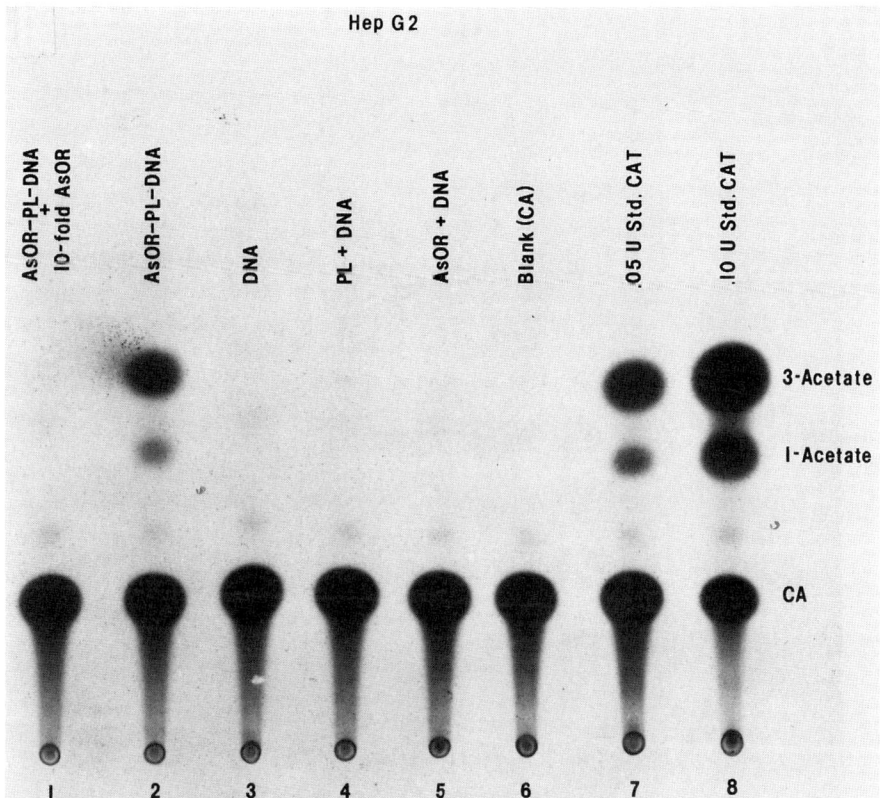

Fig. 16-2. Assay for gene transformation in Hep G2 cells. Cells were grown to one-quarter confluence and then incubated for 48 hours at 37°C under 5 percent CO_2 in the presence of filtered AsOR-poly-L-lysine complex or components of the complex. After harvesting and sonication, cells were assayed for CAT enzymatic activity by thin-layer chromatography, as described by Gorman et al.[13] The effect of incubation with AsOR-poly-L-lysine complex alone is shown in lane 2; complex plus AsOR, lane 1; DNA, lane 3; poly-L-lysine plus DNA, lane 4; AsOR plus DNA, lane 5. [^{14}C]Chloramphenicol alone is shown in lane 6; standard CAT enzyme, 0.05 units, lane 7; and CAT, 0.1 units, lane 8. PL, poly-L-lysine. (From Wu and Wu,[17] with permission.)

TARGETED GENE DELIVERY AND EXPRESSION IN VIVO

Compared with other gene transfection techniques, the asialoglycoprotein receptor-mediated system differs from most by the characteristics of solubility and targetability. These properties raised the possibility that foreign genes could be targeted in vivo by simple intravenous injection.

To test this possibility, plasmid DNA was radiolabeled with ^{32}P, complexed with carrier, and then filtered through membranes to ensure that precipitates were absent. The ^{32}P-labeled DNA complex or ^{32}P-labeled DNA alone was injected intravenously into adult rats. After 10 minutes, animals were sacrificed and organs removed and counted.[19] Injection of labeled DNA resulted in only

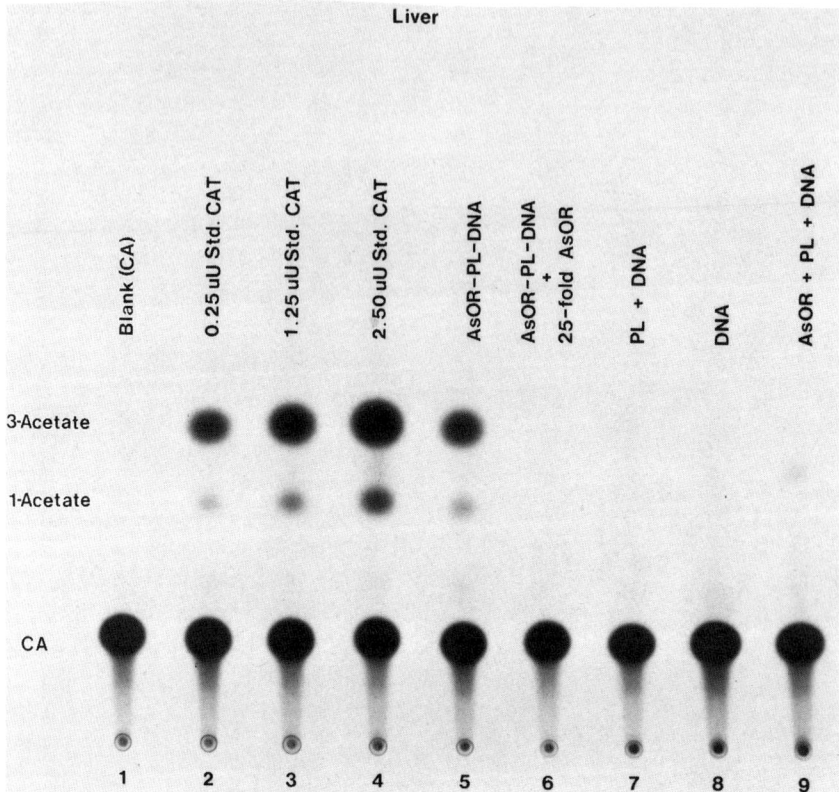

Fig. 16-3. Representative autoradiogram of a thin-layer chromatographic CAT assay of liver homogenates. Rats were injected with AsOR-PL-DNA complex or controls, all in sterile saline. After 24 hours, livers were homogenized and assayed for CAT enzymatic activity. Lane 1 contains [^{14}C]chloramphenicol alone; lanes 2–4, standard CAT enzyme; lanes 5–9, liver homogenates from rats treated with AsOR-PL-DNA complex, complex plus excess AsOR, PL plus DNA, DNA alone, or a mixture of AsOR plus PL plus DNA, respectively. (From Wu and Wu,[19] with permission.)

17 percent of the injected counts taken up by the liver. However, labeled DNA injected in the form of an AsOR-PL-DNA complex resulted in liver uptake of 85 percent of the injected counts. The organ distribution of the complex was similar to that for [^{125}I]-AsOR alone, indicating that the complex retained its ability to be recognized by asialoglycoprotein receptors in vivo.[19] Because the recognition of our targetable complex is directed by an asialoglycoprotein, receptor-mediated endocytosis of the complex would be expected to result in rapid degradation of the ligand. The radioactivity detected in liver after intravenous injection of the complex could simply have represented DNA that had been completely degraded. To determine whether any targeted CAT gene sequences could be detected substantially beyond the 10-minute postinjection period, rats were injected intravenously with unlabeled complexed plasmid containing the CAT gene. Twenty-four hours later, liver DNA was extracted,

and CAT sequences were detected by dot-blot hybridization using a ^{32}P-labeled cDNA CAT probe.[19] CAT gene sequences from transfected liver were easily detectable. Equal quantities of control liver DNA possessed no sequences hybridizable to the cDNA probe, indicating that the hybridization found in liver transfected with the complex was a result of targeted DNA and not host DNA sequences.[19]

Although CAT sequences could be detected after injection of the complex, the DNA could have been nonfunctional. To determine whether the targeted DNA could be expressed in the liver, rats were injected with complexed DNA in saline or components of the complex; DNA alone, PL plus DNA, or a mixture of AsOR, PL, and DNA all present in the same concentrations provided by the complete complex. After 24 hours, CAT enzymatic activity was assayed[19] (Fig. 16-3). None of the controls consisting of components of the complex, alone or as mixtures, resulted in gene transfection of the liver, as there was no detectable hepatic CAT activity under any of these conditions. Targeted delivery by the complete AsOR-PL-DNA complex did result, however, in CAT gene expression in liver in vivo, as seen from the presence of acetylated chloramphenicol derivatives.[19] To determine whether the foreign gene expression was

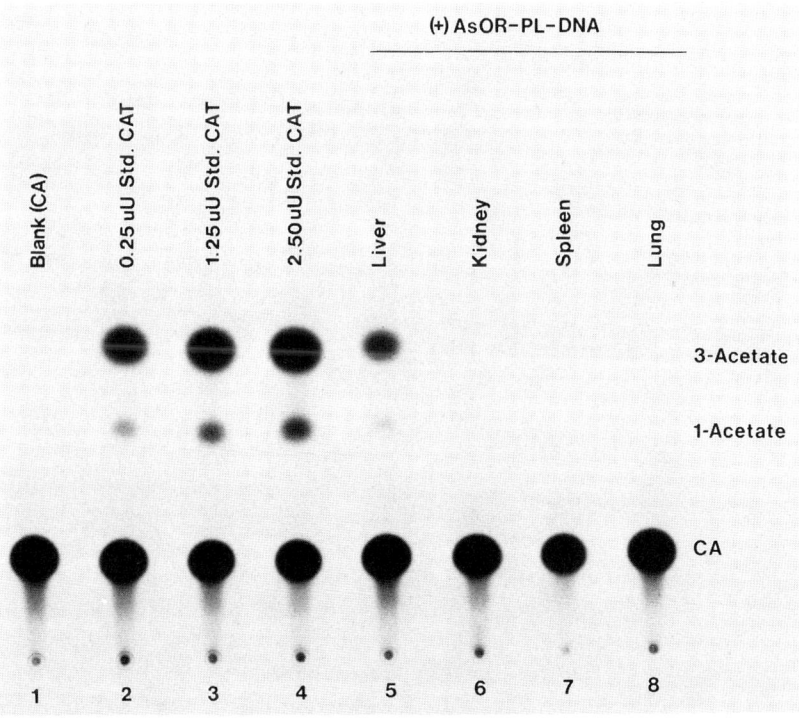

Fig. 16-4. Organ specificity of targeted CAT gene expression. Rats were injected intravenously with AsOR-PL-DNA. After 24 hours, organs were homogenized and assayed for CAT enzymatic activity. Lane 1 contains [^{14}C]chloramphenicol alone; lanes 2–4, CAT enzyme standards; lanes 5–8, homogenates of liver, kidney, spleen, and lungs, respectively. (From Wu and Wu,[19] with permission.)

targeted to the liver, the kidney, spleen, and lungs also were assayed for CAT enzymatic activity after administration of complexed DNA (Fig. 16-4). Only liver had produced detectable quantities of CAT enzymatic activity.[19]

EFFECTS OF NATURAL MAMMALIAN REGULATORY ELEMENTS

In our previous experiments, DNA consisted of a plasmid containing a foreign gene, CAT, driven by viral SV40 regulatory sequences to obtain high gene expression and enhance detectability. However, the use of this targetable gene delivery system for correction of genetic disorders would probably require regulation of the targeted gene in a natural fashion. To determine whether a foreign gene driven by natural mammalian regulatory elements could be targeted and expressed in vivo, a construct, alb-CAT, was prepared by Dr. James Wilson (University of Michigan). This plasmid contains the CAT gene driven by mouse/rat albumin regulatory elements. The carrier conjugate was then complexed to plasmid DNA and prepared as described previously and injected

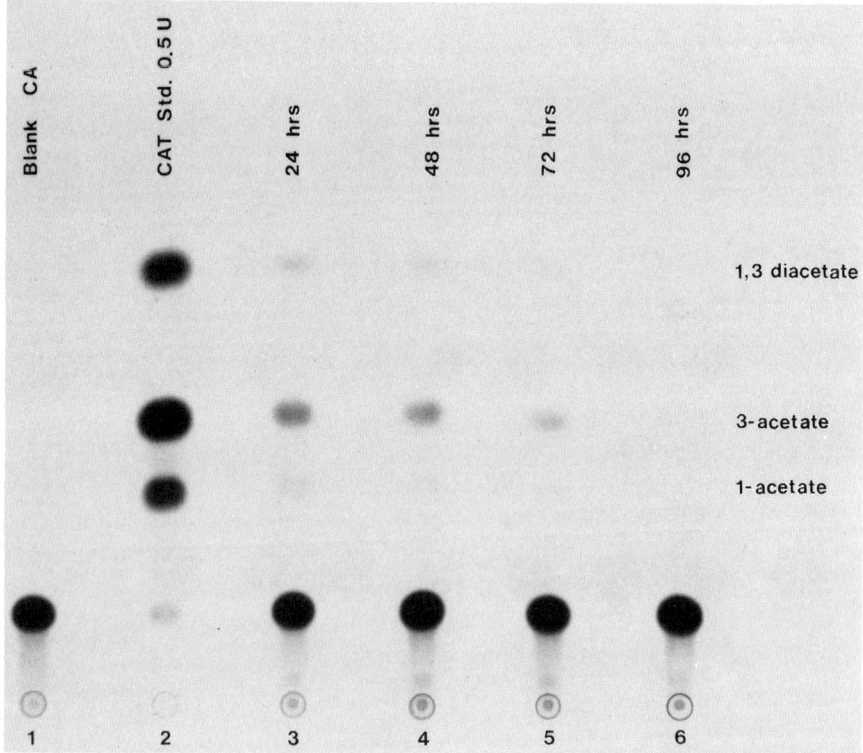

Fig. 16-5. A representative time course of targeted gene expression. Groups of rates were injected with AsOR-PL-DNA complex, and at regular intervals, animals were killed, liver samples homogenixed and equal amounts of homogenate protein assayed for CAT activity.[13] (From Wu, et al.,[16] with permission.)

intravenously into groups of rats. After 24 hours, the rats were sacrificed and livers homogenized. Equal quantities of protein were assayed for CAT enzymatic activity. Liver from the rats that received the DNA in the form of a soluble complex was successfully transfected with the foreign gene. Acetylated chloramphenicol derivatives were detected corresponding to 10 U/g liver. However, liver from control rats that received saline alone, DNA alone, or a mixture of components of the complex AsOR plus PL plus DNA, in which all the components were present in the same amounts as provided by the intact complex, did not produce detectable CAT enzymatic activity. CAT gene expression was present in liver but was absent in heart, lung, spleen, or kidney.

The time course of targeted gene expression in vivo was determined by injection of complexed DNA into groups of rats sacrificed at regular intervals, and hepatic CAT enzymatic activity was assayed (Fig. 16-5). Under these conditions, targeted gene expression was transient with a maximum at 24 hours. Thereafter, CAT enzymatic activity slowly declined, until by day 4 it was no longer detectable.[16]

PERSISTENCE OF TARGETED GENE EXPRESSION

It was previously shown that the likelihood of persistent gene expression is increased if the foreign DNA integrates into the host genome. Furthermore, the probability of integration has been found to be enhanced during host DNA replication.[20] The normal adult liver, however, is a relatively stable population containing few replicating hepatocytes.[12] This could be an important factor in the observed transient gene expression. If so, stimulation of hepatocyte replication might enhance the chances for persistent targeted gene expression. In spite of its high level of differentiation, normal liver has the remarkable capability to regenerate when injured. The classic example is regeneration of liver after injury. A well-established model for stimulation of hepatic regeneration is the two-thirds partial hepatectomy. In this procedure, a predictable response in DNA synthesis begins by 12 hours after surgery, and at least one round of cell division occurs in almost all remaining hepatocytes within 1 to 2 weeks.[12]

To examine the effect of hepatocyte stimulation on targeted gene expression, partial hepatectomies were performed 30 minutes after intravenous injection of the targetable DNA complex into groups of rats. Hepatic CAT enzymatic activity was assayed at regular intervals up to 11 weeks postinjection[16] (Fig. 16-6). CAT gene expression was not detectable 24 hours after surgery. However, by 48 hours CAT enzymatic activity was detectable, reaching a maximum level by the 8th week. Activity remained high through the 11th week postsurgery.[16]

DISCUSSION

Advances in molecular biology and genetics have resulted in the identification and cloning of a variety of structural genes. Many techniques have been developed to introduce foreign genes into cells in vitro to study how genes are

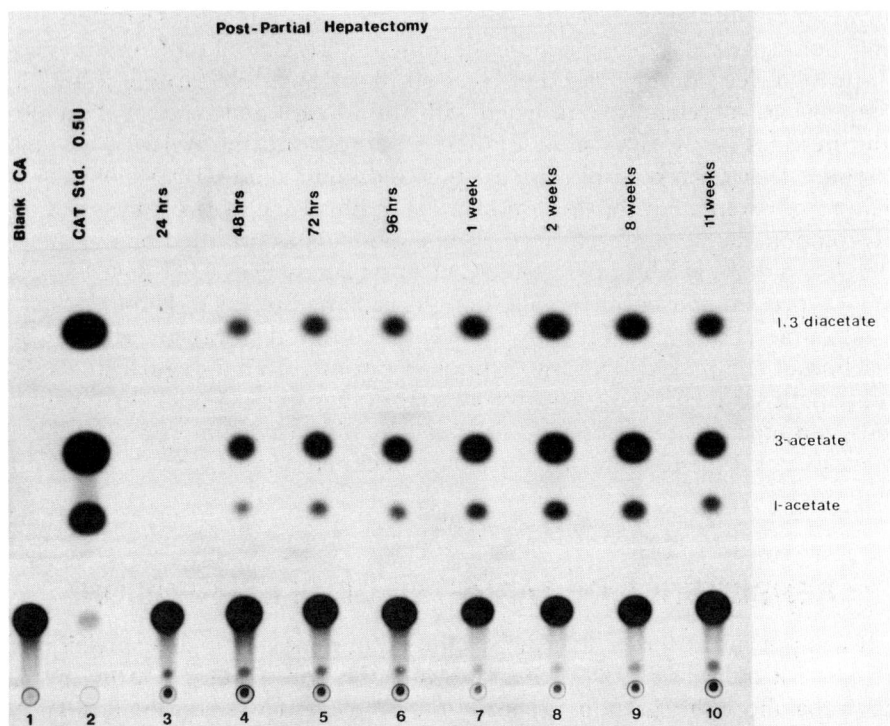

Fig. 16-6. A representative time course of targeted gene expression following partial hepatectomy. Groups of rats were injected with AsOR-PL-DNA complex immediately before 66 percent partial hepatectomy. At varying time points thereafter, livers were assayed for CAT activity.[13] (From Wu, et al.,[16] with permission).

regulated. Among the most popular are $CaPO_4$ precipitation,[21,22] electroporation,[23] microinjection,[24] scrape-loading,[25] liposomes,[26,27] and retroviruses.[28,29] These methods have been very useful biologic tools in vitro. Some of these techniques have been applied in vivo. For example, $CaPO_4$ precipitates prepared in vitro were shown to be capable of transfecting cells in living animals when administered by direct injection of the DNA into organs. Wolff et al.[30] demonstrated expression of several reporter genes after injection of DNA directly into mouse muscle in vivo. Dubensky et al.[31] demonstrated that polyoma viral DNA in the form of $CaPO_4$ precipitates injected directly into livers or spleens of adult or newborn mice resulted in viral replication and acute viral infection. In addition, the direct intraperitoneal injection of $CaPO_4$ precipitates of plasmids containing the CAT gene as a marker was found to result in CAT gene expression in liver and spleen.[32]

Liposomes have been used for hepatic transfection in vivo. Nicolau et al.[33] used liposomes containing the rat proinsulin gene and showed that intravenous injection into rats resulted in uptake primarily by liver and spleen. Levels of serum glucose were decreased and levels of serum insulin were increased as compared with control.[33] In order to enhance the specificity of delivery, liposomes have also been coated with antibody directed against a particular cell-surface antigen. For example, DNA containing a marker gene was encapsulated within liposomes coated with an antibody against a lymphoma cell antigen. Intravenous injection of these liposomes into rats with these lymphoma tumors resulted in targeted foreign gene expression in these cells.[34] The cell specificity of liposomes has also been changed by chemical modification. For example, incorporation of lactosylceramide into liposome envelopes containing foreign DNA; intravenous injection of these modified liposomes resulted in a tissue distribution enhanced in favor of liver. Delivery to hepatocytes versus nonparenchymal cells was increased as compared with control nonmodified liposomes. However, uptake by endothelial cells was still predominant.[35]

Viral vectors are particularly attractive vehicles for gene transformation because of their relatively high efficiency of transfection. Many viral vectors have been studied for transfection of foreign DNA in vitro (e.g., SV40, adenovirus, polyoma, bovine papilloma, and retroviruses).[36] However, in general, the wide spectrum of cells susceptible to viral infection has limited studies in vivo primarily to transfection of isolated target cells in vitro, followed by replacement of the transfected cells into the host animal.[37-39]

A retroviral vector containing the β-galactosidase marker gene has also been used to transfect hepatocytes in culture. This resulted in foreign gene expression with approximately 25 percent efficiency.[40] The structural gene for the human low-density lipoprotein receptor was transfected into primary cultured hepatocytes from Watanabe hyperlipidemic rabbits and was shown to correct the genetic defect in vitro.[41] Similarly, we have taken advantage of specialized cell-surface structures to deliver genes to the hepatocyte.[9,10] Although asialoglycoprotein receptors have been used to target genes specifically to liver cells as a model system, the principles may be more generally applicable. For example, other receptor-mediated endocytotic mechanisms based on the recognition of glycoproteins bearing different carbohydrate moieties have been found on other normal, nonhepatic tissues. Therefore, the concept of targeted gene delivery based on receptor-mediated endocytosis need not be limited to the asialoglycoprotein receptor and hepatocytes.

CONCLUSIONS

Our data indicate that a foreign gene driven by natural mammalian regulatory elements can be specifically delivered to hepatocytes by intravenous injection in vivo using a receptor-mediated, soluble DNA carrier system. The resultant foreign gene expression can be made to persist by stimulation of hepatocyte replication during liver regeneration. This system offers exciting prospects for the study of gene regulation in vivo and may be of value in the development of somatic cell gene-replacement therapy for inherited metabolic disorders.

ACKNOWLEDGMENTS

This work was supported in part by grants NIH CA46801 and DK 42182 from the U.S. Public Health Service; by grant CA01110, a Research Career Development Award; by an American Gastroenterology Association/Industry Research Scholar Award; and by an S. F. Wilson award. The expert assistance of Mrs. Rosemary Pavlick and Ms. Maria Lorenzetti in the preparation of this manuscript is gratefully acknowledged. We also thank Dr. James M. Wilson for his help and advice.

REFERENCES

1. Loyter A, Scangos GA, Ruddle FH: Mechanisms of DNA uptake by mammalian cells: Fate of exogenously added DNA monitored by the use of fluorescent dyes. Proc Natl Acad Sci USA 79:422, 1982
2. Ashwell G, Morell AG: The role of surface carbohydrates in the hepatic recognition and transport of circulating glycoproteins. Adv Enzymol 41:99, 1974
3. Wall DA, Wilson G, Hubbard AL: The galactose-specific recognition system of mammalian liver: The route of ligand internalization in rat hepatocytes. Cell 21:79, 1980
4. Dunn WA, Hubbard AL, Aronson NN: Low temperature selectively inhibits fusion between pinocytotic vesicles and lysosomes during heterophagy of ^{125}I-asialofetuin by the perfused rat liver. J Biol Chem 255:5971, 1980
5. Chang T-M, Kullberg DW: Studies on the mechanism of cell intoxication by diphtheria toxin fragment A-asialoglycoprotein hybrid toxins. J Biol Chem 257:12563, 1982
6. Wu GY, Wu CH, Stockert RJ: A model for the specific rescue of normal hepatocytes during methotrexate treatment of hepatic malignancy. Proc Natl Acad Sci USA 80:3078, 1983
7. Wu GY, Wu CH, Rubin MI: Acetaminophen hepatotoxicity and targeted rescue: A model for specific chemotherapy for hepatocellular carcinoma. Hepatology 5:709, 1985
8. Wu GY, Keegan-Rogers V, Franklin S, et al: Targeted antagonism of galactosamine toxicity in normal rat hepatocytes in vitro. J Biol Chem 263:4719, 1988
9. Wu GY, Wu CH: Targeted delivery and expression of foreign in hepatocytes. In Wu GY, Wu CH (eds): Liver Diseases: Targeted Diagnosis and Therapy Using Specific Receptors and Ligands. Marcel Dekker, New York (in press)
10. Wu GY, Wu CH: Delivery systems for gene therapy. Biotherapy 3:87, 1991
11. Chang C, Weiskopf M, Li HJ: Conformational studies of nucleoprotein. Circular dichroism of deoxyribonucleic acid base pairs bound by polylysine. Biochemistry 12:3028, 1973
12. Leffert HL, Koch KS, Moran T, Rubalcava B: Hormonal control of rat liver regeneration. Gastroenterology 76:1470, 1979
13. Gorman CM, Moffat LF, Howard BH: Recombinant genomes which express chloramphenicol acetyltransferase in mammalian cells. Mol Cell Biol 2:1044, 1982
14. Whitehead DH, Sammons HG: A simple technique for the isolation of orosomucoid from normal and pathological sera. Biochim Biophys Acta 124:209, 1966
15. Oka JA, Weigel PH: Recycling of the asialoglycoprotein receptor in isolated rat hepatocytes. J Biol Chem 258:10253, 1983

16. Wu GY, Wilson JM, Wu CH: Targeting genes: Delivery and persistent expression of a foreign gene driven by mammalian regulatory elements in vivo. J Biol Chem 264:16985, 1989
17. Wu GY, Wu CH: Receptor-mediated in vitro gene transformation by a soluble DNA carrier system. J Biol Chem 262:4429, 1987
18. Wu GY, Wu CH: Evidence for targeted gene delivery to Hep G2 hepatoma cells in vitro. Biochemistry 27:887, 1988
19. Wu GY, Wu CH: Receptor-mediated gene delivery and expression in vivo. J Biol Chem 263:14621, 1988
20. Varmus HE, Padgett T, Heasley S, et al: Cellular functions are required for the synthesis and integration of avian sarcoma virus-specific DNA. Cell:11, 1977
21. Graham FL, Van der Eb AJ: A new technique for the assay of infectivity of human adenovirus 5 DNA. Virology 52:456, 1973
22. Gopal TV: Gene transfer method for transient gene expression, stable transformation, and co-transformation of suspension cell cultures. Mol Cell Biol 5:1188, 1985
23. Potter H, Weir L, Leder P: Enhancer-dependent expression of human k immunoglobulin genes introduced into mouse pre-B lymphocytes by electroporation. Proc Natl Acad Sci USA 81:7161, 1984
24. Harland R, Weintraub H: Translation of mRNA injected into *Xenopus* oocytes is specifically inhibited by antisense RNA. J Cell Biol 101:1094, 1985
25. Fechheimer M, Boylan JF, Parker S, et al: Transfection of mammalian cells with plasmid DNA by scrape-loading and sonication loading. Proc Natl Acad Sci USA 84:8463, 1987
26. Nicolau C, Sene C: Liposome-mediated DNA transfer in eukaryotic cells. Biochim Biophys Acta 721:185, 1982
27. Fraley RT, Fornari CS, Kaplan S: Entrapment of a bacterial plasmid in phospholipid vesicles: Potential for gene transfer. Proc Natl Acad Sci USA 76:3348, 1979
28. Guild BC, Finer MH, Housman DE, Mulligan RC: Development of retrovirus vectors useful for expressing genes in cultured murine embryonal cells and hematopoietic cells in vivo. J Virol 62:3795, 1988
29. Zwiebel JA, Freeman SM, Kantoff PW, et al: High level recombinant gene expression in rabbit endothelial cells transduced by retroviral vectors. Science 243:220, 1989
30. Wolff JA, Malone RW, Williams P, et al: Direct gene transfer into mouse muscle in vivo. Science 247:1465, 1990
31. Dubensky TW, Campbell BA, Villarreal LP: Direct transfection of viral and plasmid DNA into liver or spleen of mice. Proc Natl Acad Sci USA 81:7529, 1984
32. Benvenisty N, Reshef L: Direct introduction of genes into rats and expression of the genes. Proc Natl Acad Sci USA 83:9551, 1986
33. Nicolau C, Le Pape A, Soriano P, et al: In vivo expression of rat insulin after intravenous administration of the liposome-entrapped gene for rat insulin I. Proc Natl Acad Sci USA 80:1068, 1983
34. Wang C-Y, Huang L: pH-Sensitive immunoliposomes mediate target-cell specific delivery and controlled expression of a foreign gene in mouse. Proc Natl Acad Sci USA 84:7851, 1987
35. Soriano P, Dijkstra J, Legrand A, et al: Targeted and nontargeted liposomes for in vivo transfer to rat liver cells of a plasmid containing the preproinsulin I gene. Proc Natl Acad Sci USA 80:7128, 1983
36. Kucherlapati R, Skoultchi AI: Introduction of purified genes into mammalian cells. CRC Crit Rev Biochem 16:349, 1984

37. Eglitis MA, Kantoff PW, Jolly JD, et al: Gene transfer into hematopoietic cells from normal and cyclic hematopoietic dogs using retroviral vectors. Blood 71:717, 1988
38. Kantoff PW, Freeman SM, Anderson WF: Prospects for gene therapy for immunodeficiency diseases. Annu Rev Immunol 6:591, 1988
39. Kasid A, Morecki S, Aebersold P, et al: Human gene transfer: Characterization of human tumor-infiltrating lymphocytes as vehicles for retroviral-mediated gene transfer in man. Proc Natl Acad Sci USA 87:473, 1990
40. Wilson JM, Jefferson DM, Roy Chowdhury J, et al: Retrovirus-mediated transduction of adult hepatocytes. Proc Natl Acad Sci USA 85:3014, 1988
41. Wilson JM, Johnston DE, Jefferson DM, Mulligan RC: Correction of the genetic defect in hepatocytes from the Watenabe heritable hyperlipidemic rabbit. Proc Natl Acad Sci USA 85:4421, 1988

17

HERPES SIMPLEX VIRUS AS A VECTOR FOR NEURONS

Xandra O. Breakefield · Neal A. DeLuca

INTRODUCTION

There are more than 80 different types of herpes viruses with varying tissue and species tropisms.[1-3] These viruses contain a linear dsDNA genome, which is transcribed and replicated and packaged in the cell nucleus, and derive their membrane by budding through the nuclear membrane. This chapter focuses on herpes simplex virus type 1 (HSV-1) as a vector for neurons. This human neurotropic virus can be maintained in a latent state in neurons over the life span of rodents and humans. In its latent state, the HSV-1 genome is believed to exist as a multicopy circular episomal element in the nucleus,[4] with one viral promoter known to be active.[5] This remarkable state, combined with the fact that neuronal physiology does not appear to be affected by viral latency, has made this virus an obvious candidate for long-term gene transfer to neurons. The virus can be taken up by nerve terminals and delivered efficiently to the nucleus over long distances by rapid retrograde transport. Furthermore, the virus can be passed transsynaptically between neurons, thereby traveling along specific neuronal pathways. Thus, neurons can become accessible to gene delivery through their endings on the skin, eye, and blood vessels, as well as through other neurons. The availability of relatively easy selections and screens for insertions in the viral genome, as well as the degree of potential variability in genome size, permit a number of "foreign" genes and promoter elements to be carried by herpes vectors.[6] Several laboratories are trying to develop HSV-1-derived vectors, which will be nontoxic to neurons and will confer regulated expression of "foreign" genes on them. Such vectors could be used for gene delivery to discrete populations of neurons in the nervous system for a number of purposes, for example (1) to supplement defective genes or provide trophic factors, (2) to alter cellular physiology, (3) to kill certain cells selectively, or (4) to test neuronal specific promoters. In fact, no other way has been found to deliver genes efficiently to postmitotic neurons, either in culture or in vivo.

Three basic issues need to be resolved to evaluate the potential uses of HSV-1 vectors.

1. *Can various viral functions associated with cellular toxicity be controlled so that the virus itself is "neutral" to cells?* Toxicity can result from productive replication of the virus, leading to cell death, but also includes shutoff of host macromolecular synthesis during infection, alterations of host cell DNA, and as yet only partially defined neurovirulence and neuroinvasive functions.
2. *Is latency specific to subsets of neurons, or can many types of neurons and even other nondividing cell types harbor herpes virus stably in this state?* Numerous studies have shown that sensory neurons can maintain latent HSV-1 in vivo,[5] and probably in culture as well.[7] Furthermore, latent viral DNA can exist in spinal cord, brain, various spinal ganglia, and even adrenal medulla.[1] Expression of latency-associated transcripts (LATs) has been noted in neurons and possibly in the glia in the brain[8] and in the motor neurons in the hypoglossal nucleus.[9] For purposes of gene transfer, it may be sufficient to define latency as a state in which the viral genome is stably maintained in a nonreplicative, nontoxic state in the cell nucleus for "long" periods; promoter(s) within it are accessible to trans-acting factors encoded in its own or in the cellular genome. For gene delivery in vivo, a "long" period might ideally be the life span of the individual, but for cells in culture, a few days to weeks might suffice for experimental purposes.
3. *Can non-HSV-1 promoters be incorporated in the viral genome and regulate gene expression for short or long periods?* Herpes vectors potentially provide a means to evaluate the neuronal specificity of promoters and to effect strong, cell-specific and/or inducible expression of genes. Herpes vectors could dramatically speed up the definition of neuronal promoters by facilitating their introduction, coupled to reporter genes, into real neurons both in culture and in vivo. Furthermore, for cell-targeted gene delivery, specificity could be conferred at the level of gene regulation, rather than at viral entry. It should be kept in mind that for some uses of HSV-1 vectors, such as evaluation of promoter activity or protein function, or in tracing neuronal pathways, it may not be necessary to achieve a state of latency, and expression of virally encoded genes for a few days may be sufficient.

This chapter briefly covers some of the properties of HSV-1 viruses and their mutant derivatives, as well as the initial studies that have been carried out toward the development of HSV-1 vectors for neurons. There is a great deal of excitement among neuroscientists and geneticists about the potential uses of these vectors. Many plans and strategies have been conceived and are now being evaluated, but only a few publications exist to date on herpes-mediated gene transfer to neurons. Our goal is to set the stage and predict the early vector models, knowing that there will be many surprises along the way and that future models will be much more sophisticated and specialized for particular uses.

VIRAL GENOME, INFECTION AND LATENCY

The HSV-1 genome (152 kb) is composed of two unique segments, U_L and U_S, each bounded by inverted repeat elements, b and c, respectively, that can invert relative to each other[10] (Fig. 17-1). DNA extracted from viruses contains four isomers varying in the position and orientation of unique segments.[11] The genome termini consist of the relatively short *a* sequence, which is also present in an inverted orientation at the joint between the long and short components of the genome (for viral sequences, see refs. 10, 12, 13). There are three origins of DNA replication in the HSV-1 genome, one in the U_L region and a duplicated one in the c region.[14] During replication, the termini join, and replication is thought to proceed as a "rolling circle," forming a concatamer of HSV-1 genomes[2] (Fig. 17-2). Packaging involves the presence of a cleavage and packaging recognition site in the *a* sequence.[15–17] The HSV-1 genome contains about 70 genes encoding transcription units with unique mRNAs and promoters. Some genes are present in two copies within the duplicated elements, and others in one copy within unique elements. Genes include those coding for regulatory factors, enzymes, and structural proteins. The structure and function of many of these genes have been determined by mutational analysis. Others are defined as containing open-reading frames or because they regulate a function, although their gene structure has not yet been completely elucidated.

The HSV-1 genome has three states within the infected nucleus of neurons. In one state, productive infection ensues, leading to viral progeny and cell death; in a second state, the viral genome is latent and relatively silent with

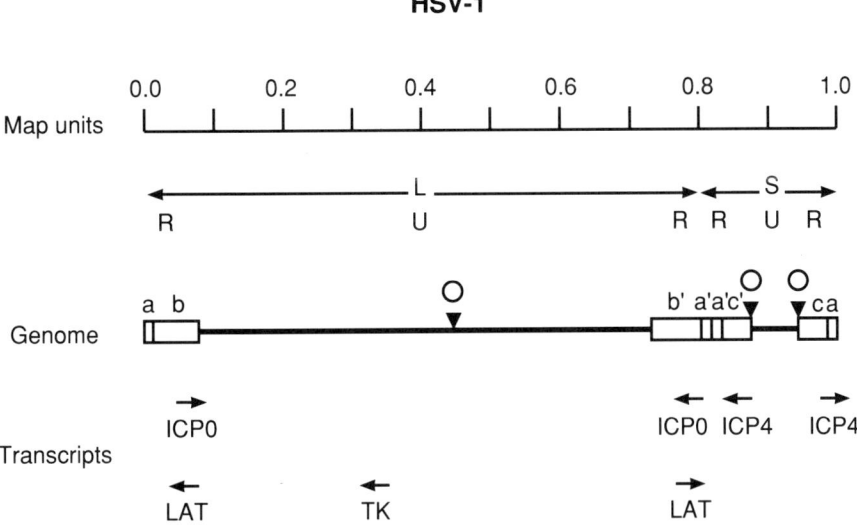

Fig. 17-1. Diagrammatic representation of HSV-1 genome. Map units (mu) are shown in the top line within an approximately 150 kb genome. Repeat elements (R_L and R_S) contain repeats a,b and a, c, respectively. Unique regions (U) are designated as long (L) and short (S). The three origins (O) of replication are designated by ∇. The position of a few of the transcripts discussed in this paper (out of the 72 total) are indicated by arrows. (Modified from McGeoch et al.,[10] with permission.)

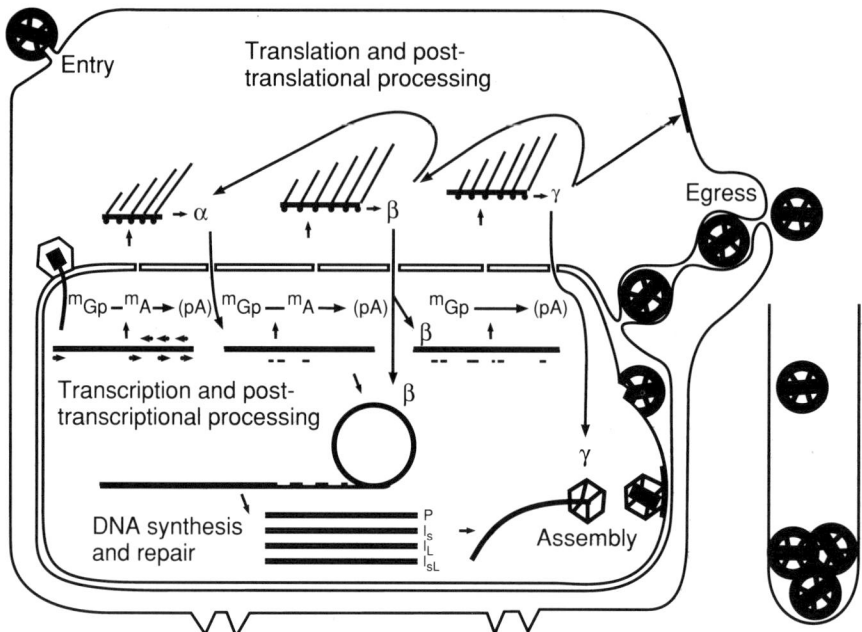

Fig. 17-2. Summary of the major features of the herpes simplex virus replicative cycle. The virus envelope fuses with the plasma membrane, and the virus capsid is transported to the nuclear membrane. Viral DNA is released into the nucleus, where it is transcribed sequentially in a cascade with immediate-early genes (α) first, then early genes (β) next, and finally late genes (γ). Viral proteins are translated in the cytoplasm. Viral DNA is replicated and packaged in the nucleus and buds out through the nuclear membrane. (From Roizman and Batterson,[2] with permission.)

respect to transcription; and in a third state, the latent genome responds to changes in neuronal physiology and reactivates viral replication. In productive infection, viral particles attach to specific receptors on the cell surface and are taken up by fusion of the viral envelope with the plasma membrane[2] (Fig. 17-2). Attachment and viral entry are thought to be mediated by glycoproteins in the viral coat, including gC, gB, and gD,[18] which interact with heparin-like glycosaminoglycans[19] and apparently, the basic fibroblast growth factor receptor.[20] Virion nucleocapsids proceed to the nucleus where the DNA is uncoated. Viral genes are transcribed by RNA polymerase II.[21]

In productive infection, the first genes to be transcribed are the viral immediate-early (IE) genes in the early stages of infection (2 to 4 hours)[22] (Table 17-1[23]). The transcription of IE genes is stimulated by a viral coat protein, VP16, also termed Vmw65 and α-TIF.[24,25] Five immediate-early viral genes encode the proteins: ICP0, ICP4, ICP22, ICP27, and ICP47. Immediate-early proteins act as transcriptional factors and are required for the expression of the second kinetic class of viral genes, the "early" genes.[26] The early proteins are maximally expressed at 4 to 7 hours postinfection. They are involved primarily in the replication of viral DNA and include thymidine kinase (TK), ribonucleotide reductase, DNA polymerase, and ssDNA binding proteins. The late genes (7 to

Table 17-1. Kinetic Classes of Herpes Genes in Productive Infection

Designations	Proteins	Functions
Immediate-early (IE, α)	ICP0 (Vmw110) ICP4 (Vmw175) ICP22 (Vmw68) ICP27 (Vmw63) ICP47 (Vmw12)	Transacting factors
Early (E, β)	TK DNA polymerase DNAase Ribonucleotide reductase UTPase	Viral DNA replication, transacting factors
Late (L, γ)	VP16 (Vmw65, ICP25, α-TIF) Glycoproteins (e.g., gC, gD) UL41	Structural proteins of virions, transacting factors

(From Breakefield and DeLuca,[23] with permission.)

15 hours) include structural proteins of the viral capsid and the glycoproteins. During productive infection, the virus alters host macromolecular processes, such as gene expression, protein translation, and cellular DNA replication and repair. These aspects of infection alone can be toxic to cells. Viral particles are assembled in the nucleus and can be seen as numerous dense capsids shortly after the onset of viral DNA synthesis (Fig. 17-2). After assembly, the virion particles bud out through the inner nuclear membrane, which now contains viral proteins, forming unique vesicular structures,[27] and move out to the plasma membrane, where enveloped virions are released into the extracellular space. If the infected cell is a neuron, viral particles will be passed anterogradely and retrogradely within the neuronal processes by high-affinity transport processes; they will be released first at synaptic endings, where they will be taken up by presynaptic or postsynaptic terminals, respectively.[28] Eventually, productively infected cells die and release virus onto adjacent cells.

In latent infection, the virus is first taken up by neurons, transmitted to the nucleus, and presumably unpackaged, as in productive infection. It is not clear whether viral DNA replicates before entry into latency, but a number of studies have shown that viral replication is not required for this transition.[5,29] Mutant viruses that are compromised or defective in their replication potential in neurons can still enter latency (e.g., ICP0⁻, TK⁻, and ICP4⁻ [9,30–32]), albeit sometimes with reduced efficiency, as compared with wild-type virus. This reduced efficiency may be because without viral replication in cells at the site of inoculation, fewer viral particles actually get taken up by neurons. It is not clear what signals steer the virus into latency and several, as yet unidentified, factors may be involved.

There is evidence that some neurally derived cells contain transacting factors that can bind to octamer motifs (TAATGARAT) in the regulatory elements of viral immediate-early genes and repress their expression,[33] and it appears that the LAT promoter is stronger in neuronal cells than in other types.[34–36] The molecular basis for the establishment and maintenance of latency is not understood. In fact, it has been difficult to detect the presence of the latent viral genome. Classically, the latent virus was detected by eliciting reactivation to a productive infection, either by stimulating or by injuring the neuron in vivo or by explanting ganglia and co-culturing them with permissive cells. Latent viral DNA in ganglia has also been demonstrated by Southern blot analysis and by amplification using the polymerase chain reaction (PCR).[8] Recently, a number of laboratories have demonstrated that latent virus is transcriptionally active and that LATs of unknown function can be detected in the nuclei of neurons harboring latency.[37,38] These transcripts are poly A$^-$ RNA found predominantly in the nucleus and are spliced versions of each other: 2.2 kb and 1.5 kb. Although they contain some sequences with open reading frames, it is not clear that they encode proteins.[39] In fact, they may be generated as introns from a much larger RNA species (8 kb) that is apparently unstable.[40]

Mutant viruses containing deletions encompassing all or part of the LATs can replicate and establish latency normally.[30,41] Their only possible deficit appears to be in reactivation from latency under some conditions.[30,42] The major LAT promoter element appears to lie about 700 bp 5' to the 5' end of the LATs and contains several SP1 binding sites and a TATAA box.[34,36,40,43] Several other potential promoters are contained within this 700-bp interval, however, that may also have transcriptional activity. In several studies, reporter genes have been placed downstream of the LAT promoter and shown to be stably expressed during latency in neurons (see below). Segments (0.5 to 1 kb) containing the major LAT promoter have also been inserted into plasmid vectors and used to drive the chloramphenicol acetyltransferase (CAT) reporter gene after transfection into different cell types in culture.[34–36] This promoter element was found to be positively regulated in sensory neurons and neuroblastoma cells and, in fact, was stronger than the SV40 or HSV-1 ICP0 promoters in these cells, as compared with the non-neuronal cell types tested.

As is well known from the recurrence of cold sores caused by HSV-1, the virus can reactivate from latency at periodic intervals and release active viral particles at the nerve terminals, resulting in productive infection of non-neuronal cells.[5] The process of reactivation can be triggered by nerve trauma (e.g., severing nerve fibers), certain drugs, and hormones. Interestingly, in cell culture, removal of nerve growth factor (NGF) from the medium will trigger a productive infection in sensory neurons, which have been harboring the virus in a nonreplicative state.[7,44] Such environmental stimuli are known to lead to changes in transcriptional activity in neurons, and it is tempting to speculate that cellular regulatory factors change the transcriptional state of the latent HSV-1 genome and serve to initiate reactivation. Since recurrent reactivations can occur in neurons in vivo without analgesia to the area innervated by their terminals, Stevens[5] has reasoned that neurons may be capable of limited viral replication and release of viral or subviral particles from the nerve terminals without dying. Others have argued that repeated activation from latency seems to decrease the number of neurons in sensory ganglia that are capable of viral replication, implying that dying neurons release virus, which then becomes latent in surrounding neurons.[45]

CELLULAR TOXICITY

Herpes virus can be toxic to cells in a number of ways. Fully productive infection of cells with wild-type HSV causes alterations in virtually all cellular macromolecular processes, culminating in cell death. However, even in the absence of extensive viral replication, viral infection can be toxic. After entry of the virus into cells, host cell macromolecular synthesis is shutoff in a multiphase process.[46,47] A rapid cessation of cellular polypeptide synthesis is accompanied by disaggregation of polyribosomes. This initial shutoff is mediated in large part by a structural component of the HSV-1 virion, UL41, which may be an RNase that degrades or destabilizes host cell mRNA.[48,49] Degradation of host mRNA is accelerated as this late viral protein, UL41, becomes synthesized by the cells. In addition to changes in mRNA stability and translation, there are widespread changes in transcription of cellular genes. For some cellular genes, a dramatic repression of transcription follows viral infection.[50,51] However, other cellular genes are activated. Virally encoded proteins, ICP4 and ICP0, can nonspecifically activate cellular genes.[52] Two cellular heat-shock proteins, *hsp*70 and *hsp*90,[50] and a growth-related protein, T156,[53] accumulate during infection as a result of increased transcription. Changes in cellular transcription may be mediated by viral transacting regulatory factors.

Infection with HSV-1 also leads to alterations in cellular DNA. These include stimulation of repair replication[54,55] and production of chromosomal aberrations,[56] which may be a response to damage to genomic DNA[57] and/or may represent viral DNA replicative enzymes acting nonspecifically on cellular DNA. Virally induced amplification of SV40 sequences integrated into the cellular genome has been observed and is presumed to occur for cellular sequences too.[58] These potentially disruptive actions of herpes on the host cell genome may compromise the effectiveness of herpes vectors for gene transfer, unless they can be controlled through mutations in the viral genome.

In addition, two other toxic features of herpes bear on its potential value as a vector for the nervous system. These relate to its neuroinvasiveness and neurovirulence. *Neuroinvasiveness* is defined as the ability of the virus to spread from the site of peripheral inoculation to the brain, which requires that it undergo productive replication in neurons and be passed transsynaptically. *Neurovirulence* refers to the capacity to kill the animal through encephalitis after inoculation of less than 100 pfu of virus into the brain.[59] Wild-type strains of HSV-1 or HSV-2 can differ remarkably in these two properties. When two non-neuroinvasive strains of HSV-1—ANG and KOS—are injected together into mouse footpads, they are able to complement each other and achieve substantially higher viral titers in the spinal cord and brains than either alone.[59] In this study, a few neuroinvasive recombinant viruses were isolated from spinal cord. Transfer of two specific sequences, 0.32 to 0.42 and 0.49 to 0.64 map units (mu), from ANG into another non-neuroinvasive strain, 17 syn+, could also confer invasiveness.[60] Thus, several regions of the herpes genome appear to be involved in spread through the nervous system, but the nature of the responsible genes is unknown. One might speculate that they encode (1) viral proteins, which are involved in either linking onto the cytoskeletal components needed for rapid transport in neurites or in entry into neurons or (2) regulatory factors, which determine the choice between productive infection and latency. Undoubtedly, these neuroinvasive genes also encode functions needed for productive infection of neurons and, as such, cross over into the realm of neurovirulence.

A critical feature of neurovirulence is the ability of HSV-1 to replicate in central nervous system (CNS) neurons. Any mutations that compromise the ability of the virus to replicate will reduce neurovirulence, as reviewed by Thompson et al.[61] Such mutations can occur in genes involved in viral DNA replication. Some of these viral gene products are essential for replication in nondividing cells, such as predominate in the nervous system, but are not necessary in dividing cells. Thus some mutant viruses can use host cell factors and enzymes, which are regulated in a division-dependent mode, to replicate viral DNA. A classic example is the non-neurovirulence of HSV-TK⁻ mutants.[31,32] The thymidine kinase gene is highly regulated as a function of cell division, such that there is a 20-fold increase in activity levels as cells pass from G1 to S phase in the cell cycle.[62] HSV-TK⁻ viruses propagate as efficiently as wild-type virus in rapidly growing cells in culture through use of cellular TK but are essentially nonreplicative in nondividing cells in the CNS. Virus-bearing mutations in other genes involved in DNA replication, including DNA polymerase,[63–65] and dUTPase,[66] are also non-neurovirulent.

At least four other, as yet unidentified, genes appear to affect specifically the ability of the virus to infect neural tissue productively, without affecting its ability to replicate in cultured, dividing cells or in non-neural tissue in vivo.[61] One or more of these genes has been localized to a defined region (0.82 to 0.832 mu) of the HSV-1 genome.[61] A nonvirulent recombinant virus between HSV-1 and HSV-2, RE6, has been developed that is replication-defective in the brain, but not in peripheral tissues, and does not cause a fatal encephatitis even after intracranial inoculation of 10^8 pfu. Interestingly, in some strains of virus, the degree of neurovirulence does not appear to involve the degree of viral replication in the CNS, as similar titers of virus in the brain can be achieved by strains that are nonpathogenic and those that cause a fatal encephalitis.[67] Again, using recombinant viruses between HSV-1 and HSV-2, a 9.1-kb region (0.079 to 0.143 mu) of HSV-1 has been identified, which in this case does not compromise viral replication in the CNS, but results in a loss of neurovirulence.[67] In this case, neurovirulence may depend on the tropism of the virus for specific cell types. Presumably, neurovirulent strains can replicate in neurons, as well as other cell types, and non-neurovirulent strains replicate predominantly in non-neuronal cells in the CNS, such as glia and ependymal and meningeal cells.

MUTANT VIRUSES USEFUL AS VECTOR "BACKBONES"

A number of mutations in the HSV-1 genome have been described that decrease the ability of the virus either to replicate in all cells or specifically in neurons, or to reduce its toxicity to cells, or both. Such mutant viruses can potentially form the "backbone" of virus vectors, such that they will not induce toxic effects in cells in culture or in vivo and will not result in pathogenic effects or death of experimental animals. Furthermore, such compromised vectors present less of a health risk for laboratory personnel. These features will be critical in determining the potential value of these vectors. For most gene-delivery schemes to neurons, one seeks to alter the physiology of these cells as discretely as possible and to follow the effects of gene delivery and expression over days to months. Although herpes is primarily a human pathogen, it can undergo productive and latent infection in essentially all mammalian species. One must be concerned, then, not only with the replication competence of the vector, but with its potential to reactivate latent wild-type virus, or to recombine with latent viral DNA or other herpes sequences. In the future, it may be advisable to use vectors that bear deletions, rather than point mutations in critical viral genes, such that they cannot revert to the wild-type phenotype, and which bear deletions in more than one critical gene, such that a single recombination event cannot yield wild-type virus. A number of mutant viruses appear to have properties compatible with their use as vectors for the nervous system. Some representative examples are given in Table 17-2.

Mutant HSV-1 defective in TK activity does not replicate in neurons but can establish latency in them, although it cannot reactivate from latency.[31,32,65] Injection of these viruses into rodent brains has been found to be less pathogenic as compared with wild-type virus. Many cells are capable of division, even in the adult brain, which could harbor replicating TK⁻ virus. Since division of some of these cells, such as astrocytes, can be stimulated by injury, the pathogenicity of TK⁻ virus may vary in different regions of the brain under various conditions. Other mutant viruses that are defective in their ability to replicate viral DNA in neurons include those with lesions in genes for viral DNA polymerase, UTPase, ribonucleotide reductase, a neurovirulence function (Table 17-2), and the US3 protein kinase.[71]

Other mutants are compromised or defective in replication because of lesions in immediate-early genes. The ICP0 gene encodes a protein that transactivates all kinetic classes of viral genes in transient assays.[72–75] Mutants in ICP0 are compromised in their replicative capacity in all cells, with a delay between viral entry and replication and a decrease in viral titers.[76,77] At low MOI (multiplicity of infection), the number of viral particles released per productively infected cell is reduced, while at high MOI they mimic the wild-type virus.[77] ICP0 mutants establish latency in sensory ganglia and reactivate from it with reduced efficiency relative to the wild type.[69] Intracerebral injection of an ICP0-deletion mutant into adult rat brain has proved relatively nonpathogenic.[78] Only 1 of 19 rats died from apparent viral encephalitis over 1 month (the longest time tested) after inoculation of 200,000 pfu.[78] While animals injected with the ICP0 mutant virus lived without behavioral abnormalities or apparent illness for 1 month, some animals showed weight loss and some brain pathology.

A deletion mutant in the immediate-early gene encoding ICP22 showed compromised replication in several rodent cell lines, as well as in one, but not another, human cell line; it also displayed dramatically reduced neuroviru-

Table 17-2. HSV-1 Mutant Viruses Tested in the Nervous System

Mutant Gene	Replication		Latency in Neurons	Reactivation from Latency	References
	In Dividing Cells in Culture	In Neural Cells In Vivo			
TK	+	−	+	−	Coen et al. (1989)[32] Leist et al. (1989)[31] Katz et al. (1990)[65]
DNA polymerase	−	−	+	−	Katz et al. (1990)[65]
UTPase	+	−	?	?	Thompson and Wagner (1988)[66]
Ribonucleotide	+	−	+	−	Goldstein & Weller (1988)[68] Katz et al. (1990)[65]
RE6[a]	+	−	+	?	Thompson et al. (1989)[61] Leib et al. (1989)[69]
ICP0	±	±	+	+	Leib et al. (1989)[69]
ICP22	±	±	+	?	Sears et al. (1985)[70]
ICP4	− (+ in ICP4+ cells)	−	±	−	Leib et al. (1989)[69] Katz et al. (1990)[65] Dobson et al. (1990)[9]
ICP27	− (+ in ICP27+ cells)	−	+	−	Leib et al. (1989)[69] Katz et al. (1990)[65]
LAT	+	+	+	±	Hill et al. (1990)[42] Leib et al. (1989)[30]
vhs-1	+	+	+	+	Kwong and Frenkel (1989)[18]
VP16	+	±	+	+	Steiner et al. (1990)[29]

[a] HSV intertypic recombinant with which wild-type sequences conferring neurovirulence have been maped to 0.82 to 0.832 map units.

lence in mice, although it was able to establish latency.[70] Sears and co-workers[70] suggest that the ICP22 protein, of as yet unknown function, may be involved in the expression of late viral gene products and may be replaceable by a host cell protein present in some, but not other, cell types.

Mutants in the immediate-early gene, ICP4, are deficient in enhanced expression of early and late viral genes and in attenuated expression of immediate early genes.[79] This protein is required for viral replication; mutants lacking it can be grown only in cultured cells transfected with the ICP4 gene.[80] Although ICP4⁻ mutants did not appear to enter latency in sensory neurons, as assessed in a latency paradigm that depends on reactivation in culture,[69] they have been shown to exist in a latent state in neurons in vivo after direct injection into nerve by in situ hybridization to probes for LATs[9] and by PCR amplification of viral DNA.[65] Intracerebral injection of ICP4⁻ virus into rat brain was completely nonpathogenic, as assessed by the health of animals and by neuroanatomic observations.[78]

Mutants in the immediate-early gene encoding ICP27 cannot replicate in any cells, except those transfected with the ICP27 gene.[81] This protein acts to down-regulate transcription of early viral genes and to up-regulate transcription of late genes. It has also been shown to be required for accumulation of a cellular protein, p40, during infection.[82] Although the function of the p40 protein is not known, it bears homology to the 90-kb heat-shock protein family and may represent a cellular defense mechanism. Mutants in ICP27 cannot replicate in vivo, but appear to establish latency.[65]

Knocking out viral genes involved in toxicity to the host cell should also prove useful for vectors. A mutant virus, vhs-1, is available in which the gene responsible for early shutoff of host cell macromolecular synthesis, UL41, is defective.[48] However, vhs-1 virus vectors still caused cytopathic effects on neurons in culture.[83] Apparently, other toxic functions must also be disarmed before the vectors can achieve their full potential for gene delivery into cultured neurons. Another interesting mutant, *in*1814, bears an insertion in the gene encoding VP16, such that its ability to induce expression of IE viral genes is reduced about 100-fold.[84] The replication of this mutant is compromised in non-neuronal cells in culture and is apparently defective in neurons.[29] It can enter latency in sensory neurons and reactivate from it, however. No intracranial inoculations of this mutant have been reported to date.

For long-term gene expression, mutants should ideally enter latency, but not reactivate from it. A number of mutant virus disrupted in the LAT transcripts appear to replicate and enter latency with equal efficiency to wild type but to be compromised in reactivation in vivo[42] and in explant cultures.[30]

A number of existing HSV-1 mutants then could be used as helper virus for plasmid-derived vectors or as a "backbone" for engineering of virus-derived vectors. For the latter, these different mutations serve two useful functions. First, by mutating different viral genes, it should be possible to control to what extent the vector is toxic, replication-competent, neuroinvasive, and neurovirulent. Second, for every gene that can be knocked out, the potential is open to insert either another gene or promoter element, or both, in its place.

Nonessential viral genes that can be disrupted and propagated in nonengineered cells include those encoding viral glycoproteins G, I, E, and C[85,86] and LAT, TK, ICP22, and ribonucleotide reductase. Genes that can be disrupted with some compromise in productive replication are those encoding ICP0 and VP16. Potentially any gene is dispensable to the extent that cell lines can be engineered to replace the missing functions and thus permit viral replication within them. Mutants in the many viral genes have been successfully propagated in such cells in culture, including genes for ICP4, ICP22, ICP27, glycoproteins B and D, DNA polymerase, and ICP8, as well as many others. Some of these mutants are listed in Table 17-2. Roizman and co-workers have estimated that at least 30 kb of foreign DNA can be carried within the HSV-1 genome.[87]

HERPES VECTORS

Strategic Considerations

An ever-increasing amount of work is being undertaken in many laboratories exploring the development and application of herpes vectors. To date, this is the only viral system described that can deliver and express foreign genes in adult neurons both in culture and in vivo. In fact, it has proved quite difficult, if not impossible, to deliver functional genes into mature neurons by other means, such as by DNA transfection or with other viral vectors (e.g., retroviruses). In addition, for many gene-transfer strategies, herpes vectors appear to have especially useful features. Four aspects of these vectors are considered here: (1) use in assessing neuronal promoters, (2) effectiveness of latency as a means of long-term gene expression, (3) a means of delivery from peripheral sites of inoculation to neurons throughout the nervous system, and (4) types of cells that can serve as potential recipients.

Neuronal Promoters

Neuronal promoters are defined initially by coupling them to a reporter gene and transfecting constructs into cultured neural and non-neural cells. However, primary neurons in culture do not take up DNA efficiently by any means, and "neural" cell lines often show atypical gene regulation. Full definition of these DNA regulatory elements requires the creation of transgenic mice to evaluate tissue-specific expression. Herpes vectors have the advantage of being able to infect neurons efficiently both in culture and in vivo, and to carry relatively large DNA fragments containing foreign promoter elements. For studies of gene regulation in cultured neurons, it may be sufficient to achieve "short-term" expression of reporter genes, abrogating the necessity for the virus to enter latency. However, transacting viral proteins could alter the normal repertoire of regulatory factors present in neurons and may need to be eliminated from these vectors.

Latency

It seems highly likely that HSV-1 has evolved a commensalistic relationship with neurons, resulting in the long-term maintenance of the viral genome in these cells, with the continued potential to undergo productive infection in other cell types. The virus and the neuron probably communicate through transacting factors and DNA elements. Transcription of the HSV-1 genome involves the cellular transcription machinery, individual components of which may, in turn, be altered or influenced by viral proteins. Some of these cellular transcription factors may be neuron specific, underlying the ability of these cells to "discourage" productive replication of the virus, to "encourage" entrance into latency, to maintain the viral genome in latent state, and to assist in reactivation through changes in neuronal physiology. Two regulatory systems that may be involved in these processes have recently been identified. Some neuronally derived cells appear to possess a repressor activity for the octamer-related regulatory element (TAATGARAT), essential to the expression of immediate-early viral genes.[33,88,89] These octamer elements are normally positively regulated by the viral coat protein, VP16, and the cellular protein, OCT1,[90] but may be negatively regulated by these repressors. Thus, neuroblastoma cells can be shown to yield low expression of the CAT reporter gene coupled to an immediate-early viral promoter containing the octamer element, as compared with a number of other cell types.[33] Thus, at a low MOI (low amounts of VP16), productive infection might be suppressed by neurons. At a high MOI, this suppression might be overcome, and productive infection would ensue.

Another regulatory interaction specific to neurons and HSV-1 appears to involve the LAT promoter. Several groups have linked the LAT regulatory element, at position −700, (see under Viral Genome, Infection and Latency) to a reporter gene and demonstrated that it is a much more potent promoter for neurons and neuroblastoma cells than for other cell types in culture. This finding suggests that it is responding to neuron-specific transacting factor(s).[34-36] Reactivation from latency can be provoked by some of the same mechanisms known to change the transcriptional activity of neurons. For example, treatment of sensory nerve terminals with catecholamine neurotransmitters leads to viral reactivation.[42] Catecholamines can raise cyclic adenosine monophosphate (cAMP) levels in some neurons through adrenergic receptors coupled to G-proteins, and lead to select changes in transcriptional activity of genes bearing cAMP-responsive elements (CRE), including genes for tyrosine hydroxylase and several neuropeptides.[91] Thus, neurons could promote entry of HSV-1 into the latent state, at least in part, by depression of immediate-early viral gene expression and stimulation of the LAT promoter, and could facilitate reactivation by changing levels of transacting factors that bind to the LAT or other viral promoters.

The latent state of HSV-1 in neurons appears to be almost ideally suited for gene transfer, although a lot more information about it is needed to exploit this potential fully. For the purposes of gene transfer to neurons in culture or in vivo, it is probably sufficient to define *latency* as a state in which the virus exists as episomal DNA elements in the nucleus and in which the LAT promoter is active.[5] Since latent HSV-1 DNA is not integrated into genomic DNA, it does not disrupt cellular genes, nor are levels of its own gene expression affected by the genomic site of insertion. By virtue of its longevity and apparent lack of toxicity in neurons, latent viral DNA can potentially effect a stable change in cellular phenotype by introducing novel gene products or by altering the regulation of cellular genes through the expression of transacting factors. To what extent latent viral DNA in the nucleus can participate in such regulatory interactions remains to be determined.

Delivery

One of the most difficult problems to resolve is how to deliver genes to the brains of adult animals. Although intracerebral injections are feasible, they cause cellular damage and require special procedures. Herpes offers a solution to this problem in two regards. First, it can be taken up at nerve terminals and be actively transported through the neurites to the cell nucleus. This transport is of the rapid type, with rates of 2 to 10 mm/h.[92] Thus, the neuron can be accessed through its endings, which can be a considerable distance from the cell body and which may be accessible through skin, cornea, footpads, blood vessels, or cerebrospinal fluid. Second, synaptic connections between neurons can be used as a "railroad" through the CNS. Studies using herpes virus as neuronal tracers have shown that viral particles can be passed both retrogradely and anterogradely and that release of virus from synaptic terminals occurs early in infection.[28] Thus, after infection of sensory neurons at their terminals, wild-type viral particles and antigens are observed first in neurons synaptically connected to the initially infected neurons and then in cells surrounding the cell body of the initially infected neuron.

Peripheral inoculation of neuroinvasive and neurovirulent strains of HSV-1 ultimately leads to widespread infection of neurons and other cells in the nervous system, and usually to fatal encephalitis. One possible way to reduce the spread would be to use replication-compromised viruses, which might be passed through only a limited number of synapses. If the numbers of viral particles passed across synapses were to decrease in secondarily affected neurons, the virus might eventually enter latency in the last set of neurons infected. If the virus were able to reactivate from latency without killing the host neuron, and be passed to synaptically connected neurons as subviral particles,[5] a "silent," nontoxic spread of the virus could possibly occur, with latency being established in a series of synaptically connected neurons. Ideally, the fate of the neuron and the virus could be controlled by genetic engineering of the vectors, and it should be possible to deliver a gene to neurons in the brain by inoculation of the HSV-1 vector at a peripheral site.

Recipient Cells

Another issue for gene delivery to the nervous system is the number and types of neurons and other cell types that can be recipients. For some purposes, gene delivery to a few neurons would be ideal; this might be achieved through inoculation anywhere along the extent of a neuron. A greater challenge is non-pathogenic delivery to a large number of neurons, such as would be necessary in gene replacement therapy for a lysosomal storage disease. The use of replication-compromised, non-neurovirulent vectors might permit a limited spread of virus beyond the site of inoculation through productive infection in some non-neuronal cells and neurons, at the same time allowing latent infection of many neurons. Wild-type herpes virus has a natural proclivity toward limbic structures in the brain (e.g., hippocampus, temporal lobe, amygdala, and medial septum).[93] By determining the neurotropic mechanisms underlying this preference, it may also be possible to steer the virus toward certain brain nuclei. Cell-specific promoters might also be useful in restricting the cell types that can express the transferred genes. It is not clear which neurons can harbor latency; in addition to sensory neurons, expression of LATs has been demonstrated in motor neurons[9] and in CNS neurons in many brain regions, and possibly in glia as well.[8] Once the essential elements of latency are understood, it may be possible to engineer vectors that can achieve an equivalent state in any nondividing cell.

General Principles of Vector Construction and Packaging

Current herpes vectors fall into two categories, which we shall term plasmid-derived and virus-derived. *Plasmid-derived vectors* were described initially by Frenkel et al.[94] and by Stow and McMonagle[95] and were termed *amplicons*. These plasmids contain sequences that permit cloning and amplification in bacteria (*Escherichia coli ori* and a selectable marker), as well as replication in mammalian cells (HSV-1 *ori*)[96,97] and packaging in herpes virions (the HSV-1 cleavage and packaging site)[98,99] (Fig. 17-3[100]). For propagation and packaging in HSV-1 virions, viral DNA is transfected into a mammalian cell infected with a helper virus[23] (Fig. 17-4). The helper HSV-1 can be a conditional mutant, such as a temperature-sensitive one (*ts*DNA), or a wild-type virus. The plasmid DNA is packaged as a concatenate in the virion with, for example, 10 tandem repeats of a 15-kb plasmid equivalent to the 150-kb herpes genome. HSV-1 will package appropriate genome, varying somewhat in length.

The use of plasmid-derived vectors for gene delivery to cultured neurons has been explored by Geller and co-workers[101] (see under Early Model Vectors for Neural Cells). Plasmid-derived vectors have the advantage that construction and packaging are relatively easy. However, several questions need to be answered to evaluate their value in different gene-transfer schemes. Can they be stably maintained in neurons, given that, by definition, the plasmid-derived vector itself cannot enter the latent state, even though the helper virus is able to do so? What types of helper virus would be most effective? Ideally, the helper virus should bear mutations that compromise its replication, that are not "leaky," and that do not readily revert. It should also not exert toxic effects on the recipient cells. Other problems include recombination between the plasmid and helper virus and difficulties in obtaining high yields of plasmid vectors to helper virus (Johnson P: personal communication). As yet there has

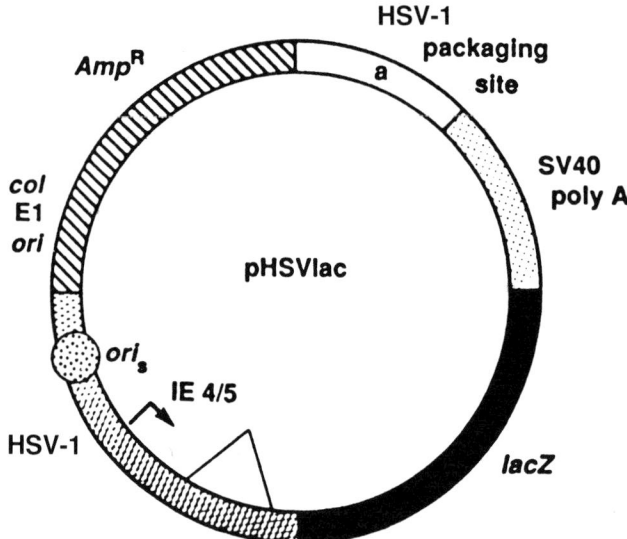

Fig. 17-3. Structure of pHSVlac plasmid. This 8.1-kb plasmid contains three kinds of genetic elements: (1) sequences that permit propagation in *Esherichia coli*-ampillicin resistance gene (Amp) and the *E. coli* origin of DNA replication (*col* El *ori*); (2) sequences that support propagation and packaging of plasmid DNA by a helper HSV-1 virus—an HSV-1 origin of DNA replication (*ori*$_s$) and a packaging site (a); and (3) a transcription unit, including the HSV-1 IE 4/5 promoter and intervening sequence, the *E. coli lacZ* gene, and the SV40 early polyadenylation site. (From Geller and Breakefield,[100] with permission.)

been no documentation of the successful use of plasmid-derived vectors for gene delivery in vivo.

HSV-1-derived vectors have been used more extensively. Construction of these vectors requires subcloning of portions of the HSV-1 genome into phage or plasmid vectors, which can hold up to 20 to 30 kb of DNA.[6] Genetic manipulation of this viral DNA is then carried out by site-directed mutagenesis or by the introduction of deletions or insertions. An example of insertion of the *lacZ* gene into the LAT region is presented in Figure 17-5.[102] The construct DNA is reintroduced into the appropriate site in the viral DNA by replacement recombination in infected cells[23] (Fig. 17-6). This is achieved by co-transfecting plasmid DNA and "infectious" viral DNA, using a molar excess of plasmid DNA into permissive cells in culture. Infectious viral DNA is intact and capable of initiating a productive infection after transfection into permissive cells. During infection, homologous recombination events occur between viral sequences in the plasmid and viral DNA. The frequency of successful recombination events varies from 10^{-1} to 10^{-4}. The higher frequencies are obtained by increasing the length of homologous sequences flanking the nonhomologous sequence. A minimum of 500 bp on either side is usually needed, but recombination has been observed with smaller extents of flanking homology. Depending on whether the targeted viral gene is present in one or two copies in the HSV-1

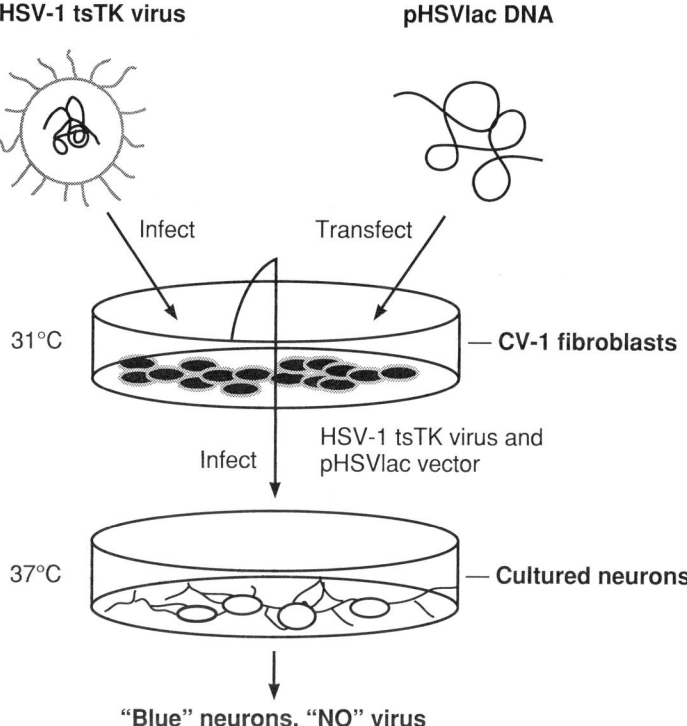

Fig. 17-4. Propagation of plasmid-derived vector. In this example, CV-1 cells permissive for HSV-1 infection are infected with a temperature-sensitive mutant of HSV-1 that can propagate at the permissive temperature (31°C) and transfected with pHSVlac DNA. During productive infection by the helper mutant virus, the plasmid DNA is also replicated and packaged into virions. Medium containing mutant virus and virions with plasmid DNA (vectors) is used to infect neurons maintained in culture at 37°C. At this restrictive temperature, the helper virus cannot propagate and the vector virions can infect confer expression of β-galactosidase on neurons. (From Breakefield and DeLuca,[23] with permission.)

genome, it may be sufficient to achieve a single successful recombination event, or subsequent growth of recombinant virus may be needed to allow for a spontaneous secondary recombination into the duplicate gene.

Some scheme is then employed to isolate the recombinant viruses. The easiest scheme is one that employs a dominant selectable marker for growth of the recombinant. Thus, if homologous recombination occurs at the site of the viral TK gene and disrupts its function, cells infected with recombinant viruses will propagate this virus in the presence of certain nucleoside analogues, such as acyclovir, while cells infected with virus bearing the intact HSV-TK gene will die.[6] Another useful scheme has been to insert a marker *lacZ* gene driven by a viral promoter that is active during productive infection into the HSV-1 genome. In this case, recombinant plaques can be identified by staining for β-galactosidase activity.[103] If no selection scheme is feasible, all plaques can be

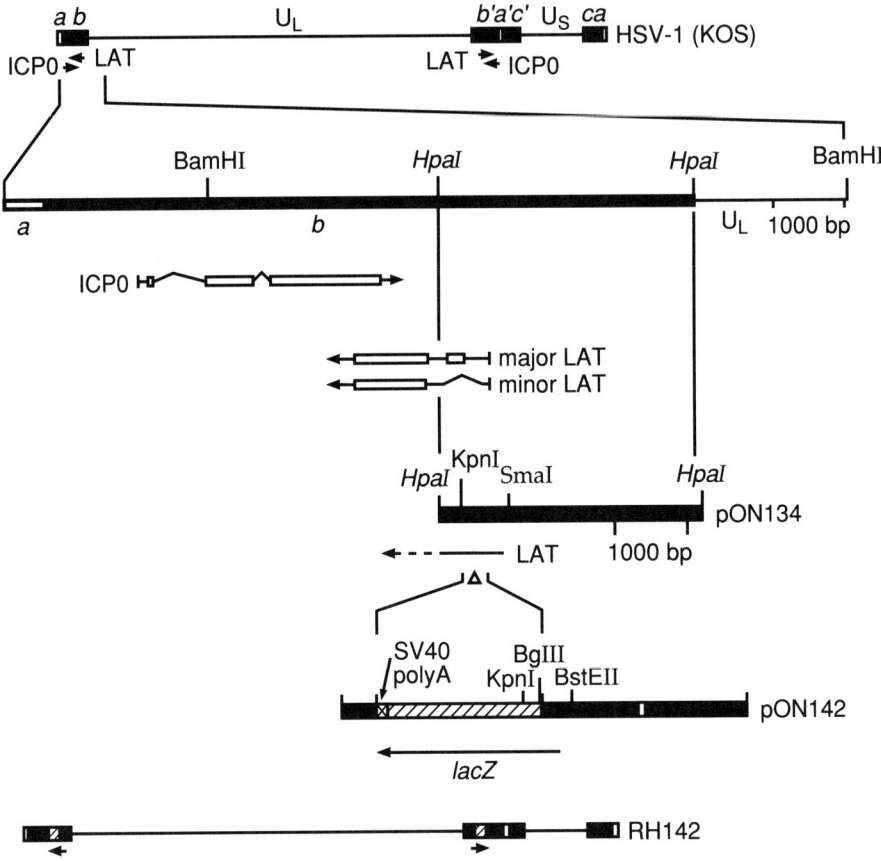

Fig. 17-5. Insertion of *lacZ* into the LAT region of the HSV-1 genome. The HSV-1 (KOS) genome is depicted on the top line. The left most repeat element is expanded in the next line showing the location and orientation of ICP0 and LAT transcripts. Open boxes represent ORFs and solid horizontal lines represent untranslated regions. One copy of the HSV-1(F) *HpaI* fragment (cloned as pON134) is shown in the expanded region with an arrow indicating the LAT transcript. The plasmid pON142 carries a *lacZ* (hatched box) gene and SV40 polyadenylation signals (cross-hatched box). The triangle represents a deletion of 373 bp introduced into the LAT gene. The recombinant virus, RH142, carries *lacZ* inserted within both copies of the LAT gene. (From Ho and Mocarski,[102] with permission.)

screened for recombination events by hybridization of plaque DNA or isolated viral DNA to inserted nonviral sequences,[90,104] by PCR amplification across nonviral and viral junctions, as has been done to assess homologous recombination in mammalian cells,[105] or by restriction digestion of viral DNA to look for altered size fragments. Recombinant virus DNA must then be analyzed by restriction digestion and sequencing, to ensure that no other recombinational or mutational changes have occurred in critical sequences.

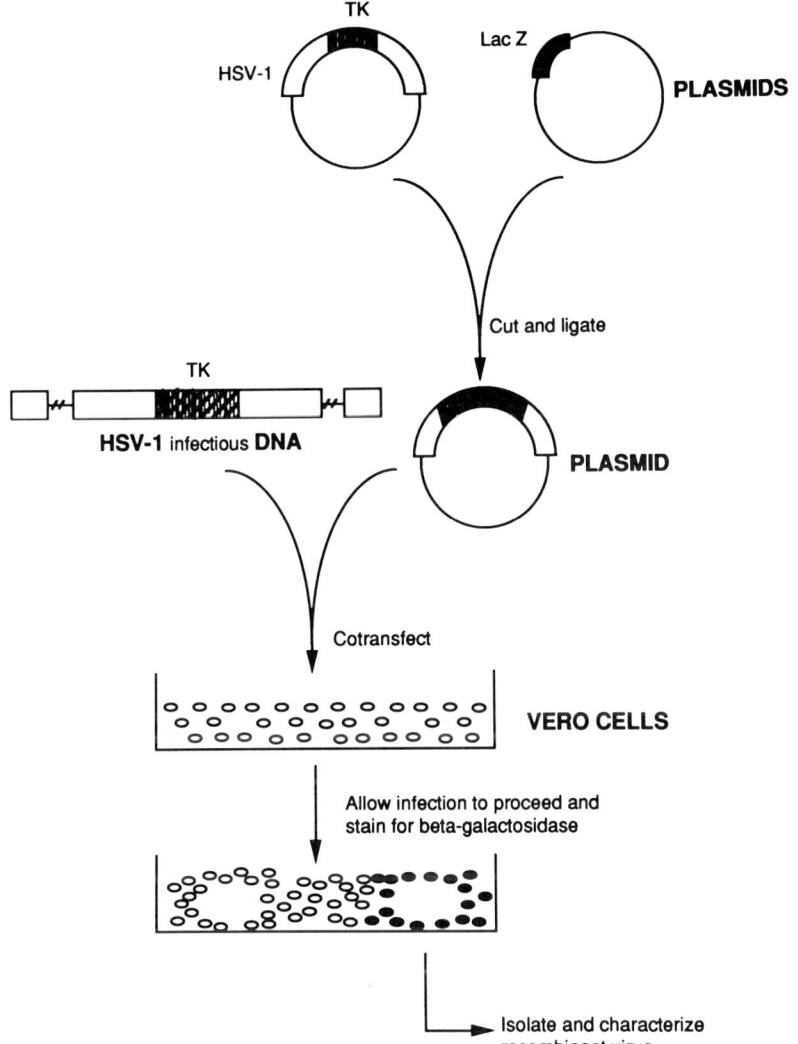

Fig. 17-6. Isolation of virus-derived vector. In this example, a portion of the HSV-1 TK gene (hatched box) and flanking sequence (open box) is carried in one plasmid and the *lacZ* gene and SV40 polyadenylation signal (solid box) in another. Plasmids are cut with appropriate restriction enzymes, such that the TK plasmid is cut once in the TK gene and the *lacZ* plasmid is cut twice, releasing the *lacZ* transcriptional element. The *lacZ* element is then ligated into the TK gene, thereby disrupting TK function, placing the *lacZ* under control of the TK promoter, and maintaining at least 500 bp of HSV-1 sequence on either side of *lacZ*. Infectious DNA from a wild-type HSV-1 is then co-transfected into permissive cells with plasmid DNA. Homologous recombination occurs between viral sequences in the viral and plasmid DNA and productive infection proceeds. Viral plaques of dead cells appear, and bordering cells are stained for β-galactosidase activity. "Blue plaques" are picked to obtain recombinant virus. (From Breakefield and DeLuca,[23] with permission.)

Early Model Vectors for Neural Cells

Much of the work that has gone into developing HSV-1 reagents useful as vectors has been undertaken by herpes virologists, not for the purpose of gene delivery to neurons, but rather to explore the functions of different viral elements in replication, latency, and reactivation. Clearly, however, there is a burgeoning interest in exploring these vectors for gene transfer. Some highlights of this work are described here in studies carried out in culture and in vivo. Using a plasmid-derived vector containing the *lacZ* marker gene driven by an immediate-early viral promoter (ICP4) and packaged by a temperature-sensitive mutant, Geller and co-workers were able to effect "stable" delivery to a variety of primary neurons[100,106] and continuous cell types,[107] for several weeks in culture.

In complementary studies, it has been shown that plasmids containing the viral LAT promoter can drive neuronal-specific expression of reporter genes, for several days, when transfected into sensory neurons in culture.[35] This neuron-active viral promoter may be used to regulate expression of foreign genes. A number of foreign genes and a few foreign promoters have been inserted into virus-derived vectors and evaluated in terms of their expression in cells in culture and/or in vivo (Table 17-3). Some principles of vector construction are emerging from this work, but more variations need to be tested before guidelines can be clearly established.

In culture, virus-derived vectors have been used to infect neural cells with mixed success. Palella et al.[118] delivered the hypoxanthine phosphoribosyltransferase (HPRT) gene to HPRT⁻ rat neuroma cells in culture. In their experiments, the human HPRT cDNA was inserted into the viral TK gene and placed under that early viral TK promoter. Because these cells expressed endogenous TK activity, hence could support viral replication, productive infection ensued, with transient expression of human HPRT activity preceding cell death. Using a similar paradigm, a large number of other genes have been delivered to other cultured cell types, including β-globin,[109,110] ovalbumin,[115] and hepatitis B virus (HBV) antigen[113] (Table 17-3). In a preliminary report, a virus vector containing *lacZ* inserted into the ICP4, immediate-early gene, and placed under the control of an immediate-early promoter from cytomegalovirus (CMV), yielded transient expression of β-galactosidase in a number of cultured cell types, including primary hepatocytes and neurons, as well as continuous lines of hepatoma and pheochromocytoma cells.[83] However, in all cases, cytopathic effects were observed within 2 weeks. Cytotoxicity was also observed with a vhs-1 mutant defective in a host cell shutoff function, but toxicity in all cases (KOS, ICP4⁻, and vhs-1) was reduced by ultraviolet (UV) irradiation of the virus, suggesting that it was mediated by early viral genes rather than by virion proteins.[83]

Several viral promoter elements have proved effective in expressing genes early in infection of cultured cells, including those for HSV-1 genes ICP0, ICP4, ICP6, gC, and TK, as well as those for other viruses, including the SV40 late region and a human CMV immediate-early protein (see Table 17-3 and references cited therein). Several experiments have assessed whether mammalian

Table 17-3. Herpes Virus-Derived Vectors Containing Nonviral Genes

Name	Viral Gene Replaced	Foreign Gene	Promoter	References
bGH-Z20	L-DNA[a]	bGH[b]	SV40L[c]	Desrosiers et al. (1985)[108]
	LAT	β-Globin	LAT	Dobson et al. (1989)[40]
RH142	LAT	lacZ	LAT	Ho and Mocarski (1989)[102]
L7/14	TK, gC	β-Globin	β-Globin	Smiley et al. (1987)[109]
				Panning and Smiley (1989)[110]
	TK	APRT	APRT	Tackney et al. (1984)[111]
HP40	TK	HPRT	HSV-TK	Palella et al. (1989)[112]
	TK	hepatitis B virus antigen	α, β	Shih et al. (1984)[113]
RH116	TK	lacZ	β-8	Ho and Mocarski (1988)[114]
RH105	TK	lacZ	ICP4	Ho and Mocarski (1988)[114]
7134	ICP0	lacZ	ICP0	Cai and Schaffer (1989)[75]
GAL4	ICP4	lacZ	ICP6	Shepard et al. (1989)[103]
Cgal	ICP4	lacZ	HCMVIE	Johnson et al. (1989)[83]
	ICP4	Ovalbumin	ICP4	Herz and Roizman (1983)[115]
8117/43	ICP4	lacZ	MoMLV LTR	Dobson et al. (1990)[9]
	UL52	lacZ	ICP6	Goldstein and Weller (1988)[116]
	UL8	lacZ	ICP6	Carmichael and Weller (1989)[117]
	gC	lacZ	gC, LAT	Fink et al. (1990)[71]

[a] First, 4.5 kb L-DNA of herpes virus saimiri at left H-L-DNA border, involved in immortalization of T cells.
[b] Bovine growth hormone.
[c] L, late region promoter.

promoters incorporated into the viral genome can effect expression of downstream genes early in productive infection. The intact rabbit β-globin gene, including about 1.2 kb of 5' flanking sequence, inserted in either of two sites in the herpes genome, gave high levels of expression of β-globin mRNA and protein early in infection of cultured Vero cells, although cellular β-globin genes were not transcriptionally active in these cells.[109,110] By contrast, incorporation of the hamster adenosine phosphoribosyltransferase (APRT) gene, with its promoter into HSV-1, yielded essentially no expression in a similar paradigm[111] and the neuronal-specific enolase (NSE) promoter inserted in a noncoding region of Us in either wild-type KOS or ICP4⁻ virus was not active early in infection of cultured neurons (Roemer K, Johnson P: personal communication).

It appears, then, that some cellular promoters incorporated into the viral genome can be used to drive expression of genes early in infection, and others cannot. One difference in the promoter elements tested to date appears to be the presence of a TATA box, which occurs in the β-globin promoter and not in the APRT[119] or NSE[120] promoters. Several other lines of evidence implicate the TATA box as important for early expression of HSV-1 genes, including the findings that HSV-induced activation of the SV40 early promoter is increased by changing sequences around the TATA box,[121] and that mutations in the TATA box of the HSV-TK promoter greatly depress its expression.[122]

The state of the art in the use of herpes vectors in achieving neuronal expression in culture can be summarized as follows. With respect to neural cells in culture, it has been possible to achieve both short-term and extended expression of *lacZ* and HPRT in a number of continuous neural cell lines and primary neurons using immediate-early and early HSV-1 promoters with both plasmid-derived and virus-derived vectors, albeit with some toxicity caused by viral host shutoff functions and/or replication, depending on the mutants used. It seems reasonable that herpes vectors, especially replication-defective or compromised ones, can be used in culture to express foreign gene products and to assess neuronal-specific promoter elements. For both uses, it will be important to minimize the ability of the virus to change cellular gene expression in a deleterious way, and to know which promoter elements in the vector are capable of being expressed during productive and nonproductive viral infections. Since neurons in culture can harbor HSV-1 over long periods in a nonreplicating state,[7] it may be possible to achieve long-term expression of foreign genes in cultured neurons. It is not clear whether this "state" is true latency, but it may provide a situation in which the genes of interest can be stably expressed off the viral genome. In fact, neurons are usually only maintained in culture for at most a month or so.

With respect to studies done in vivo, a number of virus-derived vectors have been inoculated into rodents both peripherally and intracerebrally. These studies indicate that HSV-1 vectors can be used to deliver genes to neurons in vivo and effect short-term or long-term expression of foreign genes. Short-term expression, corresponding to events of early infection, of a foreign gene, *lacZ*, has been achieved in mouse sensory neurons after corneal inoculation[9,114] and in motor neurons after inoculation of the tongue,[9] as well as in rat brain neurons after stereotactic inoculations.[70,77] In these vectors, HSV-1 genes for ICP4, TK, ICP0, or US3 were disrupted, rendering replication of the virus defective or compromised in the CNS, and thereby virtually nonpathogenic. In these vectors, *lacZ* was placed under the control of immediate-early, early, and late viral gene promoters (ICP0, ICP4, TK, gC) or the Moloney murine leukemia virus (MoMLV) long terminal repeat (LTR) element. In the case of intracerebral injections, replication-defective vectors, TK$^-$ (RH105), ICP4$^-$ (GAL4), and US3$^-$ yielded *lacZ* expression for a few days in a small number of cells near the injection site.[70,77] However, the ICP0$^-$ and US3$^-$ replication-compromised vector showed short-term *lacZ* expression in a large number of cells extending out from the injection site. Furthermore, *lacZ*-positive neurons could be demonstrated in brain regions far removed from the injection site after a few days[77] (Fig. 17-7) and up to 1 month later. This finding suggests that the virus can spread retrogradely and anterogradely through neuronal processes spanning the brain, and possibly even transsynaptically, as has been demonstrated with wild-type virus.[28] With the ICP0⁻ replication-compromised vector, there is some evidence for a smoldering infection in the CNS, although the animals survived and showed no gross behavioral abnormalities over the 1-month period in which they were evaluated.[77] Presumably, these vectors do enter latency in neurons but, since the marker gene is expressed off a viral promoter that is active only early in infection, this has not yet been assessed.

Long-term expression (corresponding to latency) of foreign genes has also been achieved in neurons in vivo. Two groups have placed foreign genes in two sites downstream of the LAT promoter and achieved long-term expression in sensory neurons in the trigeminal ganglia after corneal inoculation. Dobson et al.[40] inserted the β-globin gene 700 bp 5' to the start of the LATs and achieved stable expression for several weeks—the longest time monitored—as assessed by in situ hybridization. Ho and Mocarski[102] inserted the *lacZ* gene immediately 5' to the start of the LATs and demonstrated β-galactosidase activity by histochemical staining for up to 56 days. It is not clear which insertion site yields higher levels of gene expression, but the LAT promoter is believed to be in the −700 position.[34,36,40] Neither of these LAT⁻ vectors used was replication deficient, so it would not be possible to test them in the brain without further modification, as viral encephalitis would ensue. Fink and Gloriosa and their co-workers[70] have also reported stable expression of *lacZ* and/or LATs in CNS neurons, using a promoter element 600 bp 5' to the start of the LATs in replication defective (ICP4⁻) and compromised (US3⁻) mutants. In a recent complementary study, Dobson et al.[9] used a LAT⁻, ICP4⁻ vector in which the MoMLV LTR was used as a promoter for *lacZ*. This vector gave stable *lacZ* expression in sensory neurons in the trigeminal ganglia and fairly stable expression in the motor neurons in the hypoglossal nucleus. This latter study supports the possibilities that different types of neurons can harbor latency and that foreign promoters can confer stable gene expression during latency. Both concepts expand the potential applications of these vectors. This latter vector is replication defective and could be inoculated directly into the brain.

On the basis of preliminary studies, it is clear that virus-derived vectors can be used to confer short-term and stable gene expression onto neurons in the peripheral and central nervous systems, although stable expression of a foreign gene in the brain has yet to be documented. For peripheral inoculations, it is possible to use replication-competent virus, as it will tend to enter latency in some neurons. However, such vectors will replicate in other cells at the site of inoculation and may be transported transsynaptically into the CNS. For both peripheral and intracerebral inoculations, it seems better to use vectors, which are replication defective or compromised, as they can still enter latency, yet will cause minimal destructive damage.

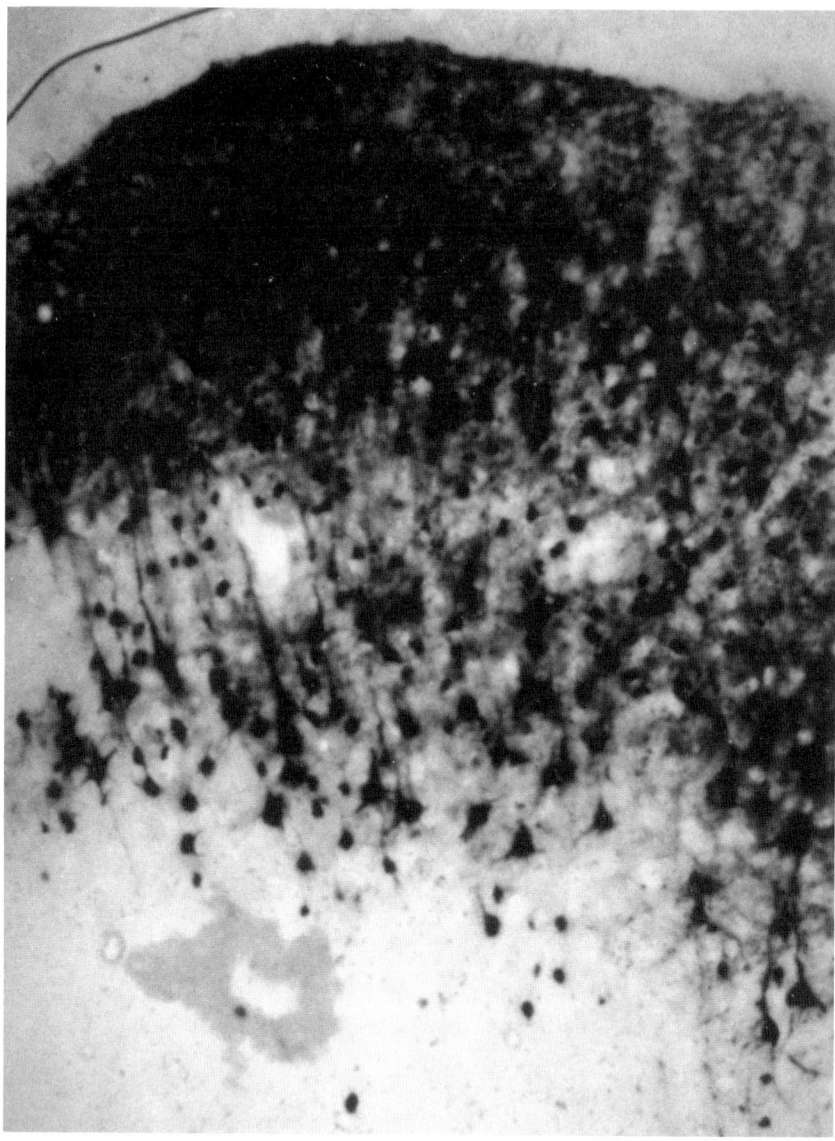

Fig. 17-7. Delivery of the *lacZ* gene to neurons in the cerebral cortex using an ICP0⁻ virus vector. A mutant HSV-1 virus (7134), containing a *lacZ* deletional substitution in both copies of the ICP0 gene, was inoculated stereotactically into the right caudate nucleus of an adult rat brain. Animals were sacrificed 3 days later. Brains were processed for histochemical staining of β-galactosidase activity and counterstained with neutral red. A coronal section of the left cerebral (cingulate) cortex is shown. Some pyramidal neurons with labeled apical dendrites are identified (arrows). Scale bar = 70 μm. (From Chiocca et al.,[77] with permission.)

SUMMARY

HSV-1 has many features that are well suited to its use as a vector for gene delivery to neurons and possibly other types of cells. Special features for use in neurons include the latent state and transneuronal and transsynaptic transport. HSV-1 vectors have been used to achieve both short-term and stable expression of foreign genes in neurons. The latter takes advantage of the ability of the virus to exist as a transcriptionally active episomal element in the cell nucleus. The viral promoter(s) that drive viral LAT expression are moderately active in latency and can be used to confer stable expression of foreign genes. It is not clear what other promoters would be active in latency and whether positional effects will limit their sites of insertion. In addition, this virus permits access to neuronal cell bodies through their terminals. This is especially important for neurons in vivo, as they are difficult to access. The virus is readily taken up by nerve terminals in the periphery (e.g., cornea and footpad) and is passed by rapid retrograde transport back to the cell nucleus. There, it may undergo productive replication or enter latency. In the case of productive infection, viral particles are then passed anterogradely and retrogradely and are preferentially passed to synaptically connected neurons. In the case of latency, reactivation may lead to subviral or viral particles, which are transported in the same way. By using replication-defective vectors, it is possible to permit latency but to block productive infection. By using replication-compromised vectors, it is possible to increase the number of neurons infected without necessarily causing pathologic viral encephalitis. A number of viral genes can be replaced so as to decrease viral toxicity or control viral replication. Probably more than 30 kb of foreign DNA can be carried in a number of different sites in the viral genome, thus providing the opportunity to deliver multiple elements simultaneously.

Two models of HSV-1 gene transfer are shown in Figure 17-8. In the first type, a replication-compromised virus (e.g., ICP0$^-$) carries a foreign gene (e.g., *lacZ*) under an immediate-early viral promoter. Infection at the nerve terminals leads to retrograde transport and expression of *lacZ* in the primary infected neuron. Viral replication proceeds, and viral particles are first passed transsynaptically to a secondary neuron, which, in turn, goes on to express β-galactosidase. The primary neuron then releases virus to surrounding cells in the ganglion, leading to the expression of β-galactosidase within them. Spread of such a virus would continue most rapidly to synaptically connected neurons, resulting in a temporal wave of "blueness" elucidating neuronal pathways. Eventually, the mutant virus might enter latency in a subset of neurons far removed from the inoculation site. If other foreign genes were controlled by promoters active in latency, they might then be stably expressed in a number of neurons. In the second type of vector model, a replication-defective virus (e.g., ICP4$^-$) carries a foreign gene (e.g., *lacZ*), under the LAT promoter. After infection at the nerve terminals, or elsewhere along the neurons, the vector would be transported to the nucleus, enter latency, and confer stable expression of a foreign gene on the primary neuron. Both types of HSV-1 vectors

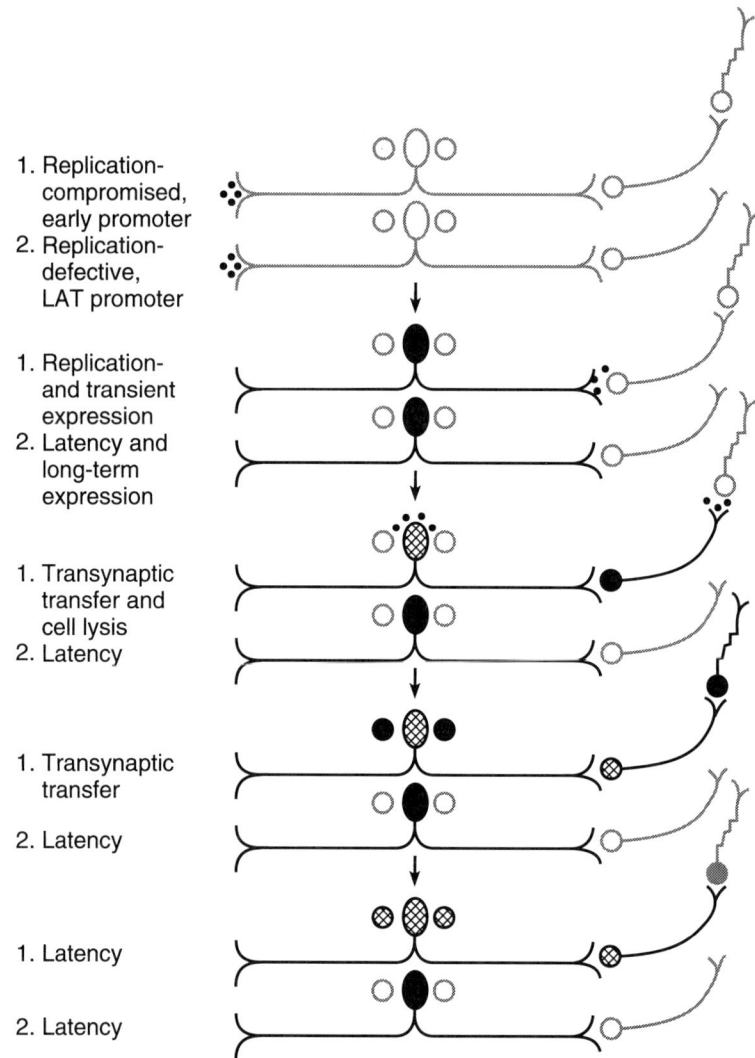

Fig. 17-8. Two modes of gene transfer with replication-deficient herpesvirus vectors. Two modes of viral gene delivery (1) and (2) are illustrated in temporal sequence from top to bottom. First, virus particles (small solid dots) are inoculated at the nerve terminals on the left. In mode 1, the vector is compromised in its replicative capacity and bears the *lacZ* gene under an immediate-early or early viral promoter. As productive viral replication proceeds in the primary infected neuron, the *lacZ* gene is expressed (cell becomes black) and virus particles are passed to synaptically connected neurons. As the primary neuron dies, virus is released to surrounding cells, which become productively infected and express *lacZ*. Infection proceeds with diminishing viral titers until the virus enters latency, at which time *lacZ* is no longer expressed. In mode 2, the vector is defective for replication but can enter latency and bears the *lacZ* gene under the LAT promoter.[5] The primarily infected neuron is not killed, but goes on to stably express *lacZ*. White, uninfected cell; black, *lacZ*-positive cell; cross-hatched, dead cell; grey, neuron harboring latent virus. (From Breakefield and DeLuca,[23] with permission.)

should prove useful in assessing cell-specific promoters, elucidating neuroanatomic pathways, and delivering functional genes, either transiently or stably, to populations of neurons and other cells.

ACKNOWLEDGMENTS

This effort was supported by grants NS24279 (XOB) and AI24306 (ND), from the National Institutes of Health. We would like to thank Ms. Suzanne McDavitt for skilled preparation of this manuscript. We are especially indebted to Jack Stevens for his intellectual generosity and to him and Paul Johnson for their sage editorial comments. We also thank the following scientists for their insights into the use of HSV-1 as vectors: Robert Dunn, Alfred Geller, Joseph Glorioso, Robert Martuza, Edward Mocarski, Michael Rosenberg, James Smiley, and Priscilla Schaffer.

REFERENCES

1. Stevens JG: Latent characteristics of selected herpesviruses. Adv Cancer Res 26:227, 1978
2. Roizman B, Batterson W: Herpes viruses and their replication. p. 497. In Fields BN, Knipe DM, Chanock RM, et al. (eds): Virology. Raven Press, New York, 1985
3. Corey L, Spear PG: Infections with herpes simplex viruses. N Engl J Med 314:686, 1986
4. Rock DL, Fraser NW: Detection of HSV-1 genome in central nervous system. Nature 302:523, 1983
5. Stevens JG: Human herpesviruses: A consideration of the latent state. Micobiol Rev 53:318, 1989
6. Roizman B, Jenkins FJ: Genetic engineering of novel genomes of large DNA viruses. Science 229:1208, 1985
7. Wilcox CL, Johnson EM Jr: Characterization of nerve growth factor-dependent herpes simplex virus latency in neurons in vitro. J Virol 62:393, 1988
8. Deatly AM, Spivack JG, Lavi E, et al: Latent herpes simplex virus type 1 transcripts in peripheral and central nervous system tissues of mice map to similar regions of the viral genome. J Virol 62:749, 1988
9. Dobson AT, Margolis TP, Sedarati F, et al: A latent, nonpathogenic HSV-1 derived vector stably expresses β-galactosidase in mouse neurons. Neuron 5:353, 1990
10. McGeoch DJ, Preston VG, Weller SK, et al: Herpes simplex virus. p. 115. In O'Brien SJ (ed): Genetic Maps Book 1 Viruses. 5th Ed. Cold Spring Harbor Laboratory, Cold Spring Harbor, New York, 1990
11. Hayward GS, Jacob RJ, Wadsworth SC, et al: Anatomy of herpes simplex virus DNA: Evidence for four populations of molecules that differ in the relative orientations of their long and short components. Proc Natl Acad Sci USA 72:4243, 1975
12. Perry LJ, McGeoch DJ: The DNA sequences of the long repeat region and adjoining parts of the long unique region in the genome of herpes simplex virus type 1. J Gen Virol 69:2831, 1988
13. McGeoch DJ, Dalrymple MA, Davison AJ, et al: The complete DNA sequence of the long unique region in the genome of herpes simplex virus type 1. J Gen Virol 69:1531, 1988

14. Hammerschmidt W, Sugden B: DNA replication of herpesviruses during the lytic phase of their life-cycles. Mol Biol Med 7:45, 1990
15. Deiss LP, Frenkel N: Herpes simplex virus amplicon: Cleavage of concatemeric DNA is linked to packaging and involves amplification of the terminally reiterated a sequence. J Virol 57:933, 1986
16. Davison AJ, Wilkie NM: Nucleotide sequences of the joint between the L and S segments of herpes simplex virus types 1 and 2. J Gen Virol 55:315, 1981
17. Spaete RR, Frenkel N: The herpes simplex virus amplicon: Analyses of cis-acting replication functions. Proc Natl Acad Sci USA 82:694, 1985
18. Kuhn JE, Kramer MD, Willenbacher W, et al: Identification of herpes simplex virus type 1 glycoproteins interacting with the cell surface. J Virol 64:2491, 1990
19. WuDunn D, Spear PG: Initial interaction of herpes simplex virus with cells is binding to heparan sulfate. J Virol 63:52, 1989
20. Kaner RJ, Baird A, Mansukhana A, et al: Fibroblast growth factor receptor is a portal of cellular entry for herpes simplex virus type 1. Science 248:1410, 1990
21. Alwine J, Stein W, Hill C: Transcription of herpes simplex type 1 DNA in nuclei isolated from infected HEp-2 and KB cells. Virology 60:302, 1974
22. Honess RW, Roizman B: Regulation of herpes virus macromolecular synthesis. I. Cascade regulation of the synthesis of three groups of viral proteins. J Virol 14:8, 1974
23. Breakefield XO, DeLuca N: Gene transfer to neurons using herpes simplex virus vectors. New Biol 3:203, 1991
24. Campbell MEM, Palfreyman JW, Preston CM: Identification of herpes simplex virus DNA sequences which encode a trans-acting polypeptide responsible for the stimulation of immediate early transcription. J Mol Biol 180:1, 1984
25. Post LE, Mackem S, Roizman B: Regulation of alpha genes of HSV: Expression of chimeric genes produced by fusion of thymidine kinase with alpha gene promoters. Cell 24:555, 1981
26. Honess RW, Roizman B: Regulation of herpes virus macromolecular synthesis: Sequential transition of polypeptide synthesis requires functional viral polypeptides. Proc Natl Acad Sci USA 72:1276, 1975
27. Lycke E, Hamark B, Johansson M, et al: Herpes simplex virus infection of the human sensory neuron. An electron microscopy study. Arch Virol 101:87, 1988
28. Kuypers HG, Ugolini G: Viruses as transneuronal tracers. Trends Neurosci 13:71, 1990
29. Steiner I, Spivack JG, Deshmane SL, et al: A herpes simplex virus type 1 mutant containing a nontransinducing Vmw65 protein establishes latent infection in vivo in the absence of viral replication and reactivates efficiently from explanted trigeminal ganglia. J Virol 64:1630, 1990
30. Leib DA, Bogard CL, Kosz-Vnenchak M, et al: A deletion mutant of the latency-associated transcript of herpes simplex virus type 1 reactivates from the latent state with reduced frequency. J Virol 63:2893, 1989
31. Leist TP, Sandri-Goldin RM, Stevens JG: Latent infections in spinal ganglia with thymidine kinase-deficient herpes simplex virus. J Virol 63:4976, 1989
32. Coen DM, Kosz-Vnenchak M, Jacobson JG, et al: Thymidine kinase-negative herpes simplex virus mutants establish latency in mouse trigeminal ganglia but do not reactivate. Proc Natl Acad Sci USA 86:4736, 1989
33. Kemp LM, Dent CL, Latchman DS: Octamer motif mediates transcriptional repression of HSV immediate-early genes and octamer-containing cellular promoters in neuronal cells. Neuron 4:215, 1990

34. Batchelor AH, O'Hare P: Regulation and cell-type-specific activity of a promoter located upstream of the latency-associated transcript of herpes simplex virus type 1. J Virol 64:3269, 1990
35. Jones C, Delhon G, Bratanich A, et al: Analysis of the transcriptional promoter which regulates the latency-related transcript of bovine herpesvirus 1. J Virol 64:1164, 1990
36. Zwaagstra JC, Ghiasi H, Slanina SM, et al: Activity of herpes simplex virus type 1 latency-associated transcript (LAT) promoter in neuron-derived cells: Evidence for neuron specificity and for a large LAT transcript. J Virol 64:5019, 1990
37. Stevens JG, Wagner EK, Devi-Rao GB, et al: RNA complementary to a herpesvirus alpha gene mRNA is prominent in latently infected neurons. Science 235:1056, 1987
38. Russell J, Stow ND, Stow EC, et al: Herpes simplex virus genes involved in latency in vitro. J Gen Virol 68:3009, 1987
39. Wechsler SL, Nesburn AB, Watson R, et al: Fine mapping of the latency-related gene of herpes simplex virus type 1: Alternative splicing produces distinct latency-related RNAs containing open reading frames. J Virol 62:4051, 1988
40. Dobson AT, Sedarati F, Devi-Rao G, et al: Identification of the latency-associated transcript promoter by expression of rabbit β-globin mRNA in mouse sensory nerve ganglia latently infected with a recombinant herpes simplex virus. J Virol 63:3844, 1989
41. Javier RT, Stevens JG, Dissette VB, et al: A herpes simplex virus transcript abundant in latently infected neurons is dispensable for establishment of the latent state. Virology 166:254, 1988b
42. Hill JM, Sedarati F, Javier RT, et al: Herpes simplex virus latent phase transcription facilitates in vivo reactivation. Virology 174:117, 1990
43. Wechsler SL, Nesburn AB, Zwaagstra J, et al: Sequence of the latency-related gene of herpes simplex virus type 1. Virology 168:168, 1989
44. Wilcox CL, Smith RL, Freed CR, et al: Nerve growth factor dependence of herpes simplex virus latency in peripheral sympathetic and sensory neurons in vitro. J Neurosci 10:1268, 1990
45. McLennan JL, Darby G: Herpes simplex virus latency: The cellular location of virus in dorsal root ganglia and the fate of the infected cell following virus activation. J Gen Virol 51:233, 1980
46. Read GS, Frenkel N: Herpes simplex virus mutants defective in the virion-associated shutoff of host polypeptide synthesis and exhibiting abnormal synthesis of alpha (immediate-early) viral polypeptides. J Virol 46:498, 1983
47. Fenwick M: The effect of herpes viruses on cellular macromolecular synthesis. Compr Virol 19:359, 1984
48. Kwong AD, Frenkel N: The herpes simplex virus virion host shutoff function. J Virol 63:4834, 1989
49. Fenwick ML, Everett RD: Transfer of UL41, the gene controlling virion-associated host cell shutoff, between different strains of herpes simplex virus. J Gen Virol 71:411, 1990
50. Latchman DS, Partidge JF, Estridge JK, et al: The different competitive abilities of viral TATTGARAT elements and cellular octamer motifs, mediate the induction of viral immediate-early genes and the repression of the histone H2B gene in herpes simplex virus infected cells. Nucleic Acids Res 17:8533, 1989
51. Stenberg RM, Pizer LI: Herpes simplex virus-induced changes in cellular and adenovirus RNA metabolism in an adenovirus type 5-transformed human cell line. J Virol 42:474, 1982

52. Everett RD, Dunlop M: Trans activation of plasmid-borne promoters by adenovirus and several herpes group viruses. Nucleic Acids Res 12:5969, 1984
53. Patel R, Chan WL, Kemp LM, et al: Isolation of cDNA clones derived from a cellular gene transcriptionally induced by herpes simplex virus. Nucleic Acids Res 14:5629, 1986
54. Nishiyama Y, Rapp F: Repair replication of viral and cellular DNA in herpes simplex virus type 2-infected human embryonic and xeroderma pigmentosum cells. Virology 110:466, 1981
55. Lorentz AK, Munk K, Darai G: DNA repair replication in human embryonic lung cells infected with herpes simplex virus. Virology 82:401, 1977
56. Waubke R, zur Hausen H, Henle W: Chromosomal and autoradiographic studies of cells infected with herpes simplex virus. J Virol 2:1047, 1968
57. Schlehofer JR, zur Hausen H: Induction of mutations within the host cell genome by partially inactivated herpes simplex virus type 1. Virology 122:471, 1982
58. Heilbronn R, zur Hausen H: A subset of herpes simplex virus replication genes induces DNA amplification within the host cell genome. J Virol 63:3683, 1989
59. Sedarati F, Javier RT, Stevens JG: Pathogenesis of a lethal mixed infection in mice with two nonneuroinvasive herpes simplex virus strains. J Virol 62:3037, 1988
60. Goodman JL, Cook ML, Sedarati F, et al: Identification, transfer, and characterization of cloned herpes simplex virus invasiveness regions. J Virol 63:1153, 1989
61. Thompson RL, Rogers SK, Zerhusen MA: Herpes simplex virus neurovirulence and productive infection of neural cells is associated with a function which maps between 0.82 and 0.832 map units on the HSV genome. Virology 172:435, 1989
62. Sherley JL, Kelly TJ: Regulation of human thymidine kinase during the cell cycle. J Biol Chem 263:8350, 1988
63. Larder BA, Lisle JJ, Darby G: Restoration of wild-type pathogenicity to an attenuated DNA polymerase mutant of herpes simplex virus type 1. J Gen Virol 67:2501, 1986
64. Day SP, Lausch RN, Oakes JE: Evidence that the gene for herpes simplex virus type 1 DNA polymerase accounts for the capacity of an intertypic recombinant to spread from eye to central nervous system. Virology 163:166, 1988
65. Katz JP, Bodin ET, Coen DM: Quantitative polymerase chain reaction analysis of herpes simplex virus DNA in ganglia of mice infected with replication-incompetent mutants. J Virol 64:4288, 1990
66. Thompson RL, Wagner EK: Partial rescue of herpes simplex virus neurovirulence with a 3.2 kb cloned DNA fragment. Virus Genes 1:261, 1988
67. Javier RT, Izumi KM, Stevens JG: Localization of a herpes simplex virus neurovirulence gene dissociated from high-titer virus replication in the brain. J Virol 62:1381, 1988
68. Goldstein DJ, Weller SK: An ICP6::lacZ insertional mutagen is used to demonstrate that the UL52 gene of herpes simplex virus type 1 is required for virus growth and DNA synthesis. J Virol 62:2970, 1988
69. Leib DA, Coen DM, Bogard CL, et al: Immediate-early regulatory gene mutants define different stages in the establishment and reactivation of herpes simplex virus latency. J Virol 63:759, 1989
70. Sears AE, Halliburton IW, Meignier B, et al: Herpes simplex virus 1 mutant deleted in the alpha22 gene: Growth and gene expression in permissive and restrictive cells and establishment of latency in mice. J Virol 33:338, 1985
71. Fink DJ, Mata M, Sternberg LR, et al: Gene transfer into brain using a herpes simplex virus vector. Soc Neurosci Abs 399.3, 1990

72. O'Hare P, Hayward GS: Evidence for a direct role for both the 175,000- and 110,000-molecular-weight immediate-early proteins of herpes simplex virus in the transactivation of delayed-early promoters. J Virol 53:751, 1985
73. O'Hare P, Hayward GS: Three trans-acting regulatory proteins of herpes simplex virus modulate immediate-early gene expression in a pathway involving positive and negative feedback regulation. J Virol 56:723, 1985
74. Everett RD: Construction and characterization of herpes simplex virus type 1 mutants with defined lesions in immediate early gene 1. Gen Virol 70:1185, 1989
75. Cai W, Schaffer PA: Herpes simplex virus type 1 ICP0 plays a critical role in the de novo synthesis of infectious virus following transfection of viral DNA. J Virol 63:4579, 1989
76. Sacks WR, Schaffer PA: Deletion mutants in the gene encoding the herpes simplex virus type 1 immediate-early protein ICP0 exhibit impaired growth in cell culture. J Virol 61:829, 1987
77. Stow ND, Stow EC: Isolation and characterization of a herpes simplex virus type 1 mutant containing a deletion within the gene encoding the immediate-early polypeptide Vmw110. J Gen Virol 67:2571, 1986
78. Chiocca EA, Choi BB, Cai W, et al: Transfer and expression of the lacZ gene in rat brain neurons mediated by herpes simplex virus insertion mutants. New Biol 2:739 1990
79. DeLuca NA, Schaffer PA: Physical and functional domains of the herpes simplex virus transcriptional regulatory protein ICP4. J Virol 62:732, 1988
80. DeLuca NA, McCarthy A, Schaffer PA: Isolation and characterization of deletion mutatns of herpes simplex virus type 1 in the gene encoding immediate-early regulatory protein ICP4. J Virol 56:558, 1985
81. McCarthy AM, McMahan L, Schaffer PA: Herpes simplex virus type 1 ICP27 deletion mutants exhibit altered patterns of transcription and are DNA deficient. J Virol 63:18, 1989
82. Estridge JK, Kemp LM, La Thangue NB, et al: The herpes simplex virus type 1 immediate-early protein ICP27 is obligately required for the accumulation of a cellular protein during viral infection. Virology 168:67, 1989
83. Johnson PA, Miyanohara A, Rosenberg MB, et al: Cytotoxicity of non-productive HSV-1 infections. Presented at the Fourteenth International Herpes Workshop, 1989 (abst)
84. Ace CI, McKee TA, Ryan JM, et al: Construction and characterization of a herpes simplex virus type 1 mutant unable to transinduce immediate-early gene expression. J Virol 63:2260, 1989
85. Longnecker R, Chatterjee S, Whitley RJ, et al: Identification of a herpes simplex virus 1 glycoprotein gene within a gene cluster dispensable for growth in cell culture. Proc Natl Acad Sci USA 84:4303, 1987
86. Schranz P, Neidhardt H, Schroder CH, et al: A viable HSV-1 mutant deleted in two nonessential major glycoproteins. Virology 170:273, 1989
87. Longnecker R, Roizman B, Meignier B: Herpes simplex viruses as vectors: Properties of a prototype vaccine strain suitable for use as a vector. p. 68. In Gluzman Y, Hughes SH (eds): Viral Vectors. Cold Spring Harbor Laboratory, Cold Spring Harbor, NY, 1988
88. Gerster T, Roeder RG: A herpesvirus trans-activating protein interacts with transcription factor OTF-1 and other cellular proteins. Proc Natl Acad Sci USA 85:6347, 1988

89. Kemp LM, Latchman DS: Regulated transcription of herpes simplex virus immediate-early genes in neuroblastoma cells. Virology 171:607, 1989
90. O'Hare P, Goding CR: Herpes simplex virus regulatory elements and the immunoglobulin octamer domain bind a common factor and are both targets for virion transactivation. Cell 52:435, 1988
91. Goodman RH: Regulation of neuropeptide gene expression. Annu Rev Neurosci 13:111, 1990
92. Price RW, Rubenstein R, Khan A: Herpes simplex virus infection of isolated autonomic neurons in culture: Viral replication and spread in a neuronal network. Arch Virol 71:127, 1982
93. Chrisp CE, Sunstrum JC, Averill DR, et al: Characterization of encephalitis in adult mice induced by intracerebral inoculation of herpes simplex virus type 1 (KOS) and comparison with mutants showing decreased virulence. Lab Invest 60:822, 1989
94. Frenkel N, Spaete RR, Vlazny DA, et al: The herpes simplex virus amplicon—A novel animal-virus cloning vector. p. 205. In Gluzman Y (ed): Eukaryotic Viral Vectors. Cold Spring Harbor Laboratory, Cold Spring Harbor, NY, 1982
95. Stow ND, McMonagle ED: Propagation of foreign DNA sequences linked to a herpes simplex virus origin of replication. p. 199. In Gluzman Y (ed): Eukaryotic Viral Vectors. Cold Spring Harbor Laboratory, Cold Spring Harbor, NY, 1982
96. Vlazny DA, Frenkel N: Replication of herpes simplex virus DNA: Localization of replication recognition signals within defective virus genomes. Proc Natl Acad Sci USA 78:742, 1981
97. Stow ND: Localization of an origin of DNA replication within the TRS/IRS repeated region of the herpes simplex virus type 1 genome. EMBO J 1:863, 1982
98. Vlazny DA, Kwong A, Frenkel N: Site-specific cleavage/packaging of herpes simplex virus DNA and the selective maturation of nucleocapsids containing full-length viral DNA. Proc Natl Acad Sci USA 79:1423, 1982
99. Stow ND, McMonagle EC, Davison AJ: Fragments from both termini of the herpes simplex virus type 1 genome contain signals required for the encapsidation of viral DNA. Nucleic Acids Res 11:8205, 1983
100. Geller AI, Breakefield XO: A defective HSV-1 vector expresses *E. coli* beta-galactosidase in cultured rat peripheral neurons. Science 241:1667, 1988
101. Breakefield XO, Geller AI: Gene transfer into the nervous system. Mol Neurobiol Rev 1:339, 1987
102. Ho DY, Mocarski ES: Herpes simplex virus latent RNA (LAT) is not required for latent infection in the mouse. Proc Natl Acad Sci USA 86:7596, 1989
103. Shepard AA, Imbalzano AN, DeLuca NA: Separation of primary structural components conferring autoregulation, transactivation, and DNA-binding properties to the herpes simplex virus transcriptional regulatory protein ICP4. J Virol 63:3714, 1989
104. Homa FL, Otal TM, Glorioso JC, et al: Transcriptional control signals of a herpes simplex virus type 1 late (gamma 2) gene lie within bases −34 to +124 relative to the 5' terminus of the mRNA. Mol Cell Biol 6:3652, 1986
105. Kim HS, Smithies O: Recombinant fragment assay for gene targeting based on the polymerase chain reaction. Nucleic Acids Res 16:8887, 1988
106. Geller AI, Freese A: Infection of cultured central nervous system neurons with a defective herpes simplex virus 1 vector results in stable expression of *Escherichia coli* beta-galactosidase. Proc Natl Acad Sci USA 87:1149, 1990

107. Boothman DA, Geller AI, Pardee AB, et al: Expression of the *E. coli lacZ* gene from a defective HSV-1 vector in various human normal, cancer-prone and tumor cells. FEBS Lett 20:159, 1989
108. Desrosiers RC, Kamine J, Bakker A, et al: Synthesis of bovine growth hormone in primates by using a herpesvirus vector. Mol Cell Biol 5:2796, 1985
109. Smiley JR, Smibert C, Everett RD: Expression of a cellular gene cloned in herpes simplex virus: Rabbit beta-globin is regulated as an early viral gene in infected fibroblasts. J Virol 61:2368, 1987
110. Panning B, Smiley JR: Regulation of cellular genes transduced by herpes simplex virus. J Virol 63:1929, 1989
111. Tackney C, Cachianes G, Silverstein S: Transduction of the Chinese hamster ovary aprt gene by herpes simplex virus. J Virol 52:606, 1984
112. Palella TD, Hidaka Y, Silverman LJ, et al: Expression of human HPRT mRNA in brains of mice infected with a recombinant herpes simplex virus-1 vector. Gene 80:137, 1989
113. Shih M-F, Arsenakis P, Tiollais P, et al: Expression of hepatitis B virus S gene by herpes simplex virus type 1 vectors carrying alpha- and beta-regulated gene chimeras. Proc Natl Acad Sci USA 81:5867, 1984
114. Ho DY, Mocarski ES: Beta-galactosidase as a marker in the peripheral and neural tissues of the herpes simplex virus-infected mouse. Virology 167:279, 1988
115. Herz C, Roizman B: The alpha promoter regulator-ovalbumin chimeric gene resident in human cells in regulated like the authentic alpha 4 gene after infection with herpes simplex virus 1 mutants in alpha 4 gene. Cell 33:145, 1983
116. Goldstein DJ, Weller SK: Herpes simplex virus type 1-induced ribonucleotide reductase activity is dispensable for virus growth and DNA synthesis: Isolation and characterization of an ICP6 *lacZ* insertion mutant. J Virol 62:196, 1988
117. Carmichael EP, Weller SK: Herpes simplex virus type 1 DNA synthesis requires the product of the UL8 gene: Isolation and characterization of an ICP6: *lacZ* insertion mutation. J Virol 63:591, 1989
118. Palella TD, Silverman LJ, Schroll CT, et al: Herpes simplex virus-mediated human hypoxanthine-guanine phosphoribosyltransferase gene transfer into neuronal cells. Mol Cell Biol 8:457, 1988
119. Park JH, Taylor MW: Analysis of signals controlling expression of the Chinese hamster ovary aprt gene. Mol Cell Biol 8:2536, 1988
120. Forss-Petter S, Danielson PE, Catsicas S, et al: Transgenic mice expressing beta-galactosidase in mature neurons under neuron-specific enolase promoter control. Neuron 5:187, 1990
121. Everett RD: Promoter sequence and cell type can dramatically affect the efficiency of transcriptional activation induced by herpes simplex virus type 1 and its immediate-early gene products Vmw175 and Vmw110. J Mol Biol 203:739, 1988
122. Coen DM, Weinheimer SP, McKnight SL: A genetic approach to promoter recognition during transinduction of viral gene expression. Science 234:53, 1986

18

IN UTERO THERAPY FOR GENETIC DISEASES

James D. Goldberg · Mitchell S. Golbus

INTRODUCTION

Recent advances in reproductive genetics have dramatically expanded the potential for in utero fetal therapy. New approaches to prenatal diagnosis have provided direct access to the fetus for administration of potential therapies. In addition, increased understanding of the mechanisms of genetic disorders have permitted therapeutic approaches using replacement of missing factors to the mother as a means of delivering in utero therapy to the fetus. Current approaches to fetal therapy include the use of intrauterine transfusions for red blood cell alloimmunization, first reported by Liley during the early 1960s.[1] This therapy has evolved into an extremely effective means of preventing hydrops and death of the affected fetus. Another effective example of fetal therapy for nongenetic conditions is the antenatal use of maternally administered steroids to enhance lung maturation in the premature infant to decrease the incidence of respiratory distress syndrome after birth.[2] This chapter reviews the attempts that have been made for the in utero treatment of genetic disease and discusses future possibilities.

MATERNAL ADMINISTRATION OF AGENTS TO TREAT THE FETUS

Congenital Adrenal Hyperplasia

The administration of appropriate compounds to the mother to treat the fetus has proved effective for certain disorders. A primary example is the therapy of congenital adrenal hyperplasia (CAH) caused by 21-hydroxylase deficiency. This autosomal-recessive disorder results from mutations in the

21-hydroxylase B genes located between the HLA-B and HLA-DR genes, more closely linked to the HLA-B gene. The frequency of the severe classic disorder varies in different populations, ranging from an incidence of 1 in 282 in the Yupik Eskimos to 1 in 20,000 in Japan.[3] The incidence in the United States is approximately 1 in 12,000 to 1 in 18,000. The classic disorder results in severe masculinization of female infants early in gestation, and the salt-wasting form results in life-threatening adrenal crisis after birth. Prenatal diagnosis can be performed by HLA-haplotyping of chorionic villus cells obtained in the first trimester of pregnancy or amniocytes from second trimester amniocentesis. In some cases, molecular techniques using probes from the HLA-B locus or 21-hydroxylase gene may be used. Elevated levels of 17-hydroxyprogesterone are also found in mid-trimester amniotic fluid in fetuses affected with the severe salt-wasting form of the disorder.

The goal of antenatal fetal therapy in this disorder has been prevention of the virilization of affected female fetuses by in utero adrenal suppression. This has been attempted by maternal administration of oral steroids. These attempts have been recently summarized by Pang and Clark.[3] Of the 16 reported cases of maternal therapy, 5 female infants with CAH had normal female genitalia at birth.[4-7] These mothers were treated with 0.5 mg of dexamethasone bid from 3 to 9 menstrual weeks until birth. Seven treated fetuses were reported to have either mild or partial virilization.[4,7-12] Four of these mothers received 0.5 mg of dexamethasone bid or tid from 3.2 to 9 weeks gestation until birth. Two mothers received 1 mg dexamethasone once daily or 0.25 mg tid to qid from 8 menstrual weeks until birth.[7,10] Another mother received 40 to 50 mg of cortisol daily from 9.4 menstrual weeks until birth.[8]

Four treated fetuses had marked virilization.[9] One received 0.5 mg of dexamethasone tid from 5 to 21 menstrual weeks. Two mothers received 1 mg of dexamethasone daily from 9 to 10 menstrual weeks until birth, and the other mother received 0.5 mg of dexamethasone daily from 5 menstrual weeks until birth.

The above findings indicate that attempts to prevent virilization of female fetuses affected with CAH have not been completely successful. While some mothers may have been on inadequate doses of dexamethasone and others may have started too late in gestation, treatment failures have occurred. At-risk couples who are contemplating this approach must be aware of these limitations.

Multiple Carboxylase Deficiency

The inability to absorb or bind biotin to the biotin-dependent mitochondrial enzymes, pyruvate carboxylase, propionyl-coenzyme A (CoA) carboxylase, and β-methylcrotonyl-CoA carboxylase results in the biotin-responsive multiple carboxylase deficiency. This disorder produces a characteristic picture of early dermatitis, severe metabolic acidosis, and organic acid excretion. All these findings can be reversed in affected persons with this autosomal-recessive disorder by biotin supplementation. Two cases of in utero treatment of this disorder have been reported.

Roth et al.[13] described a family with two affected infants that died early in neonatal life. The mother was first seen in her third pregnancy at 34 weeks of pregnancy. Prenatal diagnosis was not attempted because of the advanced stage of pregnancy. Oral administration of biotin was begun (10 mg/d) until the end of pregnancy. The pretreatment maternal urinary organic acid profile was normal. Dizygotic twins were delivered at 38 weeks gestation. Cord blood and urinary organic acid profiles were normal in both twins. Cord blood biotin concentrations were four to seven times normal. The newborn course of both infants was normal. Subsequent evaluation of the twins by analysis of cultured fibroblasts revealed one twin to be affected with the disorder.

Packman et al.[14] also reported the in utero treatment of this disorder in a couple with an affected child. In a subsequent pregnancy, amniocentesis at 17 menstrual weeks was performed. The activities of all three carboxylases were severely reduced in cultured amniocytes. Oral biotin therapy (10 mg/d) was begun at 23.5 weeks of pregnancy. At birth, the infant showed no clinical or gross chemical abnormalities. The diagnosis of an affected infant was confirmed by analysis of cultured fibroblasts.

While both studies demonstrate the successful in utero treatment of this disorder to prevent neonatal complications, the advantage over the traditional therapy of beginning biotin supplementation at birth has not been shown. Further studies are necessary to evaluate fully whether prenatal treatment has a beneficial effect.

Methylmalonic Acidemia

Some forms of methylmalonic acidemia result from the lack of the active form of vitamin B_{12}, 5'-deoxyadenosylcobalamin, or of the inability of the cofactor to bind to the apoenzyme. Administration of large doses of the appropriate form of vitamin B_{12} in these individuals may result in clinical improvement. Ampola et al.[15] reported the first attempt at prenatal therapy for the B_{12}-responsive form of this disorder in a family who had a previously affected child that died at 3 months of age. In a subsequent pregnancy, these workers performed an amniocentesis at 19 menstrual weeks. An elevated methylmalonic acid level was demonstrated in cell free amniotic fluid. Cultured amniocytes had abnormal propionate oxidation, succinate oxidation, and undetectable levels of 5'-deoxyadenosylcobalamin. An increased methylmalonic acid excretion was also found in maternal urine, all consistent with the diagnosis of methylmalonic acidemia. At 32 menstrual weeks, oral cyanocobalamin (10 mg/d in divided doses) was given to the mother. This resulted in a slight reduction in maternal urinary methylmalonic acid excretion. Because of this, at 34 weeks of pregnancy 5 mg/d of cyanocobalamin was begun intravenously, resulting in a progressive decrease in maternal urinary excretion of methymalonic acid. At delivery (41 weeks of pregnancy), the maternal urinary methylmalonate was slightly above normal. Amniotic fluid methylmalonic acid at delivery was still elevated at four times normal. The infant was confirmed to be affected but had no neonatal complications and has continued to do well on oral cobalamin supplementation without mental or motor complications.

Rosenblatt et al.[16] reported the prenatal therapy of a fetus with methylcobalamin deficiency. A prior sibling diagnosed postnatally had delayed development and a megaloblastic anemia which had responded to large doses of hydroxycobalamin. This infant's fibroblasts had a low methycobalamin level and a reduced capacity for remethylation of homocysteine by methylfolate to form methionine. In a subsequent pregnancy, an affected fetus was prenatally diagnosed, and prenatal administration of hydroxycobalamin (1 mg twice weekly) was begun at 25 menstrual weeks. The infant was normal at birth and has remained so while on therapy.

As with therapy for multiple carboxylase deficiency, the prenatal administration of vitamin B_{12} has not been demonstrated to be superior to the standard approach of beginning therapy at birth. There is some evidence that prenatal therapy may be beneficial for methylmalonic acidemia in that Nyhan[17] has reported that an increased incidence of minor anomalies may result if the disease is not treated in utero.

TREATMENT BY MATERNAL DIETARY RESTRICTION

Phenylketonuria

Since the advent of successful neonatal screening and therapy programs for infants affected with phenylketonuria (PKU), many effectively treated women with this autosomal-recessive disorder have entered into the reproductive age period. Lenke and Levy[18] reported that these women, who are no longer on dietary restriction, have a high incidence of offspring affected with significant retardation and other congenital abnormalities. This effect of the mother's disease on the fetus (an obligate heterozygote) is known as maternal PKU syndrome. Rohr et al.[19] also have shown that early pregnancy or preconception institution of dietary phenylalanine restriction can significantly reduce the incidence of this syndrome. The degree and timing of reduction of the maternal phenylalanine level to produce a maximal response are currently unknown and are the subject of a collaborative National Institute of Child Health and Human Development (NICHD) study.

Galactosemia

Galactosemia, an autosomal-recessive disorder caused by a deficiency of galactose 1-phosphate uridyl transferase, is characterized by cataracts, growth deficiency, and early ovarian failure in females. While clinical symptoms appear during the neonatal period and can be ameliorated by the elimination of galactose from the diet, irreversible damage to oocytes is thought to occur in the prenatal period.[20] Some investigators have suggested that prenatal neurologic damage also may occur in galactosemic fetuses.[21] Experiments in rats suggest that gonadal toxicity in the female fetus is most pronounced during the premeiotic stages of development, suggesting that if therapy is going to improve gonadal function, it must start very early in gestation.[22] To date, no studies have been reported using maternal galactose restriction to reduce the incidence of these complications.

TREATMENT BY FETAL TRANSPLANTATION

Hematopoietic Stem Cells

Hematopoietic stem cells (HSCs) from the bone marrow of postnatal donors and from the liver of fetal donors have been used for the treatment of certain genetic diseases (i.e., thalassemia, severe combined immunodeficiency). HSCs are hematopoietic organ-derived cells that have the ability to differentiate into a variety of progeny cells, including erythroid, thromboid, myeloid, and lymphoid lines. In addition, they can differentiate into cells of the reticuloendothelial system. When HSCs are harvested and transfered from a donor to a recipient, the donor stem cells find their way to the hematopoietic stromal environment of the recipient, where they implant, replicate, and differentiate. A genetic disorder that results from a defect in the function of cells derived from the HSCs can be reversed if stem cells capable of producing normal progeny replace existing HSCs. The use of bone marrow transplantation (BMT) as a source of stem cell replacement for the correction of certain genetic disorders is also discussed in Chapter 12.

A significant complication of postnatal bone marrow transplantation is graft-versus-host-disease (GVHD) caused by graft recognition of the host as foreign and an attempt to reject the host. Even with HLA-matching of donors, severe or mild GVHD may occur in as many as 50 percent of recipients. Another consideration of postnatal BMT is that significant disease pathology may already exist in the fetus. Accumulation of storage material in aborted fetuses with Tay-Sachs disease has been demonstrated as early as 8 to 9 weeks of pregnancy by electron microscopy.[23] Also, the necessary ablation of the recipient marrow before BMT makes the host extremely susceptible to infection. The need for long-term immunosuppression in the recipient leads to additional procedural complications of postnatal BMT.

The fetus as a potential recipient avoids many of these problems. The early fetus is ontologically prepared for engraftment, is immunotolerant, and should permit foreign grafts without rejection. A naturally occurring "experiment of nature," the freemartin cow, is an example. This female twin of a normal male co-twin is born masculinized. In 1916, Lillie[24] used a mathematical construct to argue that vascular anastomoses between male and female bovine twins permitted the exchange of humoral factors that resulted in the female's masculinization. It was 29 years later, in 1945, that Owen[25] demonstrated that these vascular connections permitted intrauterine transfusions between the bovine twins that led to blood type chimeras, an intrauterine HSC transplantation.

Jolly et al.[26] in 1976 reported that the intrauterine transfer of HSC between nonidentical co-twins altered the pathology of mannosidosis in the affected twin. Although the neurodegenerative course was not changed, histologic examination of visceral tissues revealed a significant decrease in pathology. It was argued that the transplanted hematopoietic cells that had normal mannosidase activity were unable to compensate for the absence of mannosidase in the brain tissue of the affected calf. In the human, an inadvertent experiment demonstrates the immunotolerance of the fetus. In 1973, Turner[27] reported that 5 of 65 infants who had a fetal intrauterine transfusion for Rh isoimmunization demonstrated the persistence of circulating donor white blood cells beyond 1 year of age.

Fetal immunotolerance permits the formation of these chimeras. Chimeras occur because the fetal host does not recognize the allogeneic graft as foreign and subsequently becomes tolerant of the antigenic differences of the foreign graft. From a practical standpoint, this immunotolerance would permit the use of allogeneic stem cell grafts from unmatched donors, and possibly even pooled donor cells, obviating the need for HLA-antigen typing and a specific donor search as is generally required for postnatal BMT. It would also remove the need for marrow ablation as is necessary with immunocompetent recipients who have the capacity to reject an allogeneic graft. An additional reason for marrow ablation of the recipient is to rid the bone marrow stroma of its occupant cells to make room for the implantation of the injected HSCs. Before the 20th week, the human fetus not only has an empty marrow but also is programmed for the reception of HSCs. In the human, hematopoiesis begins in the yolk sac and switches to the liver at about 6 to 7 weeks of pregnancy. The mechanism is thought to be "seeding" of the liver by HSC from the yolk sac. The bone marrow is then "seeded" by the liver at about 20 menstrual weeks. Certain factors orchestrate these events. They accomplish a variety of functions. They prepare the HSCs for reimplantation, stimulate their departure from their familiar stromal environment, allow the HSCs to move to the bone marrow by the vascular system, and permit reimplantation. Simultaneously, the bone marrow increases its available stromal space and prepares its architecture to function as a hematopoietic organ.

To avoid the problem of GVHD, fetal liver cells have been used as a source of HSC for transplantation. Post-thymic T lymphocytes (those capable of causing GVHD) are only found in the fetal liver after 18 to 20 weeks of pregnancy. Human fetal lymphocytes, however, show the capacity to proliferate in response to mitogens by 11 to 12 weeks of pregnancy and by the 14th week are capable of responding to allogeneic cells in mixed lymphocyte culture. Thus, the early fetal liver may be an ideal source for HSCs while preventing GVHD.

The fetus was not used as a recipient of fetal HSCs in an animal model until 1979, when Fleischman and Mintz[28] reported that intraplacental transfusions of fetal liver HSCs into mid-gestational fetal mice with a severe macrocytic anemia (W^v) led to the permanent cure of their anemia. Flake et al.[29] reported the transplantation of fetal liver HSCs into fetal sheep at 45 to 65 days gestation. These investigators demonstrated engraftment and stable chimerism at 6 months after birth in these lambs. There was no evidence of GVHD seen. Harrison et al.[30] reported the in utero transplantation of fetal liver HSCs in monkeys. Intraperitoneal injections of fetal liver HSCs were given at 59 to 68 days gestation. These animals exhibited stable engraftment of lymphoid (2.9 to 8.0 percent donor cells), erythroid (5.3 to 12.5 percent donor cells), and myeloid (8.5 to 15.4 percent donor cells) cell lineages for up to 2 years. There was no evidence of GVHD.

To date, there have been two reports of human in utero stem cell transplantations in the literature. Linch et al.[31] reported the in utero transplantation of maternal bone marrow in a fetus severely affected with red blood cell isoimmunization at the same time a red blood cell transfusion was being performed at 17 weeks of pregnancy. No evidence of engraftment was seen after birth of the infant. The other report is by Touraine et al.[32] of the in utero transplantation of fetal liver HSCs for a fetus affected with the bare lymphocyte syndrome. This disorder is characterized by lack of expression of HLA antigens on lymphoctes and causes an immunodeficiency syndrome. This fetus was transplanted with fetal liver cells at 30 weeks of pregnancy by the intravascular route. At 1 month of age, 10 percent of the infant's lymphocytes expressed HLA antigens. At the time of the report, 7 months after birth, the infant remains in a sterile bubble and has undergone seven further postnatal fetal liver HSC transplants.

CONCLUSION

As these cases illustrate, much work remains to be done in the area of in utero HSC transplantation. The optimum time in gestation for treatment and the best source and quantity of HSCs to transplant continue to be unanswered questions. If successful this approach would provide a significant advance in the treatment of genetic disease.

REFERENCES

1. Liley AW: Intrauterine transfusion of fetus in hemolytic disease. Br Med J 2:1107, 1963
2. Liggins GC, Howie RN: A controlled trial of antepartum glucocorticoid treatment of the respiratory distress syndrome in premature infants. Pediatrics 50:515, 1972
3. Pang S, Clark A: Newborn screening, prenatal diagnosis, and prenatal treatment of congenital adrenal hyperplasia due to 21-hydroxylase deficiency. Trends Endocrinal Metab 1:300, 1990
4. Forest MG, Betuel H, David M: Prenatal treatment in congenital adrenal hyperplasia due to 21-hydroxylase deficiency: Update 88 of the French multicentric study. Endocr Res 1989 15:277, 1989
5. Odink RJH, Boue A, Jansen M: The value of chorion villus sampling in early detection of 21 hydroxylase deficiency (21-OHD). Pediatr Res 23:131, 1988
6. Romer TE, Ginalska-Malinowska M: Successful prenatal treatment of congenital

adrenal hyperplasia resulting from the 21-hydroxylase deficiency: Is prenatal diagnosis in a mother at risk essential? Endokyrnol Pol 38:125, 1987
7. Speiser PW, Laforgia N, Kato K, et al: First trimester prenatal treatment and molecular genetic diagnosis of congenital adrenal hyperplasia. J Clin Endocrinol Metab 70:838, 1990
8. David M, Forest MG: Prenatal treatment of congenital adrenal hyperplasia resulting from 21-hydroxylase deficiency. J Pediatr 105:799, 1984
9. Dörr HG, Sippell WC, Bidlingmaier F, Knoor D: Experience with intrauterine therapy of congenital adrenal hyperplasia due to 21 hydorxylase deficiency. p. 21. Presented at the Seventieth Annual Meeting of the Endocrine Society, 1988 (abst)
10. Pang S, Kling S, Dobbins RH, Clark A: Illinois experience in newborn screening for congenital adrenal hyperplasia (CAH): A new guideline for follow-up approach. Presented at the Proceedings of the Seventh National Neonatal Screening Symposium, McLean, VA, ASTPHLD, 1990 (in press)
11. Petersen KE, Damkjaer-Nielsen M, Buus O, Couillin P: Congenital adrenal hyperplasia (CAH): Prenatal treatment. Pediatr Res 20:1201, 1986
12. Schulman DI, Mueller OT, Gallardo LA, et al: Treatment of congenital adrenal hyperplasia in utero. Pediatr Res 25:93A, 1989
13. Roth KS, Yang W, Allen L, et al: Prenatal administration of biotin: Biotin responsive multiple carboxylase deficiency. Pediatr Res 16:126, 1982
14. Packman S, Cowan MJ, Golbus MS, et al: Prenatal treatment of biotin responsive multiple carboxylase deficiency. Lancet 1:1435, 1982
15. Ampola MG, Mahoney MJ, Nakamura E, Tanaka K: Prenatal therapy of a patient with vitamin B responsive methylmalonic acidemia. N Engl J Med 293:313, 1975
16. Rosenblatt DS, Cooper BA, Schmutz SM, et al: Prenatal vitamin B_{12} therapy of a fetus with methylcobalamin deficiency. Lancet 1:1127, 1985
17. Nyhan WL: Prenatal treatment of methylmalonic aciduria. N Engl J Med 293:353, 1975
18. Lenke RR, Levy HL: Maternal phenylketonuria and hyperphenylalaninemia: An international survey of the outcome of untreated and treated pregnancies. N Engl J Med 303:1202, 1980
19. Rohr F, Doherty LB, Waisbren SE, et al: New England Maternal PKU Project: Prospective study of untreated and treated pregnancies and their outcomes. J Pediatr 110:391, 1987
20. Kaufman FR, Kogut MD, Donnel GN, et al: Hypergonadotropic hypogonadism in female patients with galactosemia. N Engl J Med 304:994, 1981
21. Segal SS: Disorders of galactose metabolism. p. 167. In Stanbury JB, Wyngarden JB, Frederickson DS (eds): The Metabolic Basis of Inherited Disease. 5th Ed. McGraw-Hill, New York, 1983
22. Chen YT, Mattison DR, Feigenbaum L, et al: Reduction in oocyte number following prenatal exposure to a high galactose diet. Science 214:1145, 1981
23. Goldberg JD, Grabowski, GA, Driscoll MC, et al: First trimester fetal diagnosis: Principles and potential pitfalls in enzymatic and molecular diagnosis. p. 218. In Fraccaro M, Simoni G, Brambat B (eds): First Trimester Fetal Diagnosis. Springer-Verlag, Berlin, 1985
24. Lillie FR: The theory of the freemartin. Science 43:611, 1916
25. Owen RD: Immunogenetic consequences of vascular anastomoses between bovine twins. Science 102:400, 1945
26. Jolly RD, Thompson CE, Murphy BW, et al: Enzyme replacement therapy—An experiment of nature in a chimeric mannosidosis calf. Pediatr Res 10:209, 1976

27. Turner JH, Hutchinson DL, Petricciani JC: Chimerism following fetal transfusion. Scand J Haematol 10:358, 1973
28. Fleischman RA, Mintz B: Prevention of genetic anemias in mice by microinjection of normal hematopoietic stem cells into the fetal placenta. Proc Natl Acad Sci USA 76:5736, 1979
29. Flake AW, Harrison MR, Adzick NS, Zanjani ED: Transplantation of fetal hematopoietic stem cells in utero: The creation of hematopoietic chimeras. Science 233:776, 1986
30. Harrison MR, Slotnick RN, Crombleholme TM, et al: In-utero transplantation of fetal liver haemopoietic stem cells in monkeys. Lancet 2:1425, 1989
31. Linch DC, Rodeck CH, Nicolaides K, et al: Attempted bone marrow transplantation in a 17 week fetus. Lancet 2:1453, 1986
32. Touraine JL, Raudrant D, Royo C, et al: In-utero transplantation of stem cells in bare lymphocyte syndrome. Lancet 1:1382, 1989

INDEX

Page numbers followed by f *indicate figures; those followed by* t *indicate tables.*

A

AAV. *See* Adeno-associated virus
Acatalasemic mice, 185
Acetoacetate, 64
Acetylcarnitine, 75
   in isovaleric acidemia, 71, 72
Acetyl CoA:α-glucosaminidase acetyltransferase deficiency, 212
N-Acetylglucosaminase 6-sulfatase deficiency, 212
α-N-Acetylglucosaminidase deficiency, 212
β-D-N-Acetylhexosaminidase A:
   and adsorptive endocytosis, 135–136
   and receptor-mediated endocytosis, 137–138
   replacement therapy, 131, 134, 135, 137–138, 141, 154
Acylcarnitines, 14t, 64, 71–73, 75
   FAB-MS detection, 70–72, 76
Acyl-coenzyme (CoA) thioesters, 71
ADA. *See* Adenosine deaminase
Adagen, 173. *See also* PEG-ADA
Addison-only adrenoleukodystrophy, 112–113, 113t, 214
   current therapeutic recommendations, 125
Adeno-associated virus (AAV):
   lytic cycle, 251
   provirus, 251
   as vector for somatic gene therapy, 250–251, 262–264, 263t, 287–319
Adenosine deaminase deficiency:
   bone marrow transplantation for, 169, 170–171, 176f, 176–177
   gene therapy for, 169, 249, 252, 253, 262f, 262–264, 266, 267t, 267–268, 268f
   partial exchange transfusion for, 171, 172f, 176
   PEG-ADA replacement therapy, 131, 169, 171–178, 176f
   future applications, 178
   immune response to, 175–176
   T cell therapy, 265–266
Adenosine phosphoribosyltransferase (APRT) gene, 307
Adrenoleukodystrophy, X-linked, 111–129
   Addison-only, 112–113, 113t, 125, 214
   bone marrow transplantation for, 122–124, 123f, 124f, 126, 214–216, 224
   carnitine effect on, 115
   clofibrate effect on, 115
   current therapeutic recommendations, 125–126
   dietary therapy for, 115–122
   glycerol trioleate oil (GTO) for, 115, 116f, 117–118, 119t
   glycerol trirucate oil (GTE) for, 116f, 116–117

Adrenoleukodystrophy, X-linked
    (Continued)
  GTE–GTO effect on, 116f, 116–117,
      120–122, 122t, 125–126
  nature and clinical spectrum of,
      112–115
  neonatal ADL, 112
  pathogenesis of, 113–114
  peripheral nerve function, 117–120,
      118t, 119t
  therapy for, 115–126
  varying phenotypes of, 112–113, 113t
  VLCFA and, 111, 112, 113, 114–116,
      121, 122, 123, 124f, 214–215
Adrenomyeloneuropathy (AMN), 111,
    112–113, 113t. See also
    Adrenoleukodystrophy,
    X-linked
  current therapeutic recommendations, 125
  GTO therapy for, 117–118
  peripheral nerve function, 117–120,
      118t, 119t
Alanine:
  anabolic effect, 60
  and MSUD therapy, 51, 52f, 53t, 55f,
      55t, 57t, 58t
  in parenteral solutions, 59t
  supplementation, 59–61
Alanine-glucose cycle, 60, 60f
Albumaid-XP, 33, 34t
Alcaptonuria:
  metabolite correction, 2–3
  nutritional therapy for, 1–2
ALD. See Adrenoleukodystrophy,
    X-linked
Alloisoleucine: and MSUD, 49, 52f, 57t,
    58t
α$_1$-Antitrypsin: gene transfer, 249, 254
α-Fucosidase deficiency, 186t, 205. See
    also Fucosidosis
α-Galactosidase A: and Fabry disease,
    134, 155
α-1,4-Glucosidase deficiency, 185, 204
α-L-Iduronidase deficiency, 186t, 205,
    209. See also
    Mucopolysaccharidosis I
α-Mannosidosis: transplantation in animal models for, 186t–193t
α-TIF protein, 290

Amino acid:
  anabolism as adjuvant, 48–52
  assay problems, 30
  branched-chained
    degradation pathways, 5f
    in medical foods, 41
  imbalance correction, 15–16
  medical sources, 91t
  nutritional therapy for disorders of,
      45–66. See also specific disorders
  organoleptic properties, 38–40, 39f
  parenteral solutions, 59t
  profile of medical foods, 38–41
  restriction, 7t
Aminogran, 33, 34t
p-Aminophenylethylamine (PAPEA):
    and Hex A endocytosis, 138
Ammonia:
  HHH syndrome
    lysine and, 61–63, 63t
    metabolic defect, 62f
  and urea cycle disorders, 7t, 14t, 87
AMN. See Adrenomyeloneuropathy
Anabolism: in MSUD therapy, 48–52, 53t
Analog XP, 32, 33, 34t, 37, 46, 48
Animal model transplantation, 183–201
  conclusions drawn from, 193–194
  for defective cells, tissues, organs,
      184–185
  as source of normal gene product,
      185–186, 186t–193t
Antenatal therapy. See Fetal therapy
Apoenzyme:
  binding, 9
  defects, 46–47
Apolipoprotein E: gene transfer, 249
Arginase deficiency, 79
  PEG-modified enzyme therapy for,
      178
Arginine:
  medical sources, 90t
  and MSUD therapy, 53t, 55t
  in parenteral solutions, 59t
  supplementation, 7t, 61, 79
  for urea cycle disorders, 87, 92–93
  and waste nitrogen synthesis, 81f
Argininosuccinase deficiency (ALD), 80,
    80f, 90, 91
  diagnosis, 88t
  dietary supplements, 87

management recommendations, 89t, 90, 93–94
Argininosuccinate (ASA): synthesis, 80, 80f
Argininosuccinate lyase deficiency: PEG-modified enzyme therapy for, 178
Argininosuccinic acid: as waste nitrogen product, 87–91, 88t
Argininosuccinic acid synthetase deficiency (ASD), 80, 87
　diagnosis, 88t
　management recommendations, 89t, 93
Aromatic amino acid hydroxylases, 10t
Arylsulfatase B deficiency: transplantation for, 212–213
　in animal models, 186t–193t
ASA. See Argininosuccinate
ASD. See Argininosuccinic acid synthetase deficiency
Asialoglycoprotein receptor, 248, 273–274, 276, 283
Asialoorosomucoid (AsOR), 248, 274
[^{125}I]-ASOR-PL conjugate, 274–275
ASOR-PL-DNA complex, 275, 276f, 278f, 278–279, 279f, 280f
L-Aspartate: in medical foods, 38–39
Aspartic acid:
　and MSUD therapy, 53t, 55t
　in parenteral solutions, 59t

B

Bare lymphocyte syndrome, 327
Benzoate:
　medical sources, 91t
　for NKH, 13, 14t, 15
　for urea cycle disorders, 87, 92–93
　and waste nitrogen excretion, 84, 85t, 86, 87
β-Galactosidase:
　defect, 186t. See also GM1 gangliosidosis
　gene transfer, 249–250, 263
　M6P/IGFIIR and, 142
β-Globin:
　as ADA promoter, 263t, 264–265
　deficiency, murine: genetic correction, 242, 242t, 262–265

as HSV-1 vector promoter, 306t, 307
β-Glucosidase:
　gene transfer, 249
　replacement therapy, 131, 140
β-Glucuronidase: gene transfer, 249
β-Glucuronidase deficiency, 186t. See also Mucopolysaccharidosis VII
　murine: genetic correction, 242, 242t
β-Hydroxy-γ-N-trimethylammonium butyrate. See Carnitine
Betaine: supplementation, 13
Betaine-homocysteine methyltransferase, 6f, 13
β-Ketothiolase deficiency, 72
β-Thalassemia mouse: genetic correction, 242, 242t
BH$_4$ deficiency, 10t, 12
Biopterin: enhancement, 10t
Biotin:
　for prenatally diagnosed multiple carboxylase deficiency, 322–323
　and propionic acidemia, 5f
　supplementation, 10t, 12, 46
Biotinidase deficiency, 12
Blood-brain barrier:
　and bone marrow transplantation, 206
　and enzyme replacement therapy, 143–146, 154
Bone marrow transplantation:
　for adenosine deaminase deficiency, 169, 170–171, 176f, 176–177
　in animal models, 183–201
　　conclusions from, 193–194
　　for lysosomal disorders, 186t–193t. See also Lysosomal disorders
　　as source of normal gene, 185–186, 186t–193t
　blood-brain barrier and, 206
　CNS microglia and, 205–206
　　turnover, 206–207
　consortium formation, 216–217
　cross-correction in vitro, 204
　cross-correction in vivo, 204–205
　effect on VLCFAs, 122–124, 124f
　enzyme replacement therapy vs., 146–147
　fetal, 325–327
　for Gaucher disease, 154

Bone marrow transplantation
    (Continued)
  graft-versus-host disease and, 206,
      207, 208, 227, 325–327
  hepatocytes and, 204–205
  intrauterine, 325–327
  Kupffer cell and, 204–205
  for leukodystrophies, 214–216
    globoid, 214, 223–236. *See also*
        Globoid cell leukodystrophy
    metachromatic, 214
    X-linked adrenoleukodystrophy,
        111, 122–124, 123f, 124f, 126,
        214–216, 224. *See also*
        Adrenoleukodystrophy
  for mucopolysaccharidoses, 207–213.
      See also specific disorders
    in animal models, 186t–193t
  for storage diseases, 203–221. *See also
      specific diseases*
    in animal models, 183–201. *See also
      animal models above*
  "time needed for migration," 235
  in utero, 325–327
Bovine papillomavirus: as vector for
    gene transfer, 251
Brain. *See* Blood-brain barrier; Central
    nervous system
Branched-chained amino acids:
  degradation pathways, 5f
  in medical foods, 41
Branched-chained α-ketoacid dehydrogenase, 10t
Butyric acid, 14t
γ-Butyrobetaine, 69
γ-Butyrobetaine aldehyde, 69
γ-Butyrobetaine hydroxylase, 69
Butyrylcarnitine, 72

C

Carbamylphosphate synthetase deficiency
    (CPSD), 79, 81f, 82, 86t, 87, 88t
  management recommendations, 89t,
      92–93
Carboxylase deficiency, 10t, 12
Carnitine:
  for adrenoleukodystrophy, X-linked,
      115
  deficiency, 63–64, 71, 73–74, 75
    and Fanconi syndrome, 64, 96, 101
  in detoxification therapy, 14t, 14–15
  diagnostic applications, 70–73
  and ketogenesis, 64, 65f
  in parenteral solutions, 59t
  supplementation, 63–65, 69–76
  synthesis and metabolism, 69–70, 70f
  transport, 75–76
  treatment with, 73–75
D-Carnitine, 69. *See also* Carnitine
L-Carnitine. *See* Carnitine
CAT enzymatic activity, 274–275, 278f,
    278–280, 279f, 280f, 281, 282f
CAT gene transfer, 250, 274, 292
Cell targeting: in enzyme replacement
    therapy, 133t, 134–139
Cell uptake: in enzyme replacement
    therapy, 133t, 134–139
Central nervous system:
  blood-brain barrier
    and bone marrow transplantation,
        206
    and enzyme replacement therapy,
        143–146, 154
    enzyme replacement delivery to,
        143–146
  microglia
    bone marrow transplantation and,
        205–206
    turnover, 206–207
Ceramidetrihexoside: and Fabry disease,
    155
Chédiak-Higashi syndrome: animal
    model, 185
Chloramphenicol aminotransferase
    (CAT):
  enzymatic activity, 250, 274, 292
  gene transfer, 250, 274, 292
Chorionic villus cells: HLA-haplotyping,
    322
Citrate: for ALD, 90
Citrulline:
  medical sources, 90t
  supplementation, 7t, 79
  synthesis, 80, 81f
  for urea cycle disorders, 87
  as waste nitrogen product, 87–91
Clofibrate: for adrenoleukodystrophy,
    X-linked, 115

CMV. *See* Cytomegalovirus
Cobalamin. *See also* Vitamin B_{12}
    enhancement, 10t, 11–12, 46
    and methylmanolic acidemia, 5f, 46
Coenzyme:
    binding, 6f, 11
    binding defects, 10t
    binding site, 9, 9f
    enhancement for metabolic errors, 10t
    production, 5f, 6f, 11–12
    synthesis defects, 10t
Combined immunodeficiency:
    animal model, 185
    severe. *See also* Adenosine deaminase deficiency
        enzyme replacement therapy for, 131, 169–182
        gene therapy for, 169, 253
Congenital adrenal hyperplasia: fetal therapy, 321–322
Corneal crystals: in cystinosis, 97, 98f, 99, 104–105
CPSD. *See* Carbamylphosphate synthetase deficiency
C26:0 very long chain fatty acid:
    synthesis blockage, 15
    and X-linked adrenoleukodystrophy, 111, 114, 115, 116f, 118, 119t, 120f, 120–121
Cyanocobalamin:
    for prenatally diagnosed methylmalonic acidemia, 323–324
    supplementation, 12
Cystathionase, 6f, 10t
Cystathionine β-synthase, 6f, 10t
    deficiency, 4, 11, 12–13. *See also* Homocystinuria
Cystathioninuria: coenzyme enhancement for, 10t, 11
Cysteamine:
    for cystinosis, 100t, 102–104, 105, 106
    eyedrops, 100t, 105
Cysteine, 95
    depletion in homocystinuria, 6
    and MSUD therapy, 53t, 55t
    in parenteral solutions, 59t
Cysteine-cysteamine: for cystinosis, 102–103
Cystic fibrosis: pancreatic enzyme supplementation, 132

Cystine:
    accumulation. *See* Cystinosis
    supplementation, 7t
Cystinosis, 95–106
    benign, 99
    and carnitine deficiency, 64
    clinical manifestations, 96f, 96–98, 98f
    cysteamine for, 100t, 102–104, 105, 106
    delayed puberty, 97, 105
    diagnosis, 98–99
    genetics, 96, 99
    hypocalcemia, 101
    hypohidrosis, 97, 105
    hypophosphatemic rickets, 100
    hypothyroidism, 97, 100t, 105
    ophthalmic, 97, 98f, 99, 100t, 104–105
    renal transplant for, 97, 100t, 102
    therapy, 99–106, 100t
        ophthalmic, 104–105
        renal Fanconi syndrome, 99–101
        renal glomerular dysfunction, 101–104
Cytomegalovirus: as vector for gene transfer, 251, 262, 262f, 263t, 307

D

dAdo, 169–170, 172f, 177–178
dATP, 169–170, 172f
D-Carnitine, 69. *See also* Carnitine
Decarboxylase: defect in MSUD, 49
2′-Deoxyadenosine (dAdo), 169–170, 172f, 177–178
Deoxyadenosylcobalamin, 11–12, 46
    methylmalonic acidemia and, 323
Dexamethasone: maternal administration in prenatal therapy, 322
DHL-9 infected lymphoblasts: ADA activity in, 263t
Dietary therapy. *See* Nutritional therapy
Dihydrofolate reductase (DHFR) gene, 265
Duchenne muscular dystrophy: mdx mouse model, 184
Dwarf little mouse: genetic correction, 242, 242t
Dystrophin deficiency, 184

E

EBV. See Epstein-Barr virus
Endocytosis in enzyme replacement
    therapy, 135–137
  acceptor-mediated, 136–137
  adsorptive, 135–136
  fluid-phase, 135
  receptor-mediated, 136, 137–138
Endothelial smooth muscle cells: retroviral gene transfer, 249–250
Energy-protein ratio of foods, 35–37, 36t, 40, 42
env gene, 244f, 244–245
Enzyme:
  availability of, 131–134, 133t
  substrate and conenzyme binding sites, 9, 9f
Enzyme replacement therapy, 131–152
  for adenosine deaminase deficiency, 131, 169–182
  availability of enzyme, 131–134, 133t
  biochemical evaluation, 139
  blood-brain barrier and, 143–146, 154
  bone marrow transplantation vs., 146–147
  CNS delivery, 143–146
  complications and side effects, 133t, 141–142
  effectiveness of, 133t, 139–141
  endocytosis
    acceptor-mediated, 136–137
    adsorptive, 135–136
    fluid-phase, 135
    receptor-mediated, 136, 137–138
  for Fabry disease, 134, 138, 140, 141, 155, 165
  functional evaluation, 140
  for Gaucher disease type I, 131, 133, 140, 141, 142, 146, 153–168. See also Gaucher disease type I
  gene replacement vs., 146–147
  plasma clearance, 133t, 134
  for Pompe disease, 138
  principles and problems, 131–143, 133t
  for severe combined immunodeficiency, 131, 169–182
  subarachnoid delivery, 144–145
  targeting and cell uptake, 133t, 134–139
  for Tay-Sachs disease, 141, 144
  unresolved basic questions on, 133t, 142
Epstein-Barr virus: as vector for gene transfer, 251
ERT. See Enzyme replacement therapy
Erythropoietin, 102
  for cystinosis, 100t, 106
Escherichia coli lacZ gene, 302f, 303, 304f, 305f, 305, 306t, 307, 308

F

FAB-MS: and acylcarnitine detection, 71–73, 76
Fabry disease, 155
  enzyme replacement therapy for, 134, 138, 140, 141, 155, 165
Factor IX: gene transfer, 249, 266–267
Fanconi syndrome, 96
  and carnitine deficiency, 64, 96, 101
  therapy for, 99–101
Fast-atom bombardment mass spectrometry (FAB-MS): and acylcarnitine detection, 70–73, 76
FDA. See U.S. Food and Drug Administration
Fetal immunotolerance, 326
Fetal liver hematopoietic stem cells, 326–327
Fetal therapy, 321–329
  for congenital adrenal hyperplasia, 321–322
  for galactosemia, 324–325
  hematopoietic stem cells transplantation, 325–327
  maternal administration of agents, 321–324
  maternal dietary restriction, 324–325
  for methylmalonic acidemia, 323–324
  for multiple carboxylase deficiency, 322–323
  for phenylketonuria, 324
  transplantation, 325–327
Flavoprotein: electron-transfer, 10t

5-Fluorouracil: stem cells treated with, 247, 263, 265
Folate: enhancement, 10t, 12
Foods:
  energy-protein ratio of, 35–37, 36t, 42
  medical. *See* Medical foods
Fructose:
  abnormalities, 7t
  hereditary intolerance: nutritional therapy for, 7t
  restriction, 7t
α-Fucosidase deficiency, 186t, 205. *See also* Fucosidosis
Fucosidosis: transplantation in animal models for, 186t–193t
Fumarylacetoacetase deficiency: PEG-modified enzyme therapy for, 178

G

gag gene, 244f, 244–245
Galactocerebrosidase deficiency, 224, 226, 226t. *See also* Globoid cell leukodystrophy
Galactocerebroside β-galactosidase deficiency, 224, 226, 226t. *See also* Globoid cell leukodystrophy
Galactokinase deficiency, 8f
Galactose:
  abnormalities, 7t
  accumulation of, 3, 7t, 8f
  degradation pathway, 8f
  restriction, 7t
    as fetal therapy, 324–325
Galactosemia:
  abnormality pathway, 8, 8f
  nutritional therapy for, 2, 3, 6, 7t, 8, 16
  substrate accumulation, 3
  in utero therapy, 324–325
Galactose-1-phosphate: accumulation, 8, 8f
Galactose-1-phosphate uridyltransferase, 8, 8f
  deficiency, 324–325
Galactosidase: PEG-modified, 178
β-Galactosidase:
  defect, 186t. *See also* GM1 gangliosidosis

  gene transfer, 249–250, 263
  M6P/IGFIIR and, 142
α-Galactosidase A: and Fabry disease, 134, 155
Galactosylceramidase deficiency, 186t, 205–206. *See also* Globoid cell leukodystrophy
γ-Butyrobetaine, 69
γ-Butyrobetaine aldehyde, 69
γ-Butyrobetaine hydroxylase, 69
Gangliosides:
  GM3, 142
  metabolism of, 142–143
Gangliosidosis:
  GM1: transplantation in animal models for, 186t–193t
  GM2: β-D-N-acetylhexosaminidase A replacement for, 131, 134, 137, 139
Gaucher disease: PEG-modified enzyme therapy for, 178
Gaucher disease type I, 153–154
  enzyme replacement therapy, 131, 133, 140, 141, 142, 146, 153–168. *See also* Enzyme replacement therapy
  early investigations, 154–157
  initial studies, 155–157, 157t
  gene therapy for, 154
  glucocerebrosidase for, 133, 157–159
    activity defect, 153
    clinical efficacy, 163–164
    dose-response study, 161–162
    effect on hepatosplenomegaly, 163
    effect on quality of life, 164
    effect on skeleton, 163–164
    hematologic effects, 163
    making effective, 158f, 158–159, 159f, 160t
    mannose-terminated trials, 159f, 159–161, 160f, 161f, 162f
  therapeutic considerations, 154
Gaucher disease type II, 153
Gaucher disease type III, 153, 155
Gene therapy, 239–286
  for adenosine deaminase deficiency, 169, 253, 262–264
  alternative vectors for somatic, 250–252

Gene therapy *(Continued)*
    approaches for, 240–246
    enzyme replacement therapy vs., 146–147
    for Gaucher disease, 154
    germline, in mice, 240–242, 241f, 242t
    into hematopoietic cells, 246–247, 261–266
    into hepatocytes, 248, 273–286. *See also* Hepatocyte gene transfer
    homologous recombination, 254–255
    positive-negative selection, 254
    prospects for, 252–253, 253f
    replacement vs. repair, 254
    retroviral, 249–250
        and hepatocytes, 283
        limitations, 250
        skin fibroblasts, 249, 266–268
        somatic cell, 243f, 243–246, 244f, 245f
    skin fibroblasts, 249, 266–268
    somatic cell, 243f, 243–246, 244f, 245f
        alternative vectors, 250–252
        disease candidates for, 252
    strategies for, 240t
    transkaryotic implantation, 249
Germline gene therapy: in mice, 240–242, 241f, 242t
Globoid cell leukodystrophy, transplantation for, 205–206, 214, 223–236
    in animal models, 186t–193t
    case history, 225–226
    discussion of, 233–236, 234t
    family case history, 226t, 226–227
    GVHD and, 227, 236
    infantile, exclusion of, 235
    juvenile, 224, 235
    results, 228–233
        achievement test percentiles, 230, 231f
        brain stem auditory evoked potentials, 229
        CSF, 230, 233
        EEG, 230
        endocrinologic, 228
        language test scores, 230, 232f
        motor nerve conduction, 229
        MRI, 230
        neurologic, 228
        neuropsychological data, 230, 231f–232f
        ophthalmologic, 228
        somatosensory evoked potentials, 229
        visual evoked potentials, 229
        WISC-R verbal performance, 230, 231f
    "time needed for migration," 235
Globoside: and enzyme replacement, 154
Glucocerebrosidase:
    for Gaucher disease type I, 133, 155–157
        activity defect, 153
        clinical efficacy, 163–164
        dose-response study, 161–162
        effect on hepatosplenomegaly, 163
        effect on quality of life, 164
        effect on skeleton, 163–164
        hematologic effects, 163
        making effective, 158f, 158–159, 159f, 160t
        mannose-terminated trials, 159f, 159–161, 160f, 161f, 162f
    gene transfer, 266
    human placental, 158, 158f, 159f
    PEG-modified, 178
Glucocerebroside:
    accumulation, 158
    clearance, 155–157, 157t
Glucose:
    abnormalities, 7t
    restriction, 7t
    supplementation, 7t
Glucose-alanine cycle, 60, 60f
β-Glucosidase:
    gene transfer, 249
    replacement therapy, 131, 140
α-1,4-Glucosidase deficiency, 185, 204
β-Glucuronidase deficiency. *See also* Mucopolysaccharidosis VII
Glucuronyltransferase deficiency, 248
L-Glutamate: in medical foods, 38–39
Glutamic acid:
    and MSUD therapy, 53t, 55t
    in parenteral solutions, 59t
    for PKU, 3
Glutamic acid decarboxylase, 10t, 11

Glutamine: in parenteral solutions, 59t
Glutarate: abnormalities, 7t
Glutaric acid, 14t
Glutaric acidemia:
  and carnitine deficiency, 64
  type I
    acetylcarnitines and, 72
    detoxification therapy, 14t
    nutritional therapy for, 7t
  type II
    coenzyme enhancement for, 10t, 12
    detoxification therapy, 14t
Glutarylcarnitine, 14t, 72
Glutaryl-CoA dehydrogenase deficiency. *See* Glutaric acidemia, type I
Glycerol trioleate oil (GTO):
  for adrenoleukodystrophy, 115, 116f, 117–118, 119t
  for adrenomyeloneuropathy, 119t
  with GTE: for adrenoleukodystrophy, 116f, 116–117, 120–122, 122t, 125–126
Glycerol trirucate oil (GTE):
  for adrenoleukodystrophy, 116f, 116–117
  with GTO: for adrenoleukodystrophy, 116f, 116–117, 120–122, 122t, 125–126
Glycine, 69
  for isovaleric acidemia, 13–14, 14t, 73
  and MSUD therapy, 53t, 55t
  and NKH, 13
  in parenteral solutions, 59t
Glycogenosis:
  enzyme replacement therapy for, 165
  type II: animal model, 185
Glycogen storage diseases: nutritional therapy for, 7t
Glycosaminoglycan: and mucopolysaccharidosis type I, 140
GM3 ganglioside, 142
GM1 gangliosidosis: transplantation in animal models for, 186t–193t
GM2 gangliosidosis: β-D-N-acetylhexosaminidase A replacement for, 131, 134, 137, 139
Gonadotropin-releasing hormone (GnRH) deficiency, 184
Graft-versus-host disease:
  and adenosine deaminase deficiency therapies, 170, 171
  and bone marrow transplantation, 206, 207, 208, 227, 325–327
  and globoid cell leukodystrophy transplantation, 227, 236
  intrauterine avoidance in transplantation, 325–327
Growth hormone:
  for cystinosis, 100t, 105–106
  murine deficiency: genetic correction, 242, 242t
GTE. *See* Glycerol trirucate oil
GTE–GTO: for adrenoleukodystrophy, X-linked, 116f, 116–117, 120–122, 122t, 125–126
  RBC microviscosity normalization, 122, 122t
GTO. *See* Glycerol trioleate oil
Gunn rats, 248
Guthrie cards, 73
GVHD. *See* Graft-versus-host disease

H

Hematopoietic cell gene therapy, 246–247, 261–266
  vectors, 262, 262f
Hematopoietic stem cell transplantation, fetal, 321–329
Heparan N-sulfatase deficiency, 212
Hepatic CAT enzymatic activity, 275, 278f, 278–280, 279f, 280f, 281, 282f
Hepatocyte gene transfer, 248, 273–286
  [^{125}I]-ASOR-PL conjugate, 274–275
  ASOR-PL-DNA complex, 275, 276f, 278f, 278–279, 279f, 280f
  CAT enzymatic activity, 275, 278f, 278–280, 279f, 280f, 281, 282f
  concept, 273–274
  conclusions, 283–284
  delivery and expression in vitro, 275–276, 276f, 277f
  delivery and expression in vivo, 277–280, 278f, 279f
  discussion, 282–283
  Hep G2 cells gene transformation, 275–276, 277f

Hepatocyte gene transfer *(Continued)*
  mammalian regulatory elements, 280f, 280–281
  persistence of expression, 281, 282f
  ^{32}P-labeled DNA complex, 277–278, 279
  SK-Hep 1 gene transformation, 275, 276f
  targetable DNA, 274–275
Hepatocytes: and bone marrow transplantation, 204–205
Hep G2 cells: gene transformation, 275–276, 277f
Hereditary cyclic hematopoiesis: dog model, 184–185
Herpes simplex virus type 1 (HSV-1):
  ANG strain, 293
  cellular toxicity, 293–294
  DNA polymerase mutant gene, 296t, 298
  genome, 289f, 289–292, 290f
    mutations, 294–298, 296t. *See also specific mutations*
  HCMVIE promoter, 306t
  ICP0 mutant gene, 295, 296t, 306t, 307, 308
  ICP4 mutant gene, 296t, 296–297, 298, 305, 306t, 307, 308, 310
  ICP22 mutant gene, 295, 296, 296t
  ICP27 mutant gene, 296t, 297, 298
  ICP6 promoter, 306t, 307
  IE genes, 250, 290
  *in*1814 mutant gene, 297
  lacZ gene insertion, 302f, 303, 304f, 305f, 305, 306t, 307, 308, 310, 310f, 312, 312f
  latency, 290, 291–292, 299–300, 311f, 312
  LAT mutant gene, 296t
  LAT promoter, 292, 305, 306, 306t, 308, 310, 312f
  LAT region: *lacZ* gene insertion, 302f, 303, 304f, 305f, 305
  mutations, 294–298
  neuroinvasiveness, 293, 300
  neurovirulence, 293–294, 300
  productive infection, 289–291, 291t
  reactivation, 292, 312

RE6 mutant gene, 296t
replicative cycle, 290, 290f
TAATGARAT binding, 291–292, 299
TK– gene, 294, 295, 296t, 306t, 307, 308
transfer models, 311f, 312
tsTK mutant gene, 302, 303f, 305
UL8 gene, 306t
UL52 gene, 306t
UL41 virion, 293, 297
US3 gene, 308
UTPase mutant gene, 296t
vhs-1 virus, 296t, 297
VP16 mutant gene, 296t, 297, 299
Herpes simplex virus type 1 (HSV-1) vector, 250, 251, 287–288, 298–310
  amplicons, 301
  bGH-Z20, 306t
  Cgal, 306t
  construction and packaging principles, 301–304, 302f, 303f, 304f, 305f
  delivery, 300
  early models for neural cells, 305–310, 306t, 310f
  8117/43, 306t
  evaluating potential use, 288
  GAL4, 306t
  HP40, 306t
  L7/14, 306t
  latency, 299–300
  neuronal promoters, 298
  neuronal-specific enolase (NSE) promoter, 307
  pHSV1ac plasmid, 250, 303f
    propagation, 303f
    structure, 302f
  plasmid-derived, 301–303
    early models, 305–310
    isolation, 304f
    propagation, 303f
  promoters, 298, 305–310, 306t
  recipient cells, 300–301
  replication-compromised model, 311f, 312
  replication-defective model, 311f, 312
  RH116, 306t
  RH142, 306t

strategic considerations, 250, 251, 298–301
summary, 312–313, 311f
Herpes simplex virus type 2 (HSV-2), 293, 294
Hex A. See β-D-N-Acetylhexosaminidase A
Hexacosanoic acid (C26:0 very long chain fatty acid):
  synthesis blockage, 15
  and X-linked adrenoleukodystrophy, 111, 114, 115, 116f, 118, 119t, 120f, 120–121
Hexanoic acid, 14t
Hexanoylcarnitine, 14t, 15, 72
HHH. See Hyperornithemia hyperammonemia homocitrullinuria syndrome
Hippurate: as waste nitrogen product, 81f, 83–87, 85t, 87t
Hippuric acid, 13, 14t
Histidine:
  and MSUD therapy, 53t, 55t
  in parenteral solutions, 59t
HLA-haplotyping: of chorionic villus cells, 322
HMG lyase, 74
Holoenzyme activity, 9
Homocitrulline, 60–63
Homocysteine:
  accumulation of, 6, 12
  reduction, 12–13
Homocystinuria:
  coenzyme enhancement for, 10t, 11
  cysteine depletion in, 6
  homocysteine reduction, 12–13
  nutritional therapy for, 4, 7t, 16
Homogentisic acid: diet and excretion of, 1, 2
Homologous recombination, 254
  hsp70, 293
  hsp90, 293
[^{3}H]thymidine, 207
Human GM2 gangliosidosis. See GM2 gangliosidosis
Human immunodeficiency virus (HIV): and enzyme replacement therapy, 133–134
Hunter syndrome, 204, 207. See also Mucopolysaccharidosis II
  transplantation for, 209–211, 217

Hurler syndrome. See Mucopolysaccharidosis I
Hydroxocobalamin:
  for prenatally diagnosed methylmalonic acidemia, 324
  supplementation, 10t, 12
21-Hydroxylase deficiency: in utero therapy, 321–322
3-Hydroxy-3-methylglutaryl-CoA lyase deficiency, 72
β-Hydroxy-γ-N-trimethylammonium butyrate. See Carnitine
Hydroxypropionic acids: urinary excretion monitoring, 47
Hydroxy-TML, 69
Hyperammonemia:
  HHH syndrome, 61–63, 63t
    metabolic defect, 62f
  and urea cycle disorders, 7t, 14t, 87, 92
  management protocol, 92
Hyperornithemia hyperammonemia homocitrullinuria syndrome:
  lysine and, 61–63, 63t
  metabolic defect, 62f
Hyperornithinemia with gyrate atrophy:
  coenzyme enhancement for, 10t
Hyperphenylalaninemia. See also Phenylketonuria
  definition of, 25–26
  dietary management of, 23–43
  malignant, 23
Hypogonadal mice, 184
  genetic correction, 242, 242t
Hypoketotic hypoglycemia syndrome:
  and carnitine deficiency, 64
Hypomethionemia, 11
Hypophosphatemic rickets: in cystinosis, 100
Hypoxanthine phosphoribosyltransferase (HPRT) gene, 306

I

ICP0 mutant gene, 295, 296t, 306t, 307, 308
ICP0 protein, 290, 291t, 292, 293

ICP4 mutant gene, 296t, 296–297, 298, 305, 306t, 307, 308, 310
ICP4 protein, 290, 291, 291t, 293
ICP6 promoter, 306t, 307
ICP22 mutant gene, 295, 296, 296t
ICP22 protein, 290, 291t
ICP27 mutant gene, 296t, 297, 298
ICP27 protein, 290, 291t
ICP47 protein, 290, 291t
Iduronate sulfatase deficiency. *See also* Mucopolysaccharidosis II
transplantation for, 209–211
α-L-Iduronidase deficiency, 186t, 205, 209. *See also* Mucopolysaccharidosis I
IE genes of HSV-1, 250, 290
Immune response: PEG-ADA replacement therapy, 175–176
Immunodeficiency:
HIV. *See* Human immunodeficiency virus
severe combined. *See* Severe combined immunodeficiency
Immunosuppression: and enzyme replacement therapy, 141
Indomethacin: for cystinosis, 100t, 106
Influenza virus: as vector for gene transfer, 251
Insulin-like growth factor II receptor (IGFIIR): and receptor mediated endocytosis, 137–138, 142
Interleukin-3: synergism in PHSC division, 247
Interleukin-6: synergism in PHSC division, 247
Intermediate-early gene promoter (1E415), 250
In utero therapy, 321–329
for congenital adrenal hyperplasia, 321–322
for galactosemia, 324–325
hematopoietic stem cells transplantation, 325–327
maternal administration of agents, 321–324
maternal dietary restriction, 324–325
for methylmalonic acidemia, 323–324
for multiple carboxylase deficiency, 322–323

for phenylketonuria, 324
transplantation, 325–327
Isobutyrylcarnitine, 72
ISOLEU, 48
Isoleucine:
degradation pathway, 5f
in medical foods, 41, 59t
and MSUD, 5f, 7t, 49, 50f, 51, 52f, 53t, 55f, 55t, 57t, 58t
restriction, 7t, 46
Isovaleric acid, 13–14, 14t
Isovaleric acidemia (IVA):
acetylcarnitine in, 71, 72
alanine for, 60
and carnitine deficiency, 64, 73
carnitine supplementation, 73–74, 75
degradation pathway blockage, 5f
glycine therapy, 13–14t, 14, 73
parenteral solutions, 58, 59t
Isovalerylcarnitine, 14, 71, 72
Isovaleryl-CoA, 5f, 13–14
Isovalerylglycine, 14, 14t
IVA. *See* Isovaleric acidemia

K

Keratinocytes: retroviral gene transfer into, 249
Ketogenesis: carnitine and, 64, 65f
β-Ketothiolase deficiency, 72
3-Ketothiolase deficiency, 74
Krabbe disease. *See* Globoid cell leukodystrophy
Kupffer cell: and bone marrow transplantation, 204–205
Kynureninase, 10t

L

Lactate abnormalities, 7t. *See also* Galactosemia
Lactose: restriction, 7t
lacZ gene, 250, 302f, 303, 304f, 305f, 305, 306t, 307, 308, 310, 310f, 312, 312f
LASN virus vector, 262, 262f, 263, 263t, 267, 268

LAT mutant gene, 296t
LAT promoter: and HSV-1, 292, 305, 306t, 308, 310, 311f
LAT region: *lacZ* gene insertion, 302f, 303, 304f, 305f, 305
L-Carnitine. *See* Carnitine
Leucine:
  degradation pathway, 5f
  in medical foods, 41, 59t
  and MSUD, 5f, 7t, 49, 50f, 51, 52f, 53t, 55f, 55t, 56, 57t, 58t
  restriction, 7t, 13–14
Leukodystrophies: transplantation for, 214–216
  globoid cell, 214, 223–236. *See also* Globoid cell leukodystrophy
  metachromatic, 214, 224
  X-linked adrenoleukodystrophy, 122–124, 123f, 124f, 126, 214–216. *See also* Adrenoleukodystrophy
α-L-Iduronidase deficiency, 186t, 205, 209. *See also* Mucopolysaccharidosis I
Liposomes: for hepatic transfection in vivo, 283
LNBBA virus vector, 262, 262f, 263t
LNCA virus vector, 262, 262f, 263t
LNLA virus vector, 262, 262f, 263t
LNSA virus vector, 262, 262f, 263t
Lofenalac, 32, 33, 34t, 46
Low-density lipoprotein receptors: gene transfer, 266
Lymphotropic papovavirus promoter (LNLA), 262, 262f, 263t
Lysine, 69
  and HHH syndrome, 61–63, 63t
  and MSUD therapy, 53t, 55t
  in parenteral solutions, 59t
  restriction, 7t
Lysinuric protein intolerance: nutritional therapy for, 7t
Lysosomal disorders. *See also* Cystinosis *and specific disorders*
  enzyme replacement therapy for, 131–132, 138–140. *See also* Enzyme replacement therapy
  therapy for, 95

transplantation, 203–221. *See also* Bone marrow transplantation
  animal models, 183–201, 186t–193t
  background, 204–207
  blood-brain barrier, 206
  cross-correction in vitro, 204
  cross-correction in vivo, 204–205
  microglial cell turnover, 206–207
  microglia of CNS, 205–206
  mucopolysaccharidoses, 186t–193t, 207–213
transplantation in animal models, 186t–187t
  complications, 188t
  conclusions, 193–194
  donor-derived peripheral blood cells, 188t
  effect on CNS enzymatic activity, 191t
  effect on CNS lesion morphology, 192t
  effect on CNS substrate accumulation, 192t
  effect on corneal disease, 189t
  effect on kidney morphology, 190t
  effect on liver enzyme, 190t
  effect on longevity, 189t
  effect on peripheral nervous system, 191t
  effect on skeletal disease, 190t
  effect on urinary substrate excretion, 189t
  genotype of donor, 187t
  overall effect, 193t
  procedural mortality, 187t

M

MADD. *See* Multiple acyl-CoA dehydrogenase deficiency
Malignant melanoma: tumor infiltrating lymphocytes for, 252–253, 265–266
Mannose 6-phosphate-specific receptor (M6PR): and Hex A endocytosis, 137–138, 142

344   Index

Mannose-terminated glucocerebrosidase:
  carbohydrate units, 159f
  for Gaucher disease type I, 159f, 159–161, 160f, 161f, 162f
α-Mannosidosis: transplantation in animal models for, 186t–193t
Maple syrup urine disease (MSUD), 71
  amino acid imbalance correction, 15–16, 49–42
  anabolism as adjuvant to therapy, 48–52, 53t
  coenzyme enhancement for, 10t
  degradation pathway blockage, 5f
  isoleucine and, 5f, 7t, 49, 50f, 51, 52f, 53t, 55f, 55t, 57t, 58t
  leucine and, 5f, 7t, 49, 50f, 51, 52f, 53t, 55f, 55t, 56, 57t, 58t
  nutritional therapy for, 4, 7t, 11, 15, 16, 49–59
  parenteral solution for, 52f, 53t, 54f, 55f, 55t, 57t, 58t, 59t
Maroteaux-Lamy syndrome: *See* Mucopolysaccharidosis VI
Maternal administration of agents for fetal therapy, 321–324
Maternal dietary restriction, 324–325
Maxamaid X-P, 33, 34t, 36t, 38, 39f, 48
Maxamum XP, 33, 34t
MBP deficiency, murine: genetic correction, 242, 242t
MCAD. *See* Medium-chain acyl-CoA dehydrogenase (MCAD) deficiency
mdx mice, 184
Medical foods. *See also* Nutritional therapy; Parenteral solution
  altering role in diet, 431–42
  amino acid profile of, 38–41
  characteristics of, 24
  current products for PKU, 32–35, 34t, 36t, 46. *See also specific products*
  definition of, 32
  energy/protein ratios of, 35–37, 36t, 42
  organoleptic properties, 24, 35, 37, 38
  tyrosine in, 40
Medium-chain acyl-CoA dehydrogenase (MCAD) deficiency, 15, 64, 71, 72, 73
  carnitine supplementation, 73, 74, 75
Mental retardation:
  nutritional therapy and, 1, 16
  phenylketonuria and, 1, 3, 4, 16, 46
Metabolite:
  alternative pathways stimulation, 6f, 12–13
  blocking substrate production, 15
  coenzyme binding, 6f, 11
  coenzyme (precursor) enhancement, 10t
  coenzyme production, 5f, 6f, 11–12
  correction of secondary imbalances, 15–16
  detoxification of toxic, 5f, 7t, 13–15, 14t
  dietary correction of imbalances, 2–8
  dietary reduction, 1–2
  nontoxic conjugate binding, 13–15
  precursor administration, 1–2
  product supplementation, 2f, 4–8, 7t
  relief of block, 9f, 9–12, 10t
  substrate accumulation, 2f, 3f, 3–4, 5f, 6f, 7t
Metachromatic leukodystrophy, 206, 207
  transplantation for, 214, 224
Methionine:
  metabolism pathway, 6f
  and MSUD therapy, 53t, 55t
  in parenteral solutions, 59t
  restriction, 7t, 12, 46
Methionine adenosyltransferase, 6f
Methionine synthase, 6f, 10t, 11
Methylcitric acid: urinary excretion monitoring, 47
Methylcobalamin:
  deficiency
    detoxification therapy, 14t
    prenatal diagnosis, 323–324
  supplementation, 10t, 12
Methylenetetrahydrofolate reductase, 6f
  deficiency: coenzyme enhancement for, 10t, 12
3-Methylglutarylcarnitine, 72
Methylmalonic acid, 14t
  urinary excretion monitoring, 47, 48f
Methylmalonic acidemia:
  acylcarnitines and, 72

alanine supplementation for, 60, 61f
and carnitine deficiency, 64
carnitine supplementation, 73–74
coenzyme enhancement for, 10t, 11
degradation pathway blockage, 5f
detoxification therapy, 14t, 15
monitoring, 47, 48f
nutritional therapy for, 7t, 46, 58, 59t, 60
in utero therapy, 323–324
Methylmalonylcarnitine, 72
Methylmalonyl CoA mutase, 10t, 11, 46
Mice:
    acatalasemic, 185
    ADA, 263–264, 264f
    genetic correction, 242, 242t
    germline gene therapy in, 240–242, 241f, 242t
    hematopoietic cell gene therapy in, 263f, 263–264
    hypogonadal, 184, 242, 242t
    mdx, 184
    transgenic, 240–242, 241f, 242t
    twitcher, 205, 224
    W/W^v, 263
Milk withdrawal: for galactosemia, 3, 6
Minafen, 33, 34t
MMA. See Methylmalonic acidemia
Moloney murine leukemia virus, 308, 310
Morquio disease: transplantation for, 212. See also Mucopolysaccharidosis IV
M6PR. See Mannose 6-phosphate-specific receptor
MSUD. See Maple syrup urine disease
Mucopolysaccharidoses:
    enzyme replacement therapy for, 165
    transplantation for, 203, 204, 207–213
        in animal models, 186t–193t
Mucopolysaccharidosis I (Hurler syndrome), 204, 207
    glycosaminoglycan and, 140
    transplantation for, 208, 209, 217
        in animal models, 186t–193t
Mucopolysaccharidosis II (Hunter syndrome), 204, 207
    transplantation for, 209–211, 217
Mucopolysaccharidosis III (Sanfilippo syndromes): transplantation for, 211–212, 217
Mucopolysaccharidosis IV (Morquio disease): transplantation for, 212
Mucopolysaccharidosis VI (Maroteaux-Lamy syndrome): transplantation for, 212–213
    animal models, 186t–193t
Mucopolysaccharidosis VII:
    murine genetic correction, 242, 242t
    transplantation for: in animal models, 186t–193t
Multiple acyl-CoA dehydrogenase deficiency (MADD), 71, 72
Multiple carboxylase deficiency:
    coenzyme enhancement for, 10t
    fetal therapy, 322–323
Murine disease. See also Mice
    genetic correction, 242, 242t
Muscle cells: retroviral gene transfer into, 249

N

N-Acetylglucosaminase 6-sulfatase deficiency, 212
N-Acetylglucosamine-galactosamine-4-sulfatase deficiency. See Arylsulfatase B deficiency
neo gene, 246, 247, 252, 263, 265
Nephropathic cystinosis. See Cystinosis
Niemann-Pick disease:
    transplantation in animal models for, 186t–193t
    type B: enzyme replacement therapy for, 165
Nitrogen:
    dietary, 79
    stoichiometry between dietary intake and waste excretion, 82–83, 83f–85f, 85t, 86t
    waste, 79
        arginosuccinic acid and, 87–91
        citrulline and, 87–91
        hippurate and, 83–87
        phenylacetylglutamine and, 83–87
        synthesis pathway in urea cycle disorders, 80f, 80–82, 81f

NKH. *See* Nonketotic hyperglycinemia
N-methyl-D-aspartate receptors, 13
Nonketotic hyperglycinemia (NKH):
  metabolite detoxification, 13, 14t
Nutritional therapy, 1–17
  for adrenoleukodystrophy, X-linked, 115–122
  for amino acid disorders, 45–66. *See also specific disorders*
  anabolism as adjuvant, 48–52
  background on, 1–2
  blocking substrate production, 15
  carnitine, 69–76
  characteristics of medical foods, 24. *See also* Medical foods
  coenzyme binding, 6f, 11
  coenzyme production, 5f, 6f, 11–12
  conceptual basis for, 1–2, 2f
  correction of metabolite imbalances, 2–8
  correction of secondary imbalances, 15–16
  for cystinosis, 99–106. *See also* Cystinosis
  detoxification of toxic metabolites, 5f, 7t, 13–15, 14t
  education and, 24, 42
  for MSUD, 4, 7t, 11, 15, 16, 49–59, 52f, 53t, 54f, 55f, 55t, 57t, 58t, 59t
  for organic acid disorders, 45–66. *See also specific disorders*
  outcomes, 16
  for phenylketonuria, 1–2, 4, 5, 7t, 16, 23–43. *See also* Phenylketonuria
  principles and applications, 2–16
  product supplementation, 4–8, 7t. *See also specific products*
  relief of metabolic block, 9f, 9–12, 10t
  stimulation of alternative pathways, 6f, 12–13
  substrate accumulation, 3f, 3–4, 5f, 6f, 7t
  for urea cycle disorders, 79–80, 87–91. *See also* Urea cycle disorders

O

Octanoic acid, 14t
Octanoylcarnitine, 14t, 15, 72, 75
Octenoyl, 72
OCT1 protein, 299
Oleic acid: for peroxisomal disorders, 15
Ophthalmic cystinosis, 97, 98f, 99
Organic acid disorders: nutritional therapy for, 45–66. *See also specific disorders*
Organoids, 248
Organ targeting: in enzyme replacement therapy, 133t, 134–139
Organ transplantation: for Gaucher disease, 154
Ornithine:
  supplementation, 61
  transport defect, 61
Ornithine α-amino transferase, 10t
Ornithine transcarbamylase deficiency (OTCD), 60–61, 79, 80, 81f, 82, 87, 88t
  management recommendations, 89t, 92–93
  murine: genetic correction, 242, 242t
Orphan Drug Act, 134
OS1, 48
OS2, 48
Osteogenesis imperfecta: gene defect in, 254
Osteopetrosis: animal model, 184
OTCD. *See* Ornithine transcarbamylase deficiency
Oxaloacetate deficiency, 91

P

PA. *See* Propionic acidemia
PAG. *See* Phenylacetylglutamine
p-Aminophenylethylamine (PAPEA):
  and Hex A endocytosis, 138
Parenteral solution. *See also* Medical foods; Nutritional therapy
  for MSUD, 51–59, 52f, 53t, 54f, 55f, 55t, 57t, 58t, 59t
PEG-ADA replacement therapy, 169, 171–178, 176f
  immune response, 175–176

Pegademase, 173. *See also* PEG-ADA
PEG-modified enzymes: future applications, 178
Pentamannosyl 6-phosphate (PMP): and Hex A endocytosis, 138
Periflex, 33
Peroxisomal disorders: oleic acid for, 15
Phenylacetate:
　for urea cycle disorders, 14t, 15, 87, 92–93
　and waste nitrogen excretion, 81f, 84, 85t, 86, 86t, 87
Phenylacetic acid: and PKU, 45
Phenylacetylglutamine (PAG), 14t
　as waste nitrogen product, 81f, 83–87, 85t, 86t
Phenylacetylglutamine nitrogen (PAG-N), 86, 86t, 88t
Phenylalanine:
　accumulation in PKU, 2, 3, 3f, 5, 7t, 23–25, 26–27, 45–46. *See also* Phenylketonuria
　breast-feeding and, 31–32
　in dietary proteins, 36
　dietary restriction, 2, 3, 7t, 26–29, 45–46
　　current products, 32–35
　　degree of, 26–27, 29
　　duration of, 27–29
　　maternal, 324
　and MSUD therapy, 53t, 55t
　optimum blood level by age, 30–31
　in parenteral solutions, 59t
Phenylalanine hydroxylase deficiency, 26–27, 45. *See also* Hyperphenylalaninemia; Phenylketonuria
Phenylbutyrate:
　medical sources, 91t
　for urea cycle disorders, 14t, 15
　and waste nitrogen excretion, 81f, 84, 85t, 86, 86t
Phenyl-Free, 32, 33, 34t, 36t, 37, 37t, 38, 39, 39f, 40
Phenylketonuria, 71
　amino acid imbalance correction, 15–16
　breast-feeding and, 31–32
　coenzyme enhancement for, 10t, 12
　cognitive and psychological deficits in, 16

　current management principles, 30–31
　definition of, 25–26
　failure of diet adherence, 23–24
　historical background on, 24–25
　maternal syndrome, 29, 32
　medical foods for, 32–35, 34t, 46, 58, 59t
　　altering dietary role, 41–42
　　amino acid profile, 38–41, 39f
　　energy/protein ratios, 35–37, 36t, 42
　　organoleptic properties, 38
　mental retardation in, 1, 3, 4, 16, 46
　metabolic derangement in, 3f, 3–4
　nutritional therapy, 1–2, 4, 5, 7t, 16, 23–43, 45
　　current products, 32–35, 46. *See also* medical foods *above*
　　education and, 24, 42
　phenylalanine dietary restriction, 2, 3, 7t, 26–29, 45–46
　　degree of, 26–27, 29
　　duration of, 27–29
　　maternal, 324
　protein and, 23, 24, 26
　screening programs, 28
　tyrosine levels in, 5
　in utero therapy, 324
Phenylpyruvic acid: and PKU, 2, 3, 3f, 45
Phosphocysteamine: for cystinosis, 104
PHSC. *See* Pluripotent hematopoietic stem cells
PK Aid I, 34t
PKU. *See* Phenylketonuria
Pluripotent hematopoietic stem cells: gene transfer into, 246–247, 261–266
PNP. *See* Purine nucleoside phosphorylase
pol gene, 244f, 244–245
Polycitra, 99
Polyethylene glycol (PEG)-enzyme replacement, 169, 171–178, 176f
　future applications, 178
　immune response, 175–176
Poly-L-lysine:
　[^{125}I]-ASOR conjugation to, 274
　and Hex A, 135–136, 144

Polyvinylpyrrolidone (PVP)-derived enzyme, 141
Pompe disease, 204
    enzyme replacement therapy for, 138
Prenatal diagnosis. *See* Fetal therapy
Product 80056, 33, 92
Proline:
    and MSUD therapy, 53t, 55t
    in parenteral solutions, 59t
Propionate: metabolic disorders, 60
Propionic acid, 14t
Propionic acidemia:
    acetylcarnitine in, 71, 72
    alanine supplementation for, 60, 61f
    and carnitine deficiency, 64, 65f
    carnitine supplementation, 73–74, 75
    degradation pathway blockage, 5f
    detoxification therapy, 14, 14t
    monitoring, 47
    nutritional therapy for, 7t, 46, 58, 59t, 60
Propionylcarnitine, 14t, 15, 71, 72
Propionyl-CoA, 15, 46
Protein:
    energy ratio in foods, 35–37, 36t, 42
    intake, and urea nitrogen excretion, 82, 83f
    and methylmalonic acidemia, 47
    natural low, 35
    and PKU diet, 23, 24, 26
Protein-free diet powder: sources, 91t
Pterin defects: coenzyme enhancement for, 10t
Purine nucleoside phosphorylase:
    and ADA activity in infected DHL-9 lymphoblasts, 263t
    PEG-modified enzyme therapy for, 178
    skin fibroblast gene transfer, 266, 267, 267t
Pyridoxine:
    enhancement, 10t, 11
    seizures: coenzyme enhancement for, 10t, 11
Pyruvate:
    abnormalities, 7t
    and alanine formation, 60
Pyruvic acidemia:
    coenzyme enhancement for, 10t
    nutritional therapy for, 7t

R

Rat model: UDP-glucuronyltransferase deficiency, 184
Renal failure: in cystinosis, 100t, 101–104
Renal glomerular dysfunction: in cystinosis, 101–104
Renal transplant: and cystinosis, 97, 100t, 102
Reticular endothelial system: and enzyme replacement therapy, 133
Retroviral gene therapy, 249–250
    limitations, 250
    skin fibroblasts, 249, 266–268
    somatic cell, 243f, 243–246, 244f, 245f
Retroviral genome structure, 244, 244f
Retroviral life cycle, 243, 243f
Retroviral long terminal repeat, 262
Riboflavin: enhancement, 10t

S

Sandhoff variant of Tay-Sachs disease, 154
Sanfilippo syndromes: transplantation for, 211–212, 217
Selenium deficiency: in MSUD, 15
Serine:
    and MSUD therapy, 53t, 55t
    in parenteral solutions, 59t
Severe combined immunodeficiency. *See also* Adenosine deaminase deficiency
    enzyme replacement therapy for, 131, 169–182
    gene therapy for, 169, 253
Shiverer mouse: genetic correction, 242, 242t
Simian virus 40: as vector for gene transfer, 251, 262, 262f, 263t, 280, 283, 306t, 307
Skeletal muscle cells: retroviral gene transfer, 249–250
SK-Hep 1: gene transformation, 275, 276f
Skin fibroblasts: retroviral gene transfer into, 249, 266–268
Somatic cell gene therapy, 243f, 243–246, 244f, 245f

alternative vectors, 250–252
disease candidates for, 252
HSV-1 vector, 250, 251, 287–319.
See also Herpes simplex virus
type 1
retroviruses and, 243f, 243–244, 244f
Sparce mouse: genetic correction, 242, 242t
Spherocytosis, hereditary: gene defect in, 254
Sphingomyelinase defect, 186t. See also Niemann-Pick disease
Steroids: maternal administration in prenatal therapy, 322
Subarachnoid delivery of enzyme replacement therapy, 144–145
Sudden infant death syndrome (SIDS), 15
SV40. See Simian virus 40
Sweaty foot syndrome, 13

T

Taurine:
and MSUD therapy, 53t, 55t
in parenteral solutions, 59t
Tay-Sachs disease:
enzyme replacement therapy for, 141, 144
Sandhoff variant, 154
Tetanus toxin (TT):
and acceptor-mediated endocytosis, 136–137
fragment C (TTC), 137
Tetrahydrobiopterin: supplementation, 10t, 12
β-Thalassemia mouse: genetic correction, 242, 242t
Thiamine:
and MSUD, 5f
supplementation, 10t, 11
THRE, 48
Threonine:
and MSUD therapy, 53t, 55t
in parenteral solutions, 59t
restriction, 7t, 46
Thyroxine: for cystinosis, 100t, 105
α-TIF protein, 290
TIL. See Tumor infiltrating lymphocytes

T lymphopenia: and adenosine deaminase deficiency, 170, 174
T156 protein, 293
Transfusion, partial exchange: for adenosine deaminase deficiency, 171, 172f, 176
Transgenic mice, 240–242, 241f, 242t
Transkaryotic implantation, 249
Transplantation. See Bone marrow transplantation
Tricarboxylic acid (TCA), 90
Triglylcarnitine, 72
Trihexosylceramide: and Fabry disease, 134
Trimethylamine, 72
Trophamine, 58
Tryptophan:
in parenteral solutions, 59t
restriction, 7t
Tryptophan hydroxylase: activation, 12
TT. See Tetanus toxin
Tumor infiltrating lymphocytes (TILs):
for malignant melanoma, 252–253, 265–266
Tumor necrosis factor: tumor infiltrating lymphocyte delivery of, 252–253, 266
Twitcher mouse, 205, 224
Tyrosine:
for alcaptonuria, 1–2
and BH_4, 12
levels in PKU, 5, 27
in medical foods, 40, 59t
phenylalanine conversion to, 3
restriction, 7t
supplementation, 5–7, 27
Tyrosinemia:
nutritional therapy for, 4, 58, 59t
type I (hereditary), 4, 7t
PEG-modified enzyme therapy for, 178

U

UDPgalactose-4-epimerase deficiency, 8f
UL41 protein, 293
Urea:
nitrogen excretion, 82–83, 83f–85f
synthesis pathways, 80f, 80–82

Urea cycle disorders:
  detoxification therapy, 14t, 15
  management recommendations, 89t, 92
  nutritional therapy for, 7t, 79–80, 89t, 90t–91t
  stoichiometry between nitrogen intake and waste excretion, 82–83, 83f–85f
  treatment of, 79–94
Uridine: supplementation, 7t, 8
Uridine diphosphate (UDP) galactose deficiency, 8, 8f
U.S. Food and Drug Administration (FDA):
  and human gene therapy trials, 252, 253f
  "medical food" classification, 32

V

Vaccinia virus: as vector for gene transfer, 251
VAL, 48
Valine:
  degradation pathway, 5f
  in medical foods, 41, 59t
  and MSUD, 5f, 7t, 49, 50f, 51, 52f, 53t, 55f, 55t, 57t, 58t
  restriction, 7t, 46
Very long chain fatty acids (VLCFAs):
  and adrenoleukodystrophy, X-linked, 111, 112, 113, 114–116, 121, 122, 123, 124f, 214–215
  carnitine effect on, 115
  clofibrate effect on, 115
  dietary therapy effect on, 115–122

  glycerol trioleate oil (GTO) effect on, 115, 116f, 117–118, 119t
  glycerol trirucate oil (GTE) effect on, 116f, 116–117
  GTE–GTO effect on, 116f, 116–117, 120–122, 122t, 125–126
  source in ALD, 114–115
  synthesis blockage, 15
Vitamin B_{12}. *See also* Cobalamin
  for methylmalonic acidemia, 46
  in utero, 323–324
Vitamin B_6. *See* Pyridoxine
Vitamin responsive disorders, 9
VLCFAs. *See* Very long chain fatty acids
Vmw65 protein, 290
VP16 mutant gene, 296t, 297, 299
VP16 protein, 290, 291t

W

Waste nitrogen. *See* Nitrogen, waste
W/W^v mice, 263

X

Xanthurenic aciduria: coenzyme enhancement for, 10t, 11
X-linked adrenoleukodystrophy. *See* Adrenoleukodystrophy, X-linked

Z

Zellweger cerebrohepatorenal syndrome, 112
Zinc deficiency: in MSUD, 15